BIODEGRADATION OF NITROAROMATIC COMPOUNDS

ENVIRONMENTAL SCIENCE RESEARCH

Series Editor:

Herbert S. Rosenkranz
Department of Environmental and Occupational Health
Graduate School of Public Health
University of Pittsburgh
130 DeSoto Street
Pittsburgh, Pennsylvania

Founding Editor:

Alexander Hollaender

Recent Volumes in this Series

Volume 39 — GENETIC TOXICOLOGY OF COMPLEX MIXTURES
Edited by Michael D. Waters, F. Bernard Daniel, Joellen Lewtas, Martha M. Moore, and Stephen Nesnow

Volume 40 — NITROARENES: Occurrence, Metabolism, and Biological Impact
Edited by Paul C. Howard, Stephen S. Hecht, and Frederick A. Beland

Volume 41 — ENVIRONMENTAL BIOTECHNOLOGY FOR WASTE TREATMENT
Edited by Gary S. Sayler, Robert Fox, and James W. Blackburn

Volume 42 — CHEMISTRY FOR THE PROTECTION OF THE ENVIRONMENT
Edited by L. Pawlowski, W. J. Lacy, and J. J. Dlugosz

Volume 43 — PRIMARY PRODUCTIVITY AND BIOGEOCHEMICAL CYCLES IN THE SEA
Edited by Paul G. Falkowski and Avril D. Woodhead

Volume 44 — SECONDARY-METABOLITE BIOSYNTHESIS AND METABOLISM
Edited by Richard J. Petroski and Susan P. McCormick

Volume 45 — GLOBAL CLIMATE CHANGE: Linking Energy, Environment, Economy, and Equity
Edited by James C. White

Volume 46 — PRODUCED WATER: Technological/Environmental Issues and Solutions
Edited by James P. Ray and F. Rainer Engelhardt

Volume 47 — GLOBAL ENERGY STRATEGIES: Living with Restricted Greenhouse Gas Emissions
Edited by James C. White

Volume 48 — GLOBAL ATMOSPHERIC–BIOSPHERIC CHEMISTRY
Ronald G. Prinn

Volume 49 — BIODEGRADATION OF NITROAROMATIC COMPOUNDS
Edited by Jim C. Spain

A Continuation Order Plan is available for this series. A continuation order will bring delivery of each new volume immediately upon publication. Volumes are billed only upon actual shipment. For further information please contact the publisher.

BIODEGRADATION OF NITROAROMATIC COMPOUNDS

Edited by

Jim C. Spain
Armstrong Laboratory
Tyndall Air Force Base, Florida

Plenum Press • New York and London

Library of Congress Cataloging-in-Publication Data

On file

Sponsored by U.S. Air Force Office of Scientific Research

Based in part on the symposium on Biodegradation of Nitroaromatic Compounds, held May 22–23, 1994, in Las Vegas, Nevada

ISBN 0-306-45014-3

A Division of Plenum Publishing Corporation
233 Spring Street, New York, N.Y. 10013

10 9 8 7 6 5 4 3 2 1

Printed in the United States of America

PREFACE

During the past five years increased awareness of environmental contamination by nitroaromatic compounds has led to a dramatic increase in research on their biodegradation. The resulting discoveries have markedly extended our understanding of degradation mechanisms and pathways in bacteria and fungi. Futhermore, this new basic knowledge promises the development of field applications of biodegradation systems for nitroaromatic compounds.

In May of 1994, an International Symposium on the Biodegradation of Nitroaromatic Compounds was held in Las Vegas, Nevada. This symposium brought together the scientists at the frontiers of research into the biodegradation of nitroaromatic compounds. The invited speakers were asked to review their area of expertise and write a critical, comprehensive synthesis of their work and related work by others. This book is the result of their efforts.

The emphasis of the reviews is on basic research in biodegradation and biotransformation. Therefore, the reactions of nitroaromatic compounds in plants, animals, bacteria, fungi, soil, and even nonbiological systems are considered. The goal of the work is to provide the reader with an appreciation of the tremendous range of possibilities for metabolism of aromatic nitro compounds and the experimental approaches used to understand them. This volume should be of interest to biochemists, microbiologists, engineers, toxicologists, and anyone interested in the behavior of synthetic chemicals in the environment or in living systems. Furthermore, a variety of commercial applications can be envisioned for some of the reactions described here.

The reviews are arranged into four main sections and an introductory overview. Each of the chapters is designed to stand alone, therefore, some key research may be described by different authors to place specific work in context with recent developments in their areas of expertise. Chapter 1 provides an overview of microbial metabolism of nitroaromatic compounds and a discussion of the areas of uncertainty that must be overcome before biodegradation can be used to remove nitroaromatic contaminants from the environment.

The section on aerobic biodegradation comprises three chapters. The first (Chapter 2) describes metabolic pathways used by aerobic bacteria during growth with nitroaromatic compounds as the sole source of carbon and energy. The strategies described in this chapter are noteworthy because most of the transformations described in other chapters are fortuitous and do not benefit the responsible organisms. Chapter 3 describes efforts to use novel catabolic reactions discovered in aerobic nitrobenzoate degraders to produce synthetic intermediates such as catechols for industrial applications. This is a little explored area that may become more important as additional unusual enzyme reactions are discovered in bacteria able to transform the nitro group. The final chapter in this section discusses the use of molecular biology to construct and extend the substrate range of strains able to degrade nitrotoluenes. This approach has been pursued extensively for development of strains able to degrade other synthetic organic compounds, but these authors are the first to apply it to nitroaromatic compounds.

Substitution by more than two nitro groups seems to render the aromatic ring very resistant to attack by electrophiles such as oxygen. The section on anaerobic processes contains two chapters describing bacteria that reduce the nitro group of TNT as the first step in its metabolism. Chapter 5 provides an overview of reduction reactions and then focuses on metabolism by sulfate reducing bacteria that degrade TNT. The discovery of consortia that mineralize TNT led the authors of Chapter 6 to isolate strains of *Clostridium* able to degrade TNT in pure culture. They also describe recent efforts to develop the system for practical applications.

The biotransformation of polycyclic nitroaromatic compounds by the intestinal microflora of mammals is important because of the potential toxicity of such compounds and their metabolites. Chapter 7 is focused mainly on anaerobic metabolism by intestinal microflora and related model systems. It also includes an extensive comparison of nitroreductase enzymes from a variety of sources.

Fungi, particularly the *Phanerochaete* can degrade a wide range of synthetic organic compounds including the isomeric nitrotoluenes. The current understanding of the process and potential application strategies are described in Chapter 8. One of the major impediments to the use of fungi for biodegradation of TNT is its toxicity to the fungus. The mechanism of toxicity has been discovered recently and is described in Chapter 9 along with insight into the pathway of TNT degradation by the fungus.

The final section contains three reviews by authors working in areas that provide new perspectives on the chemistry and biochemistry of nitroaromatic compounds. Chapter 10 describes the bioorganic chemistry of the reactions involved in the interconversion of aromatic nitro groups and amino groups in a variety of systems. The understanding derived from studies of drug metabolism in eucaryotes provides an appreciation of the range of reactions possible in biological systems.

At first glance it would seem that the biochemistry of inorganic nitrogen metabolism would be analogous to the metabolism of nitroaromatic compounds. Chapter 11 dispels such naive notions with an insightful review of the current understanding and mysteries of inorganic nitrogen metabolism. However, the author suggests that metalloenzymes which are crucial to the interconversion of nitrate and ammonia may also play a role in catalyzing reactions of organic nitrogen compounds.

Much of the interest in biodegradation of nitroaromatic compounds stems from concerns about their environmental fate and transport. Studies of biodegradation in natural systems are difficult or impossible without a good understanding of the many nonbiological reactions to which the nitro group is susceptible. The final chapter explains not only abiotic transformations of nitroaromatic compounds in the environment, but also the relationships between geomicrobiology and the origins of the materials that reduce the nitro group.

Shirley Nishino was a major contributor to the project through her unfailing work in organizing the symposium and formatting and editing the manuscripts. I am grateful to Chuck Somerville for invaluable help with organizing the symposium and Cathy Vogel for help with administration. Funding for this project was provided by the Air Force Office of Scientific Research under the guidance of Walter Kozumbo.

CONTENTS

Section I. Overview

Section II. Aerobic Bacteria

Section III. Anaerobic Bacteria

Section IV. Fungi

Section V. Related Biochemistry

1

BASIC KNOWLEDGE AND PERSPECTIVES ON BIODEGRADATION OF 2,4,6-TRINITROTOLUENE AND RELATED NITROAROMATIC COMPOUNDS IN CONTAMINATED SOIL

Paul-Gerhard Rieger[1] and Hans-Joachim Knackmuss[1,2]

[1] Institut für Mikrobiologie
Universität Stuttgart
Allmandring 31
70569 Stuttgart
[2] Fraunhofer-Institut für Grenzflächen- und Bioverfahrenstechnik
Nobelstrasse 12
70569 Stuttgart

INTRODUCTION

Although a few aromatic compounds bearing one nitro group as a substituent are produced as secondary metabolites by microorganisms (31, 44, 45, 49) the majority of nitroaromatic compounds in the environment are due to anthropogenic activities. Nitrations are important reactions for the large-scale production of aminoaromatic structures that are synthons for pesticides, dyes, polymers, and pharmaceuticals. Nitroaromatic compounds such as nitrobenzene are used as solvents, whereas polynitroaromatic compounds serve as explosives. According to Hartter (16) 2,4,6-trinitrotoluene (TNT) is produced in amounts of 2 million pounds per year. Nitroaromatic compounds are therefore abundantly present in industrial waste streams and surface waters. 2,4,6-Trinitrotoluene is commonly found as the main contaminant of soil and ground water originating from facilities for manufacturing, processing, and disposing of explosives. Often these contaminants have leached from disposal lagoons into the surrounding soil, and in the case of military burdens of World War I and II, have contaminated the groundwater (13). Consequently, in Germany large areas of highly contaminated soils at former production plants must be remediated. TNT, its metabolites, and related compounds represent an environmental hazard because they exhibit considerable toxicity to humans, fish, algae, and microorganisms (39, 43, 50). Since incineration, the only proven technology for the destruction of explosives, is prohibitively costly, bioremediation represents an important alternative approach, which deserves to be considered.

Biodegradation of Nitroaromatic Compounds
Edited by J.C. Spain, Plenum Press, New York, 1995

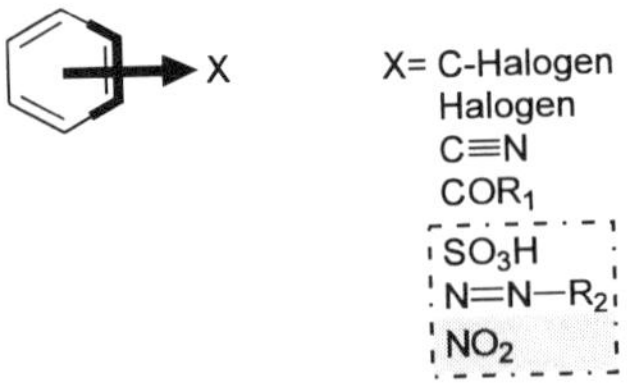

Figure 1. Electron-withdrawing groups X (R_1=alkyl or aryl; R_2=aryl) deactivate compounds for electrophilic (oxidative) attack by aerobic microorganisms.

This article will present a critical overview of the major known catabolic mechanisms and will discuss several new insights in microbial degradation of nitro-substituted aromatic compounds. These data will be weighed with respect to their applicability for bioremediation of contaminated soil. Based on the current status of knowledge, two alternative processes will be discussed in more detail.

Due to the strong electron-withdrawing character of the nitro groups, nitroaromatic compounds are electron deficient. Besides the xenobiotic character of certain structural elements such as halogen, SO_3H, N=N-R, and NO_2 (Fig. 1), it is the electron-withdrawing character of these groups and thus the electron deficiency of the aromatic ring, which impedes electrophilic attack by oxygenases of aerobic bacteria. As has been shown for substituted benzoates, the log V_{rel} values of dioxygenation can readily be correlated with the Hammett substituent constant (32). This renders aromatic compounds with electron-withdrawing substituents less susceptible to electrophilic oxygenation, particularly if several groups of this character are present on the aromatic ring.

The structure of TNT can be pictured (Fig. 2) as the resonance hybrid of several canonical forms, which include Kekulé structures such as formula (Fig. 2a) and ionic

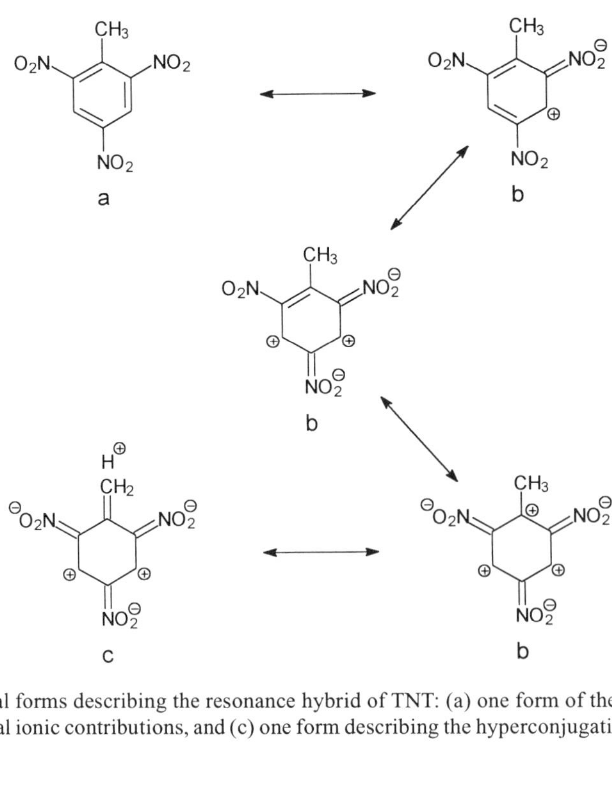

Figure 2. Canonical forms describing the resonance hybrid of TNT: (a) one form of the Kekulé structure, (b) some of the potential ionic contributions, and (c) one form describing the hyperconjugative effect of the methyl group.

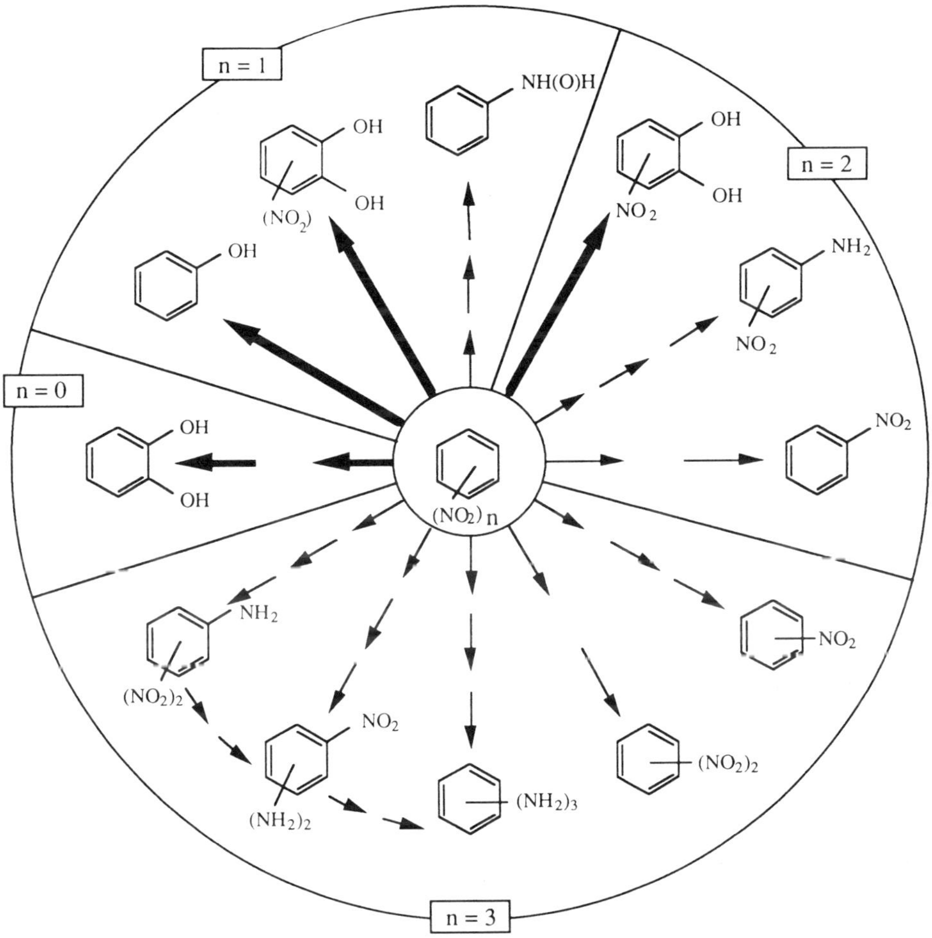

Figure 3. Initial biochemical transformations of nitroaromatic compounds: whereas with $n \leq 2$ oxidative mechanisms prevail (heavy arrows), reductive initial reactions (light arrows) were observed with increasing numbers of nitro substituents.

contributions (Fig. 2b). The figure indicates that part of the π-electrons of the aromatic nucleus are withdrawn by the nitro groups, making the nucleus electrophilic and thus prohibiting electrophilic attack by oxygenases. According to theoretical examinations by Wagniere (47), the nitro substituent of nitrobenzene withdraws 0.06 electrons from the benzene ring in the electronic ground state. The methyl group of TNT is an electron-donating group which on the one hand weakens the electron deficiency of the benzene ring, but on the other hand itself becomes less susceptible to oxygenases, i.e., to side chain hydroxylation, by contributing a so-called hyperconjugative form (structure *c* in Fig. 2) to the actual structure of TNT.

The known reactions of the degradation of aromatic compounds by aerobic bacteria through initial oxygenation (n=0 in Fig. 3) applies to mononitroaromatic compounds (n=1) and to a limited degree also to dinitroaromatic compounds. With increasing numbers of nitro substituents on the aromatic ring, initial reductive reactions are observed. Characteristically, oxygenase reactions are unknown for trinitro compounds such as TNT or picric acid. Instead, nitro groups or the aromatic nucleus are reduced (Fig. 3, n=3), which in the latter case gives rise to the replacement of a nitro group by hydrogen.

Numerous reductive transformations of nitroaromatic compounds have been described in the literature (Preuß and Rieger, this volume). These include gratuitous microbial reactions or chemical reductions by reducing agents in the medium. Thus, in iron-reducing cultures, 4-chloronitrobenzene was rapidly converted into 4-chloroaniline (17). These gratuitous reductions proceed via nitroso and hydroxylamino intermediates and are favored under microaerophilic and even more under anaerobic conditions.

INITIAL OXYGENATION OF NITROAROMATIC COMPOUNDS

Aerobic degradation of aromatic nitro compounds may be initiated by mono- or dioxygenases (Fig. 4a and b). The mechanism of oxygenolytic removal of nitro groups was first identified in 4- and 2-nitrophenol-utilizing bacteria. This initial reaction was shown to be catalyzed by monooxygenases (37, 51), as illustrated for 4-nitrophenol ($X=NO_2$) in Figure 4a. It involves the introduction of a hydroxyl group followed by spontaneous elimination of nitrite. In the case of 2- or 4-nitrophenol, the elimination of nitrite generates quinoid structures. Since ring cleavage by dioxygenases, however, requires a hydroquinone (1,4- or 1,2-dihydroxyaromatic ring system), reduction must precede the dioxygenolytic ring cleavage (Fig. 4a). Spanggord et al. (40) described nitrite elimination from 2,4-dinitrotoluene as the result of a dioxygenation reaction. Dioxygenolytic nitrite removal (Fig. 4b) was also described for 1,3-dinitrobenzene and 2,6-dinitrophenol degradation, generating 4-nitrocatechol and 4-nitropyrogallol, respectively (5, 8). Interestingly, an additional mechanism exists for the elimination of the second nitro group as nitrite. In the case of 2,6-dinitrophenol, evidence has accumulated which demonstrates that the second nitrite elimination must occur after ring cleavage (Fig. 5).

Nitrotoluenes may be subject to ring dioxygenation as well as to methyl group oxygenation. In the case of 4-nitrotoluene, the initial oxidation is catalyzed by a monooxygenase, which generates 4-nitrobenzylalcohol. This is further oxidized to 4-nitrobenzoic acid and finally metabolized to protocatechuate (see Spain, this volume) (15, 33).

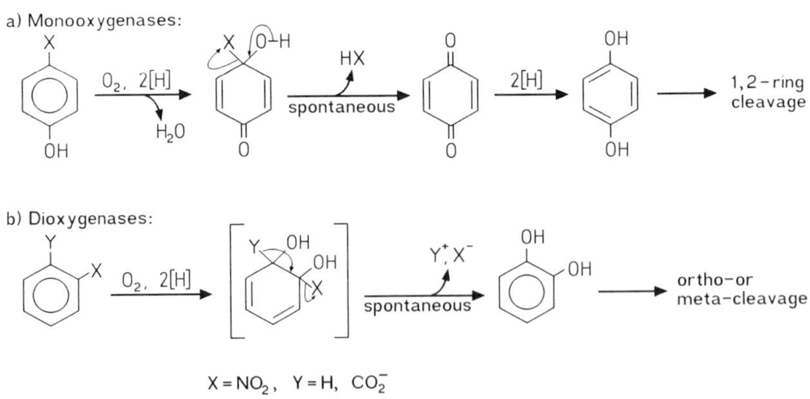

Figure 4. The known anionic elimination of substituents like X=Cl, SO_3, O-Alkyl or O-Aryl through (a) monooxygenases or (b) dioxygenases give rise to the release of nitrite in the case of nitroaromatics ($X=NO_2$). Hydrogen or COO^- in the *ortho*-position were eliminated as cations or CO_2.

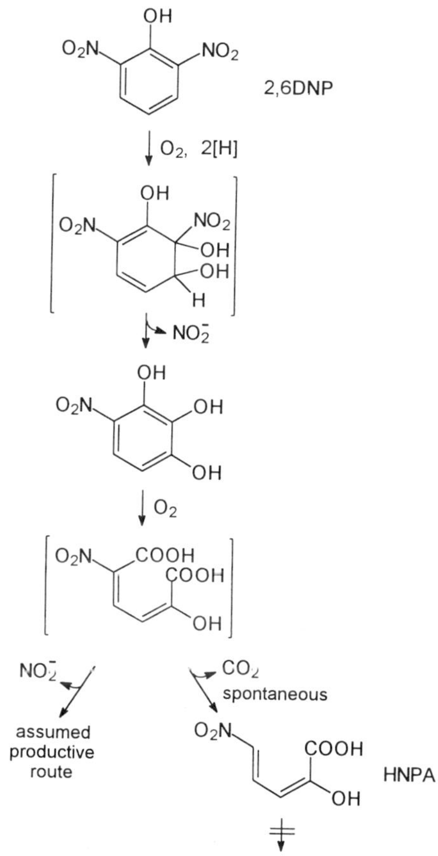

Figure 5. Initial catabolic steps of 2,6-dinitrophenol (2,6-DNP) degradation by *Alcaligenes eutrophus* JMP222 and JMP134 (8).

Another mechanism described by Haigler and Spain (14) showed that oxidation of nitroaromatics by mono- and dioxygenases can give rise to nitrocatechols which were generated from nitrobenzene by several *Pseudomonas* strains grown on toluene.

Initial denitration has been demonstrated to occur in various bacteria with different nitroaromatic structures. Characteristically, initial removal of nitro groups as nitrite may confer selective advantage to a broad range of bacteria that do not necessarily mineralize nitroaromatic compounds completely (3). A single enzyme rather than a complete catabolic sequence is sufficient to confer selective advantage on microorganisms which utilize nitroaromatic compounds as their sole source of nitrogen. Surprisingly, some of these bacteria, originally isolated by use of a readily degradable carbon source (e.g. succinate), could also mineralize these compounds completely (3, 6). From the very beginning of growth the nitroaromatic compound was degraded and nitrite was released, irrespective of whether ammonia was present in the growth medium.

REDUCTIVE ATTACK BY AEROBIC BACTERIA

The electrophilic character of the nitro group favors its reduction by microbes. Therefore even aerobic organisms harbor the ability to reduce nitroaromatic compounds. Whereas complete reduction of nitro groups to amino groups has been described in the literature, particularly as gratuitous reactions, partial reduction of nitroaromatics to hydroxylamino-derivatives has recently been identified as a key reaction of productive catabolic routes. Groenewegen and de Bont (12) and Groenewegen et al. (11) (see also de Bont et al., this volume) have described a *Comamonas* strain that utilizes 4-nitrobenzoate. Characteristically, these cells were simultaneously induced for turnover of 4-nitroso- and 4-hydroxylaminobenzoic acid but not of 4-aminobenzoic acid. A NAD(P)H-dependent nitroreductase was demonstrated *in vitro*. This enzyme and a hydroxylamino lyase, catalyze the conversion of nitrobenzoate into protocatechuate. The same catabolic mechanism has also been discovered in two *Pseudomonas* strains during degradation of 4-nitrotoluene via 4-nitrobenzoate and protocatechuate (15, 33).

A related catabolic mechanism appears to be involved in the pathway for degradation of nitrobenzene described by Nishino and Spain (29). *Pseudomonas pseudoalcaligenes* reduces nitrobenzene to hydroxylaminobenzene. The latter metabolite is rearranged, yielding 2-aminophenol. Although lacking a second hydroxyl group in the *ortho* or *para* position for ring cleavage, 2-aminophenol is directly subject to an oxygenolytic reaction at the 2,3-position. Obviously the amino group as an electron-donating substituent can also activate the aromatic nucleus for an electrophilic attack by a dioxygenase. This corresponds to the ring cleavage mechanisms of 3-hydroxyanthranilate and 5-aminosalicylate (19, 42).

The enzyme-mediated transformation of a hydroxylamino arene into an *ortho*-aminophenol corresponds to a chemical rearrangement known as the Bamberger reaction (25), which generates the 4-aminophenol from hydroxylaminobenzene under acidic conditions. It is likely that this intermolecular rearrangement proceeds via an arylnitrene, which by the additional function of a hydroxylamino lyase can generate protocatechuate plus ammonia rather than 4-amino-3-hydroxybenzoate from 4-hydroxylaminobenzoate.

A similar mechanism is discussed by Schenzle (Diploma thesis, University of Stuttgart, 1993) for the utilization of 3-nitrophenol (3NP) by *Alcaligenes eutrophus* JMP134. Cell-free extracts of 3NP-grown strain JMP134 converted 3NP into 3-hydroxylaminophenol. This is further converted into a metabolite which was extremely sensitive to autoxidation and identified as aminohydroquinone. Resting cells of 3-nitrophenol-grown JMP134 also readily cometabolized nitrobenzene into a mixture of 2- and 4-aminophenol.

Incomplete reduction of a nitro group yielding a hydroxylaminoaromatic compound and the rearrangement of the latter into an aminophenol was first described in the yeast *Rhodosporidium* sp. as part of cometabolic transformation of 4-chloronitrobenzene (4). Characteristically, reductive pathways have been postulated for the mineralization of a number of nitroaromatics (de Bont, this volume). In many cases, however, aminoaromatic structures could not be identified as intermediates. Therefore, the above novel catabolic pathway with a hydroxylaminoaromatic compound as a key metabolite could be of more general importance in aerobic mineralization of nitroaromatic compounds.

REDUCTION OF THE AROMATIC SYSTEM

Another unconventional reductive mechanism has been recently discovered by Lenke et al. in *Rhodococcus* strains which were selected with 2,4-dinitrophenol as the sole source of nitrogen (21, 22). Although succinate was supplied as a carbon and energy source during selection, the isolates could also grow with 2,4-dinitrophenol as the sole source of carbon. The accumulation of 4,6-dinitrohexanoate in the culture fluid indicated that degradation of the nitrophenol might be initiated by hydrogenation of the aromatic ring (4 hydrogens are required for 4,6-dinitrohexanoate formation) (22). Support for this mechanism came from the isolation of a spontaneous mutant, which also recognized picric acid (2,4,6-trinitrophenol) as a substrate and source of nitrogen. This picrate-adapted culture generated an orange-red metabolite which was proposed to be a H^--Meisenheimer-σ-complex of picrate

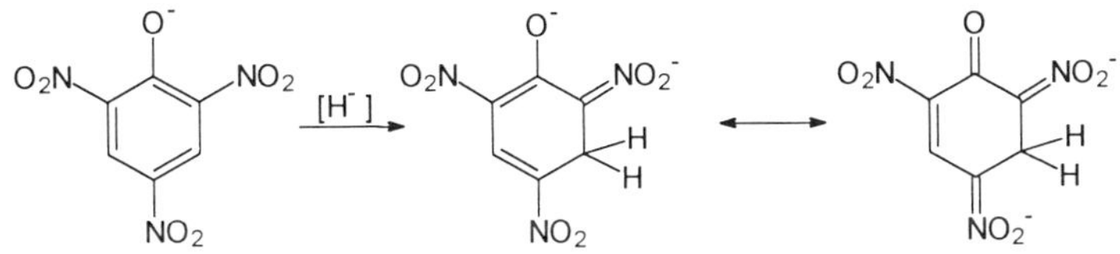

Figure 6. Initial ring hydrogenation of picrate by *Rhodococcus erythropolis* HLPM-1, yielding a H⁻-σ-complex.

(21). Unequivocal evidence for the structure of the complex given in Figure 6 came from very recent NMR-studies by Rieger et al. (34). Remarkably, these strains exclusively reduce the aromatic ring and do not attack the nitro substituents. As shown in Figure 7, the initial hydroaromatic structure is enzymatically rearomatized by generating nitrite and 2,4-dinitrophenol. When 2-chloro-4,6-dinitrophenol was utilized as an analogous substrate, 2,4-dinitrophenol was also generated and consequently chloride rather than nitrite eliminated.

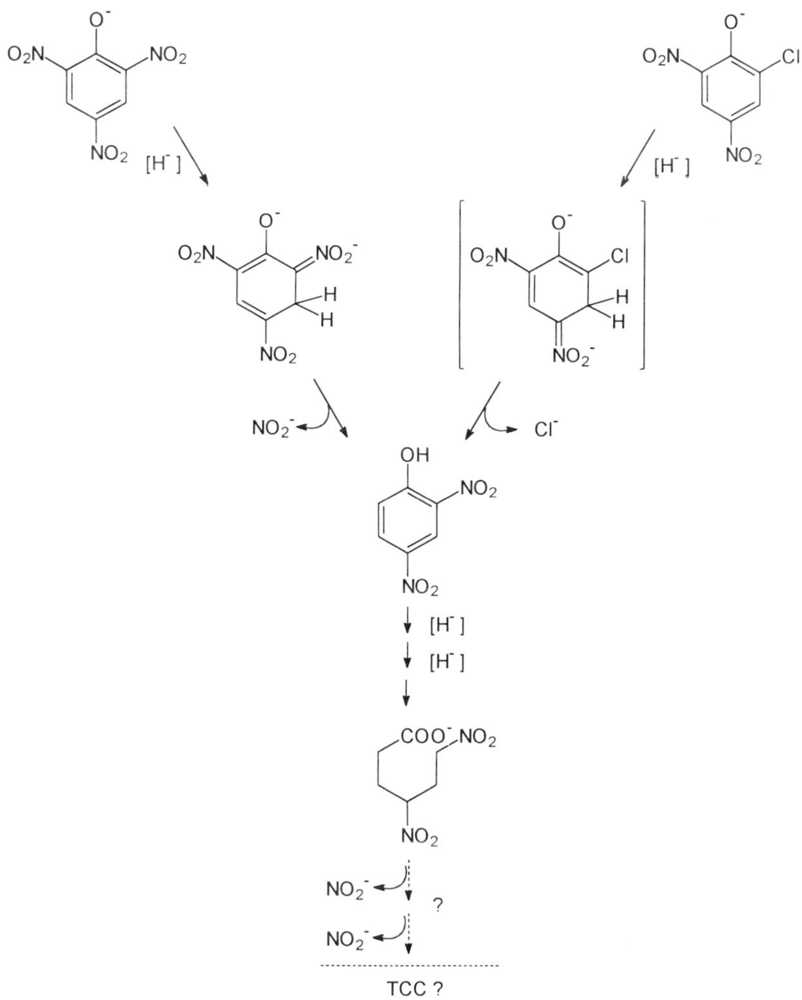

Figure 7. Transformation of picrate, 2-chloro-4,6-dinitrophenol, and 2,4-dinitrophenol by cell-free extracts of *Rhodococcus erythropolis* HLPM-1.

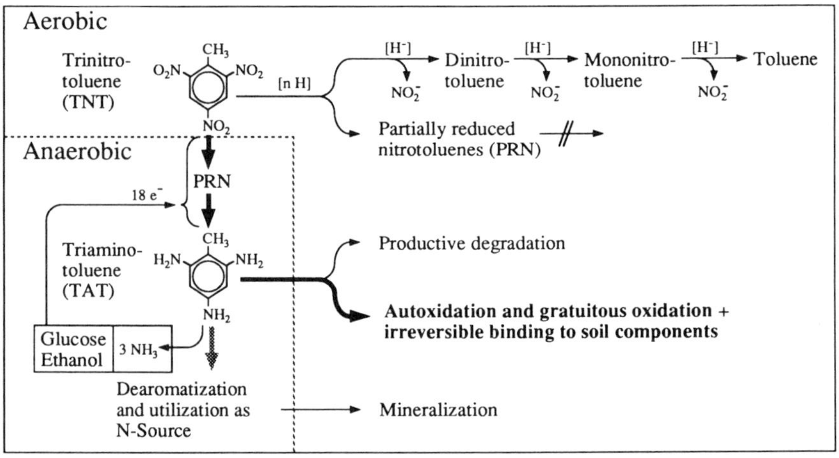

Figure 8. Perspectives on the biodegradation of TNT. Heavy arrows indicate a remediation concept which is suggested on the basis of the current knowledge. Intermediates of nitro group reduction give rise to partially reduced nitrotoluenes (PRN), which are depicted in Fig. 10. Other reactions of denitration of TNT are discussed in the text.

2,4-Dinitrophenol must be subject to further ring hydrogenation because 4,6-dinitrohexanoate could also be identified as a metabolite during turnover with cell-free extracts (Fig. 7) (34).

Replacement of NO_2 by H seems to be also applicable to trinitrotoluene, since Duque et al. (7) have implicated dinitrotoluenes, mononitrotoluenes and even toluene as metabolites of TNT (Fig. 8, horizontal sequence of reactions). This catabolic route is supported by recent observations of Vorbeck et al. (46), who actually identified the H^--σ-complex during cometabolic transformation of TNT by a 4-nitrotoluene-degrading bacterium (Fig. 9). Hydride addition and nitrite elimination either by a sequence as suggested by Duque et al. (7) or one analogous to that of picrate (21), could allow mineralization and productive metabolism of TNT. The use of such a system in mineralization processes is limited, however, by the reduction of nitro groups, which generates aminonitrotoluenes as major by-products. These metabolites are dead-end metabolites and are generated gratuitously not only by indigenous microorganisms but also by the strains described by Duque et al. (7). Although a clone was described which by aquisition of the TOL plasmid pWW0-Km reportedly grows on TNT as the sole carbon and nitrogen source, 2-amino-4,6- and 4-amino-2,6-dinitrotoluene as well as azoxy binuclear compounds were abundant products.

To our present knowledge such partially reduced nitrotoluenes and azoxy compounds are not utilized by aerobic microorganisms. The great number of potential products are illustrated in Figure 10. Those, which were actually identified are marked by shaded aromatic rings. The insert indicates the formation of secondary products such as azoxy compounds,

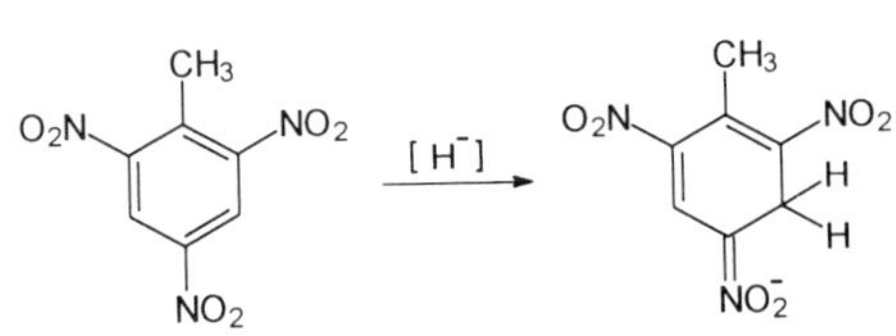

Figure 9. Gratuitous reduction of TNT by a 4-nitrotoluene-utilizing bacterial strain yielding a H^--σ-complex (46).

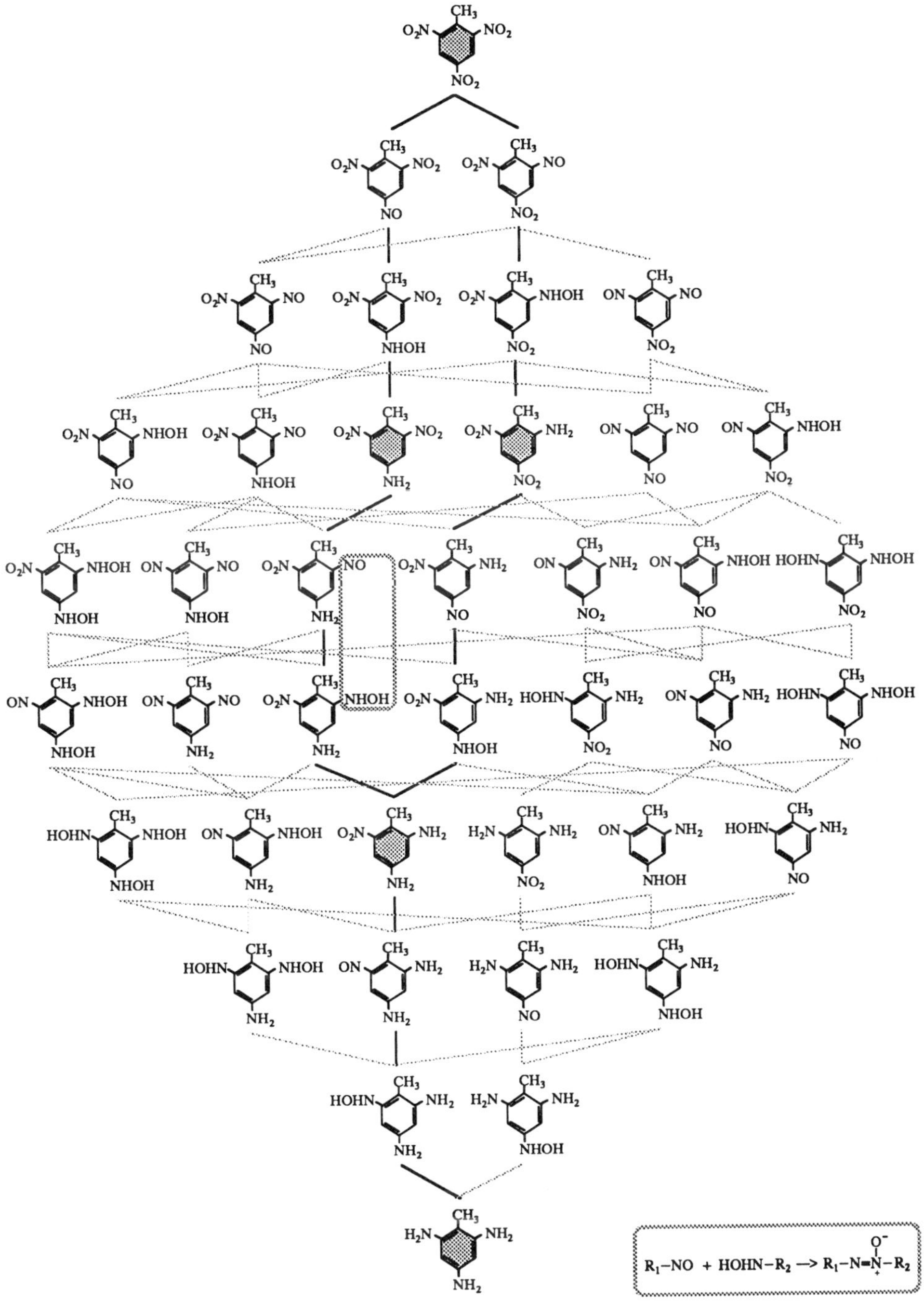

Figure 10. Potential intermediates and condensation products originating from gratuitous reduction of TNT. Identified metabolites are indicated by shading of the aromatic rings. Heavy lines connecting these compounds indicate the most likely reaction sequence between TNT and triaminotoluene (TAT).

which may be generated by spontaneous intermolecular condensation of nitroso- and hydroxylamino-intermediates.

TRANSFORMATIONS UNDER ANAEROBIC CONDITIONS

The reduction of nitro substituents by anaerobic microorganisms has been described for a number of mono-, di-, and trinitroaromatic compounds including TNT (Preuß and Rieger, this volume). McCormick et al. (26) first described the sequential reduction of the nitro groups of TNT. The electron-donating character of amino substituents weakens the electron deficiency originally present in TNT. Therefore reaction velocities decrease with the number of nitro groups being converted into amino groups. As a consequence, complete conversion of TNT into 2,4,6-triaminotoluene (TAT) requires strict anaerobic conditions ($E_h \leq -200$ mV) (10, 23). The intermediates of sequential reduction of TNT are the above-mentioned partially reduced nitrotoluenes shown in Figure 10. A likely reaction sequence is indicated by the heavy lines connecting those aminonitro-intermediates which have actually been identified. The reduction of 2,4-diamino-6-nitrotoluene to TAT via the 2,4-diamino-6-hydroxylaminotoluene intermediate was shown to be the rate-limiting step in sulfate-reducing bacteria (30; for detailed discussion see Preuß and Rieger, this volume).

Obviously, the potential of reduction of TNT is based on gratuitous transformation activities which are ubiquitous in anaerobic populations. This potential has been used for the degradation of nitrobenzene by a continuous two-step anaerobic-aerobic process (Fig. 11). In contrast to the hydrogenolytic mechanisms for nitrite elimination from nitroaromatic compounds by aerobic bacteria, gratuitous reduction of the nitro group does not require the establishment of special non-indigenous microorganisms. Substrates such as glucose or components of a synthetic sewage like alcohols or acetone are sufficient for activating the anaerobic biomass and supplying the reducing equivalents for the reduction of the nitro group of nitrobenzene. After adaptation of the microbial population, nitrobenzene is stoichiometrically reduced to aniline. In the subsequent aerobic process aniline is mineralized and utilized by the indigenous aniline-degrading bacteria of activated sludge, as shown by Dickel et al. (6).

Compared to the aerobic breakdown of nitrobenzene by a *Pseudomonas* strain (29), the two-step concept described above offers another major advantage for a bioremediation process. Reduction to aniline in the anaerobic system drastically reduces the vapor pressure of nitrobenzene and avoids the volatilization that occurs from the aeration tank of an aerobic sewage treatment system.

Like nitrobenzene,TNT is completely reduced by anaerobic sludge provided the redox potential is lowered by the addition of an electron-donating substrate ($E_h \leq -200$ mV) (10, 23). Under strongly reducing conditions all known partially reduced nitrotoluenes including the binuclear azoxy and other congeneric compounds such as mono- and dinitroaromatics are also completely reduced. In addition, abiotic reduction must be taken into account for the generation of partially reduced nitrotoluenes. Preuß et al. (30) observed the formation of 2,4-diamino-6-nitrotoluene in the presence of Fe^{2+} and HS^-. Therefore, the

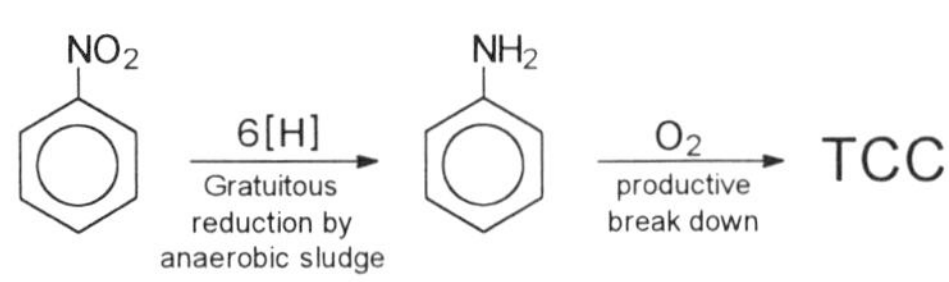

Figure 11. Mineralization of nitrobenzene by an anaerobic-aerobic process (6).

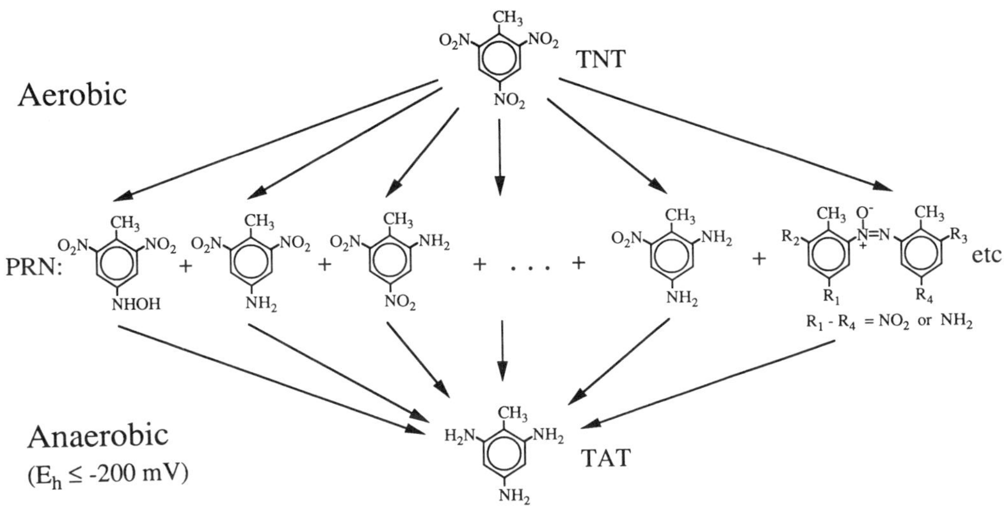

Figure 12. Under aerobic or limited oxygen conditions partially reduced nitrotoluenes (PRN) and secondary condensation products are generated. These are completely reduced to TAT under strict anaerobic conditions.

reduction products of TNT and condensation compounds also occur as pollutants of soil and groundwater at sites of former ammunition factories (13).

As illustrated in Figure 12, TNT and its reduced congeners converge during anaerobic treatment to TAT, which is chemically unstable. It is further degraded by an unknown mechanism (Preuß and Rieger, this volume). Compared to the high rates of reduction of TNT, the rates of successive disappearance of aminodinitrotoluene (ADNT), 2,4-diamino-6-nitrotoluene (DANT), and TAT decrease with increasing numbers of nitro groups reduced (Fig. 13). Since the degradation of TAT is slow and unproductive according to our experience, the elimination of TNT and its congeners via anaerobic pathways (Fig. 8, vertical reaction sequence) to yield harmless products, preferably CO_2, cannot yet be used for bioremediation

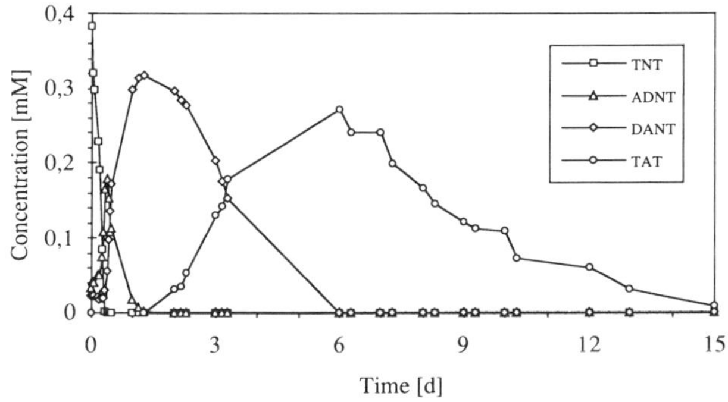

Figure 13. Anaerobic reduction of 2,4,6-trinitrotoluene (TNT) by a mixed bacterial culture (A_{546}=1-2) originating from anaerobic sludge. The process was operated in a fed batch reactor (10 liters) supplied with fresh mineral medium. Glucose (20 mM) was added daily. Initial concentration of TNT (d=0) was 0.4 mM. During the entire process the pH was maintained at 7.0-7.5 and the redox potential at $E_h^0 \leq$ -200 mV. Concentrations of amino-dinitrotoluenes (ADNT), 2,4-diamino-6-nitrotoluene (DANT), and triaminotoluene (TAT) were determined by HPLC (B. Wagener, diploma thesis, University of Stuttgart-Hohenheim, 1994).

of polluted sites. In contrast, Funk et al. (10) suggest further degradation of TAT involving *p*-cresol and organic acids.

Like aniline, TAT is an electron-rich compound and should be readily oxidized by aerobic microorganisms. In the presence of O_2, particularly when heavy metal ions (e.g. Mn^{2+}) are present, TAT undergoes rapid autooxidation to generate dark polymers. These abiotic reactions prevent productive metabolism of TAT by aerobic microorganisms. A promising yet unexploited possibility is the degradation of TAT under denitrifying conditions (30), because TAT is chemically not oxidized by nitrate (O. Dickel, personal communication).

DEGRADATION OF TNT AND ITS CONGENERS IN SOIL

Complete reduction of TNT and its partially reduced derivatives can also be accomplished when these compounds are present in soil under strict anaerobic conditions (Fig. 14). Notably, after 4 days TNT and partially reduced nitrotoluenes have completely disappeared from the aqueous phase, whereas considerable amounts of these contaminants (approx. 180 mg TNT and 40 mg ADNT per kg) are still present in soil. Obviously, at these lower levels of contamination the process of desorption becomes slower than the reduction reaction. The reduction of these compounds in heaps or compost piles, where aerobic or semiaerobic conditions prevail, remains incomplete. Products of partial reduction such as mononitro-diamino- and dinitro-monoamino-toluenes and azoxy compounds are extractable from soil. In contrast to the observations made in the absence of soil (Fig. 13), TAT as a product of complete reduction of TNT is irreversibly bound to the soil matrix and thus remains undetectable in the aqueous phase (23). Interestingly, this irreversible binding does not require the participation of O_2. Therefore, primary adsorption of TAT corresponds to the well-known mechanism of intercalation of diamines into clay minerals such as montmorillonite (18, 48).

The strongly hydrated polyanion layers (Fig. 15) exhibit a large binding and exchange capacity for cations such as organic ammonium compounds. Montmorillonite loaded with TAT turns dark on exposure to oxygen, which indicates that the primary process of interlamellar

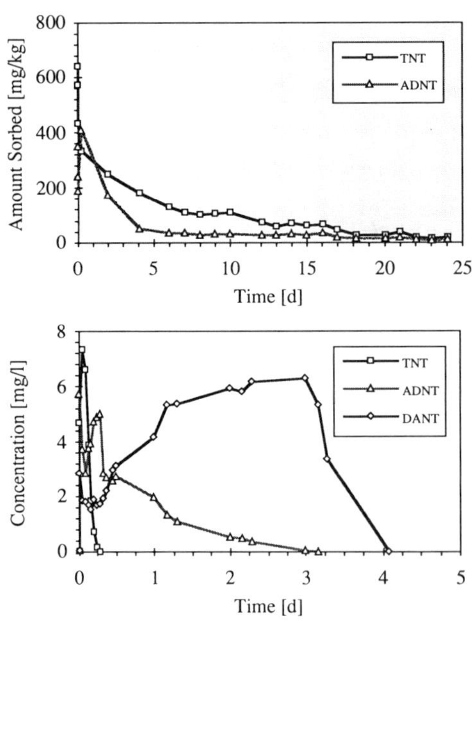

Figure 14. Anaerobic reduction of TNT in a soil slurry contaminated with TNT, aminodinitro- and diaminonitrotoluenes (ADNT and DANT). The upper graph shows the disappearance of TNT (initial contamination 640 mg/kg soil) and ADNT (approx. 200 mg/kg) from solid material. The successive reduction of the contaminants in the aqueous phase (lower graph) was followed by HPLC analysis. Composition of the aqueous medium, supply of glucose, pH, and E_h^0 were the same as described in Figure 13 (B. Wagener, diploma thesis, University of Stuttgart-Hohenheim, 1994).

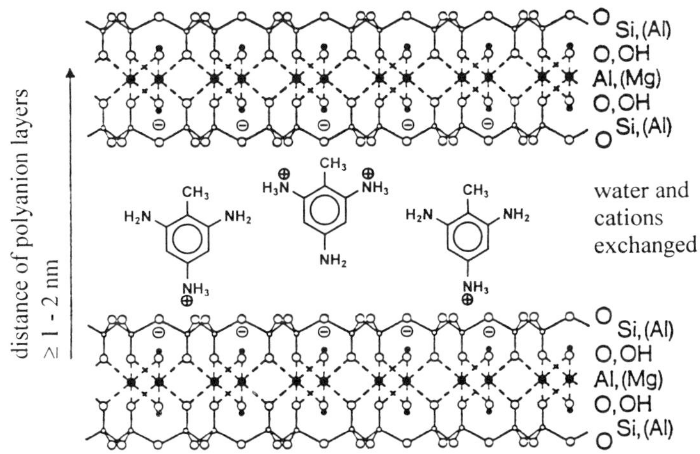

Figure 15. Intercalation of triaminotoluene into the clay mineral montmorillonite in analogy to Weiss (48). Structure of montmorillonite from Hofmann (18).

adsorption must be followed by oxidative polymerization (B. Wagener, diploma thesis, University of Stuttgart-Hohenheim, 1994). This is clearly an oxygen-initiated reaction and is likely to follow the mechanism of aniline black formation (28). In contrast to TAT, aniline contains only one amino group and therefore requires a strong oxidizing agent for polymerization. These agents generate reactive intermediates like aminoaryl radical cations through an initial one electron transfer step as the rate-limiting step (28). Little is known about the subsequent steps and the complex mechanism of polymer formation except that it must be fast and involves radical as well as quinoid intermediates. TAT harbors three amino groups so that O_2 is sufficient as an oxidant and an even more complex mechanism of polymerization must proceed (Fig. 16). At least in the final stage this must include the formation of extremely stable phenazine structures and cross links between the linear polymers (20).

Soil with a high content of humic substances harbors an additional binding capacity for basic molecules like TAT. Again, primary cationic sorption is favored by the polyanion character of the natural polymer (Fig. 17). Phenolic structural elements derived from ligninolytic processes are good reactants for covalent binding of TAT through radical induced carbon-carbon and carbon-oxygen bond formation. Quinoid structures are well known for their ability to react with amino compounds through 1,4-additions or Schiff base (quinone arylimine) formation (1, 9, 36).

Therefore, in natural soil diverse mechanisms of binding of TAT must exist. These range from ionic interaction of the TAT monomer with the polyanion structures of clay minerals and humic substances up to very complex subsequent O_2-dependent reactions that give rise to the formation of polymers. This latter process in particular is highly irreversible and resembles the formation of humic substances from monocyclic aromatic precursors (24).

The process of interlamellar adsorption to clay and final "humification" can also be applied to diaminotoluenes. These are formed by reduction of the respective dinitrotoluenes that are cocontaminants of TNT. The diaminotoluenes are eliminated somewhat more slowly than TAT but in the long run they completely and irreversibly disappear from the aqueous phase (E. Herkner, unpublished).

Mononitrotoluenes are also by-products of TNT production and hence accompany TNT as contaminants in soil. They are also completely converted into the corresponding toluidines (E. Herkner, unpublished). Binding to the soil matrix, however, is slow and

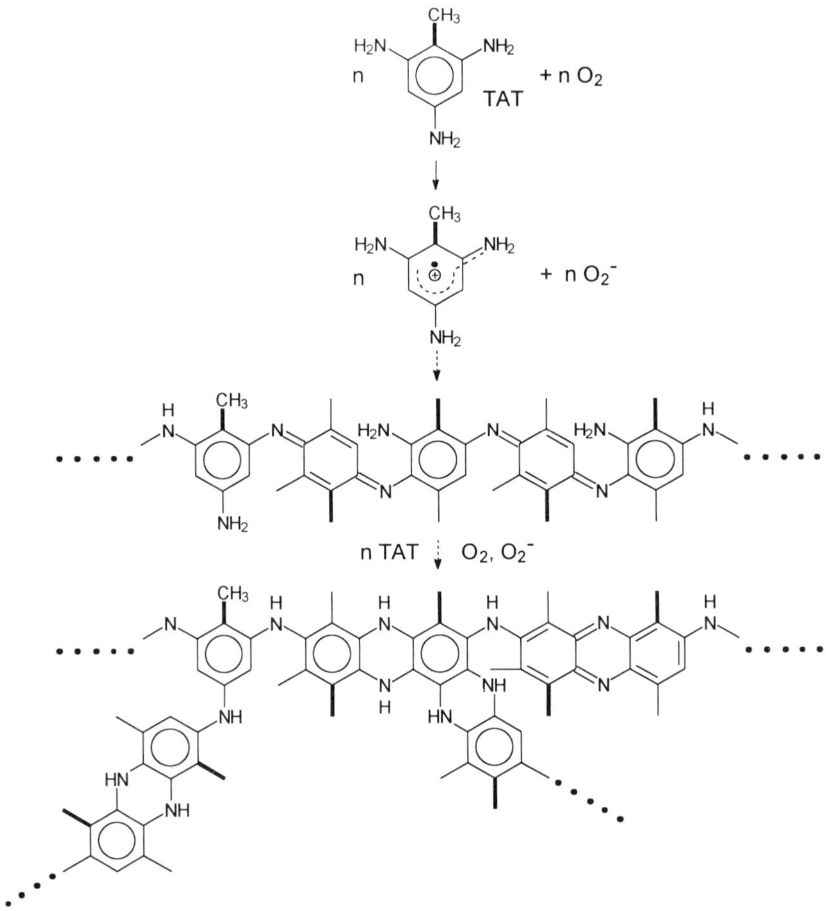

Figure 16. Potential mechanism of oxidative polymer formation from triaminotoluene.

incomplete and the ability to generate reactive species for oxygen-initiated polymerization is not pronounced. In contrast, monoaminoaromatic compounds, including toluidines, are readily mineralized by indigenous aerobic microorganisms (E. Herkner, unpublished).

PRACTICAL CONSIDERATIONS

At present, the only proven technology for elimination of TNT and its congeners from soil is incineration. Several recent observations have suggested a variety of alternative biological techniques for treating TNT contaminated soil. This includes composting, the use of white-rot fungi as agents for cooxidation, and several bio-slurry technologies (10). Although mass balances and biochemical mechanisms of conversion of TNT are largely missing, one of these processes has already been brought to a developmental stage that may offer a realistic, cost-effective alternative for clean up of TNT-contaminated sites (Crawford, this volume).

During composting of contaminated soil the ultimate fate of nitroaromatic compounds has not been clarified, and from our observations (see above) it is rather unlikely that under aerobic or limiting oxygen conditions transformation of TNT passes the amino-

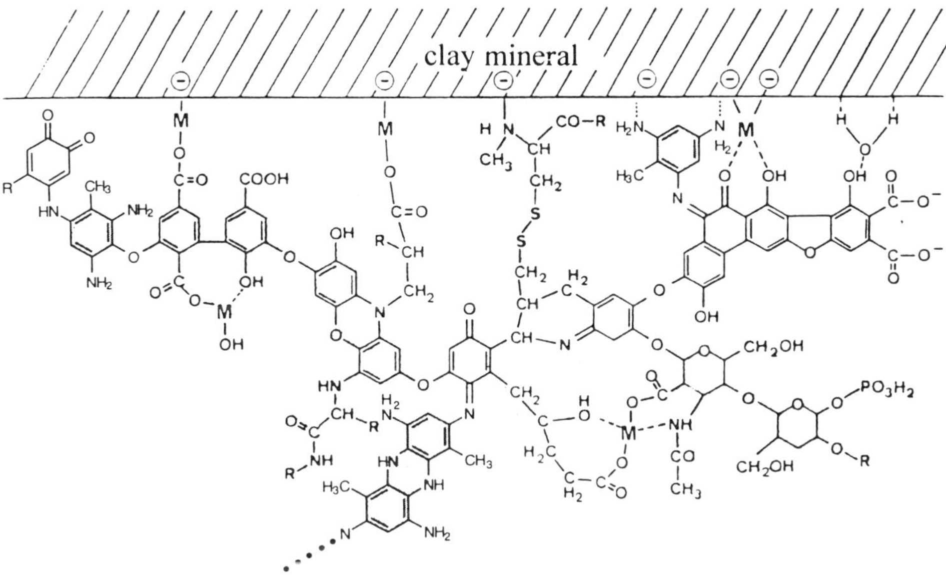

Figure 17. Potential binding of TAT and its oxidation products to humic substances (modified structure from Scheffer and Schachtschabel (35)).

dinitro- or diaminonitro-intermediate stage. Although *Phanaerochaete chrysosporium* has been shown to mineralize TNT to some extent (Michels and Gottschalk, and Stahl and Aust, this volume), its use at heavily contaminated sites has been questioned because this organism will not grow at concentrations $\geq$ 66 μM (41) or 88 μM (27).

The anaerobic-aerobic biotreatment process described previously should be a suitable method for bioremediation of TNT-contaminated soil because the highly electrophilic TNT and its partially reduced nitrotoluenes are readily converted to TAT under strictly anaerobic conditions. TAT is highly reduced and reactive for further degradation. On the one hand, under anaerobic conditions TAT is further degraded to compounds which are potential substrates for aerobic mineralization (23). On the other hand, TAT rapidly reacts with soil. Exposure to O_2 in particular gives rise to polymer formation and covalent binding to soil components, as outlined above. Although not firmly established, novel mechanisms of reductive TNT degradation have been proposed. After reduction of the nitro groups the proposed mechanisms involve hydrolytic displacements of amino groups to yield polyphenols. Reductive dehydroxylations were discussed as a mechanism for generating simple phenols such as 4-methylphenol (Crawford, this volume). Similarly, toluene has been postulated to be a reductive deamination product of TAT (2).

The process of direct anaerobic-aerobic treatment resulting in humification of TNT and its congeners offers a convincing and simple concept (Fig. 18 process II) for the remediation of soil contaminated with TNT and congeneric compounds. This is evident by comparison with the alternative process I (Fig. 18), which requires the desorption of the contaminants by innocuous solvents such as methanol or ethanol. Desorption plus the separation of particles must precede the actual anaerobic process of polynitrotoluene reduction. Furthermore, the anaerobic stage requires a microbial system which degrades TAT and diaminotoluenes productively or at least to such an extent that subsequent aerobic treatment results in complete mineralization.

Remediation of TNT-contaminated soil by humification of di- and trinitrotoluenes through an anaerobic-aerobic treatment process will require careful analysis of the treated

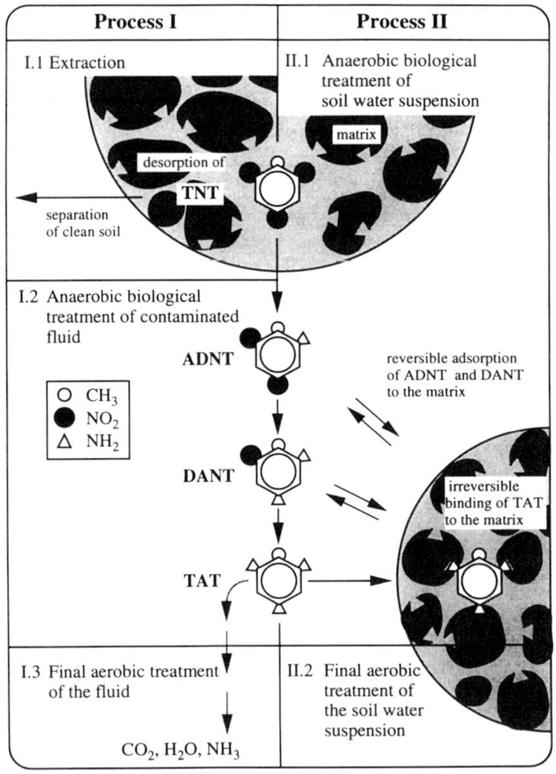

Figure 18. Alternative processes for biological treatment of TNT-contaminated soil.

soil. First of all, irreversible binding to the soil matrix must be validated by use of ^{14}C-labeled compounds. In addition, the products of oxidative polymerization have to be assessed with respect to their long-term biological stability and ecotoxicological quality (43). Finally, it will be of interest to elucidate the structure of the TAT-derived polymer in connection to and comparison with naturally occurring humic substances. This is the subject of current investigations in our laboratory.

ACKNOWLEDGEMENT

Part of the work has been supported by a grant of Hessische Industriemüll GmbH, Bereich Altlastensanierung, the Bundesministerium für Forschung und Technologie, FuE-Vorhaben 325-4003-031055 I A and the Air Force Office of Scientific Research, Air Force Systems Command USAF, under grant AFOSR-91-0237.

We thank G. Daun and W. Haug for their assistence in preparing the artwork.

REFERENCES

1. Beyer, H., and W. Walter. 1991. Lehrbuch der Organischen Chemie, 22th ed. S. Hirzel, Stuttgart.
2. Boopathy, R., C. F. Kulpa, and M. Wilson. 1993. Metabolism of 2,4,6-trinitrotoluene (TNT) by *Desulfovibrio* sp. (B strain). Appl. Microbiol. Biotechnol. 39:270-275.

3. Bruhn, C., H. Lenke, and H.-J. Knackmuss. 1987. Nitrosubstituted aromatic compounds as nitrogen source for bacteria. Appl. Environ. Microbiol. 53:208-210.
4. Corbett M. D., and B. R. Corbett. 1981. Metabolism of 4-chloronitrobenzene by the yeast *Rhodosporidium* sp. Appl. Environ. Microbiol. 41:942-949.
5. Dickel, O., and H.-J. Knackmuss. 1991. Catabolism of 1,3-dinitrobenzene by *Rhodococcus* sp. QT-1. Arch. Microbiol. 157:76-79.
6. Dickel, O., W. Haug, and H.-J. Knackmuss. 1993. Biodegradation of nitrobenzene by a sequential anaerobic-aerobic process. Biodegradation 4:187-194.
7. Duque, E., A. Haïdour, F. Godoy, and J. L. Ramos. 1993. Construction of a *Pseudomonas* hybrid strain that mineralizes 2,4,6-trinitrotoluene. J. Bacteriol. 175:2278-2283.
8. Ecker, S., T. Widmann, H. Lenke, O. Dickel, P. Fischer, C. Bruhn, and H.-J. Knackmuss. 1992. Catabolism of 2,6-dinitrophenol by *Alcaligenes eutrophus* JMP134 and JMP222. Arch. Microbiol. 158:149-154.
9. Finley, K. T. 1974. The addition and substitution chemistry of quinones. *In* S. Patai (ed.) The chemistry of the quinonoid compounds. Pt. 2. J. Wiley & Sons, New York.
10. Funk, S. B., D. J. Roberts, D. L. Crawford, and R. L. Crawford. 1993. Initial phase optimization for bioremediation of munition compound-contaminated soils. Appl. Environ. Microbiol. 59:2171-2177.
11. Groenewegen, P. E. J., P. Breeuwen, J. M. L. M. v. Helvoort, A. A. M. Langenhoff, F. P. de Vries, and J. A. M. de Bont. 1992. Novel degradative pathway of 4-nitrobenzoate *Comamonas acidovorans* NBA-10. J. Gen. Microbiol. 13:1599-1605.
12. Groenewegen, P. E. J., and J. A. M. de Bont. 1992. Degradation of 4-nitrobenzoate via 4-hydroxylaminobenzoate and 3,4-dihydroxybenzoate in *Comamonas acidovorans* NBA-10. Arch. Microbiol. 158:381-386.
13. Haas, R., and E. v. Löw. 1986. Grundwasserbelastung durch eine Altlast. Die Folgen einer ehemaligen Sprengstoffproduktion für die heutige Trinkwassergewinnung. Forum Städte-Hygiene 37:33-43.
14. Haigler, B. E., and J. C. Spain. 1991. Biotransformation of nitrobenzene by bacteria containing toluene degradative pathways. Appl. Environ. Microbiol. 57:3156-3162.
15. Haigler, B. E., and J. C. Spain. 1993. Biodegradation of 4-nitrotoluene by *Pseudomonas* sp. strain 4NT. Appl. Environ. Microbiol. 59:2239-2243.
16. Hartter, D. R. 1985. The use and importance of nitroaromatic chemicals in the chemical industry, p. 1-14. *In* D. E. Rickert (ed.) Toxicity of nitroaromatic compounds. Chemical Industry Institute of Toxicology Series. Hemisphere Publishing Corp., New York.
17. Heijman, C. G., C. Holliger, M. A. Glaus, R. P. Schwarzenbach, and J. Zeyer. 1993. Abiotic reduction of 4-chloronitrobenzene to 4-chloroaniline in a dissimilatory iron-reducing enrichment culture. Appl. Environ. Microbiol. 59:4350-4353.
18. Hofmann, U. 1968. Aus der Chemie des Tons. Angew. Chem. I. E. 7:681.
19. Keys, L. D., and G. A. Hamilton. 1987. The mechanism for the conversion of α-amino-β-carboxymuconate-ε-semialdehyde to quinolinate, an apparent nonenzymic step in the biosynthesis of the nicotinamide coenzymes from tryptophan. J. Am. Chem. Soc. 109:2156-2163.
20. Lantz, R., and J. Gascon. 1965. Condensation oxydante de "l'éméraldine de Willstätter" avec l'aniline. Remarques concernant les réactions analogues génératrices de noirs d'aniline et la constitution de ces colorants. Bull. Soc. Chim. Fr. 816-821.
21. Lenke, H., and H.-J. Knackmuss. 1992. Initial hydrogenation during catabolism of picric acid by *Rhodococcus erythropolis* HL24-2. Appl. Environ. Microbiol. 58:2933-2937.
22. Lenke, H., D. H. Pieper, C. Bruhn, and H.-J. Knackmuss. 1992. Degradation of 2,4-dinitrophenol by two *Rhodococcus erythropolis* strain HL24-1 and HL24-2. Appl. Environ. Microbiol. 58:2928-2932.
23. Lenke, H., B. Wagener, G. Daun, and H.-J. Knackmuss. 1994. TNT-contaminated soil: a sequential anaerobic/aerobic process for bioremediation, abstr. Q-383, p. 456. Abstr. 94th Annu. Meet. Am. Soc. Microbiol. 1994.
24. Lewis, W. G., and E. Jamamoto. 1990. Lignin: occurrence, biogenesis and biodegradation. Ann. Rev. Plant Physiol. Plant Mol. Biol. 41:455-496.
25. March, J. 1985. Advanced Organic Chemistry, 3rd ed. J. Wiley & Sons, New York.
26. McCormick, N. G., F. E. Feeherry, and H. S. Levinson. 1976. Microbial transformation of 2,4,6-trinitrotoluene and other nitroaromatic compounds. Appl. Environ. Microbiol. 31:949-958.
27. Michels, J., and G. Gottschalk. 1994. Inhibition of the lignin peroxidase of *Phanerochaete chrysosporium* by hydroxylamino-dinitrotoluene, an early intermediate in the degradation of 2,4,6-trinitrotoluene. Appl. Environ. Microbiol. 60:187-194.
28. Mohilner, D. M., R. N. Adams, and W. J. Argasinger. 1962. Investigation of the kinetics and mechanism of the anodic oxidation of aniline in aqueous sulfuric acid solution at a platinum electrode. J. Am. Chem. Soc. 84:3618-3622.

29. Nishino, S. F., and J. C. Spain. 1993. Degradation of nitrobenzene by a *Pseudomonas pseudoalcaligenes*. Appl. Environ. Microbiol. 59:2520-2525.
30. Preuß, A., J. Fimpel, and G. Diekert. 1993. Anaerobic transformation of 2,4,6-trinitrotoluene (TNT). Arch. Microbiol. 159:345-353.
31. Raistrick, H. 1949. Chloromycetin: its structure and synthesis. Nature 163:5553-554.
32. Reineke, W., and H.-J. Knackmuss. 1978. Chemical structure and biodegradability of halogenated aromatic compounds. Substituent effects on 1,2-dioxygenation of benzoic acid. Biochim. Biophys. Acta 542:412-423.
33. Rhys-Williams, W., S. C. Taylor, and P. A. Williams. 1993. A novel pathway for the catabolism of 4-nitrotoluene by *Pseudomonas*. J. Gen. Microbiol. 139:1967-1972.
34. Rieger, P.-G., A. Preuss, V. Sinnwell, W. Francke, H. Lenke, and H.-J. Knackmuss. 1994. H^--additions as initial steps of aerobic bacterial degradation of 2,4,6-trinitrophenol (picric acid), abstr. Q-120, p. 409. Abstr. 94th Annu. Meet. Am. Soc. Microbiol. 1994.
35. Scheffer, F., and P. Schachtschabel. 1989. Lehrbuch der Bodenkunde, p. 57, 12th ed. F. Enke, Stuttgart.
36. Sollenberger, P. Y., and R. B. Martin. 1968. Carbon-nitrogen and nitrogen-nitrogen double bond condensation reactions. *In* S. Patai (ed.) The chemistry of the amino group. J. Wiley & Sons, New York.
37. Spain, J. C., O. Wyss, and D. T. Gibson. 1979. Enzymatic oxidation of *p*-nitrophenol. Biochem. Biophys. Res. Commun. 88:634-641.
38. Spain, J. C., and D. T. Gibson. 1991. Pathway for biodegradation of *p*-nitrophenol in a *Moraxella* sp. Appl. Environ. Microbiol. 57:812-819.
39. Spanggord, R. J., K. E. Mortelmans, A. F. Griffing, and V. F. Simmon. 1982. Mutagenicity in *Salmonella typhimurium* and structure activity relationships of waste water components emanating from the manufacture of trinitrotoluene. Environ. Mutagen 4:163-179.
40. Spanggord, R. J., J. C. Spain, S. F. Nishino, and K. E. Mortelmans. 1991. Biodegradation of 2,4-dinitrotoluene by a *Pseudomonas* sp. Appl. Environ. Microbiol. 57:3200-3205.
41. Spiker, J. K., D. L. Crawford, and R. L. Crawford. 1992. Influence of 2,4,6-trinitrotoluene (TNT) concentrations on the degradation of TNT in explosive-contaminted soils by the white rot fungus *Phanaerochaete chrysosporium* . Appl. Environ. Microbiol. 58:3199-3202.
42. Stolz, A., B. Nörtemann, and H.-J. Knackmuss. 1992. Bacterial metabolism of 5-aminosalicylic acid. Initial ring cleavage. Biochem. J. 282:675-680.
43. Tan, E. L., C. H. Ho, W. H. Griest, and R. L. Tyndall. 1992. Mutagenicity of trinitrotoluene and its metabolites formed during composting. J. Toxicol. Environ. Health 36:165-175.
44. Van Pee, K. H., O. Salcher, and F. Lingens. 1981. Synthese von 7-Chlor-L- und 7-Chlor-D-tryptophan; Biosynthese von Pyrrolnitrin. Liebigs Ann. Chem. 2:233-239.
45. Vanulet J., and R. L. Van Etten. 1981. Biochemistry and pharmacology of the nitro and nitroso groups, p. 201-289. *In* H. Feuer (ed.) The chemistry of the nitro and nitroso groups. Krieger Publishing Company, New York.
46. Vorbeck, C., H. Lenke, P. Fischer, and H.-J. Knackmuss. 1994. Identification of ahydride Meisenheimer complex as a metabolite of 2,4,6-trinitrotoluene by a *Mycobacterium* sp. J. Bacteriol. 176:932-934.
47. Wagniere, G. H. 1981. Theoretical aspects of C-NO and C-NO_2 bonds. *In* H. Feuer (ed.) The chemistry of the nitro and nitroso groups. Krieger Publishing Company, New York.
48. Weiss, A. 1963. Organische Derivate der glimmerartigen Schichtsilikate. Angew. Chem. 75:113-148.
49. Williams, M. C., and R. C. Barneby. 1977. The occurence of nitro-toxins in North American *A. stragalus* (Fabaceae). Brittonia 29:310-326.
50. Won, W. D., L. H. Di Salvo, and J. Ng. 1976. Toxicity and mutagenicity of 2,4,6-trinitrotoluene and its microbial metabolites. Appl. Environ. Microbiol. 31:576-580.
51. Zeyer, J., and P. C. Kearney. 1984. Degradation of *o*-nitrophenol and *m*-nitrophenol by a *Pseudomonas putida*. J. Agric. Food Chem. 32:238-242.

2

BACTERIAL DEGRADATION OF NITROAROMATIC COMPOUNDS UNDER AEROBIC CONDITIONS

Jim C. Spain

US Air Force Armstrong Laboratory
139 Barnes Dr., Ste. 2
Tyndall AFB, Florida 32403

INTRODUCTION

During the past 20 years there have been numerous reports of biotransformation of nitroaromatic compounds. However, surprisingly few instances of complete degradation or utilization as growth substrates by microorganisms have been reported. This seems to be primarily due to the ability of many microorganisms to catalyze the nonspecific reduction of the nitro group and formation of amino derivatives or condensation products that are resistant to further microbial attack (26). Recently, several strains of aerobic bacteria have been isolated by selection for the ability to use nitroaromatic compounds as a source of carbon or nitrogen. This chapter will focus on the metabolic strategies used by these strains for degradation of nitroaromatic compounds.

Bacteria able to grow at the expense of nitroaromatic compounds have been isolated by selective enrichment from soil or wastewater contaminated with the nitro compounds of interest. Preliminary observations indicate that such strains are not widely distributed in ecosystems that have not been exposed to the nitro compounds. No work on the microbial ecology or evolutionary history of these strains has been published, so it is not clear whether they arise de novo after contamination of an ecosystem or whether they are present at very low numbers in uncontaminated soil. Ongoing studies on the molecular biology of a few isolates may shed some light on their origins and relatedness.

Bacteria studied to date use one of four approaches for aerobic metabolism of the aromatic nitro group as a prelude to growth at the expense of the parent molecule. The most widely studied approach involves the initial oxidative removal of the nitro group as nitrite in a reaction catalyzed by a monooxygenase enzyme. This type of reaction has been known for many years (46), yet details of the mechanism are still unknown. Recently, dioxygenase-catalyzed reactions have also been reported to eliminate nitro groups from the aromatic ring with the formation of the corresponding catechols and nitrite (48). Very recent reports (13, 14, 31) have revealed a third mechanism used by bacteria to initiate complete degradation of nitroaromatic compounds. The pathway involves a partial reduction of the nitro group to the hydroxylamino derivative and subsequent reactions that release the nitrogen as ammonia. A fourth mechanism of initial attack has been described for the degradation of trinitroaromatic compounds and results in the replacement of a nitro group by hydrogen (24, 53, Rieger and Knackmuss, this volume). These unusual reactions, which will be discussed in detail below, avoid the production of the amino analogs which can potentially be more toxic than the parent molecules.

OXIDATIVE REMOVAL OF THE NITRO GROUP

Monooxygenase-Catalyzed Conversion of Nitrophenols

The biodegradation of nitrophenols has been studied extensively because of the wide use of pesticides that contain nitrophenyl groups and because the nitrophenols are used as preservatives for leather. Simpson and Evans (42) isolated a *Pseudomonas* from soil that was able to grow on 4-nitrophenol with concomitant release of nitrite. Based on evidence from simultaneous adaptation studies they proposed that 4-nitrophenol was converted to hydroquinone. Subsequently, Munnecke and Hsieh (28) detected traces of hydroquinone in the medium when cultures of a soil pseudomonad were grown on 4-nitrophenol. They proposed that hydroquinone was hydroxylated to 1,2,4-benzenetriol as the next step in the metabolic pathway but provided no experimental evidence. Nitrite release was detected during the degradation of a wide range of nitrophenols and nitrobenzoates (20) over the next 20 years, but the mechanism of the reaction remained obscure because the enzyme responsible for the initial reaction could not be studied in cell extracts.

Twenty five years after the initial observation of nitrite release from 4-nitrophenol the enzyme responsible for the reaction was partially purified from a *Moraxella* isolated from activated sludge (46). The nitrophenol oxygenase requires oxygen and reduced pyridine nucleotides for activity and is stimulated by the addition of flavin adenine dinucleotide. Experiments with $^{18}O_2$ revealed that the reaction is a monooxygenation. The reaction requires the oxidation of two moles of NADPH for each mole of 4-nitrophenol converted to hydroquinone. The stoichiometry of the reaction suggests that the initial reaction produces 1,4-benzoquinone which is subsequently reduced to hydroquinone by an inducible quinone reductase (Fig. 1). However, rigorous proof that 1,4-benzoquinone is the product of the initial reaction was not possible due to the high specific activity of the quinone reductase which could not be separated from the monooxygenase enzyme. The hydroquinone thus formed was later shown to undergo a dioxygenase-catalyzed ring-fission reaction which produces γ-hydroxymuconic semialdehyde (45). The ring-fission product is subsequently converted to ß-ketoadipic acid via maleylacetic acid by a reaction sequence requiring only catalytic amounts of NAD^+.

Enzymes for the metabolism of 1,2,4-benzenetriol are also induced in cells of the *Moraxella* grown on 4-nitrophenol, but there is no evidence that hydroquinone can be converted to 1,2,4-benzenetriol. However, the induction of enzymes for catabolism of 1,2,4-benzenetriol might be evidence for the existence of an alternate pathway in which 4-nitrophenol is converted to 1,2,4-benzenetriol via 4-nitrocatechol. Indeed, Raymond and Alexander (36) suggested the participation of 4-nitrocatechol in the pathway after they detected very slow accumulation of 4-nitrocatechol from 4-nitrophenol in toluene-treated cells of a *Flavobacterium* grown on 4-nitrophenol. However, they did not provide evidence that 4-nitrocatechol was further degraded. 4-Nitrocatechol is slowly degraded with release of nitrite by 4-nitrophenol-grown cells of the *Moraxella*, but the rates are slow compared to the rate of degradation of 4-nitrophenol. Therefore, it is possible that some of the 4-nitrophe-

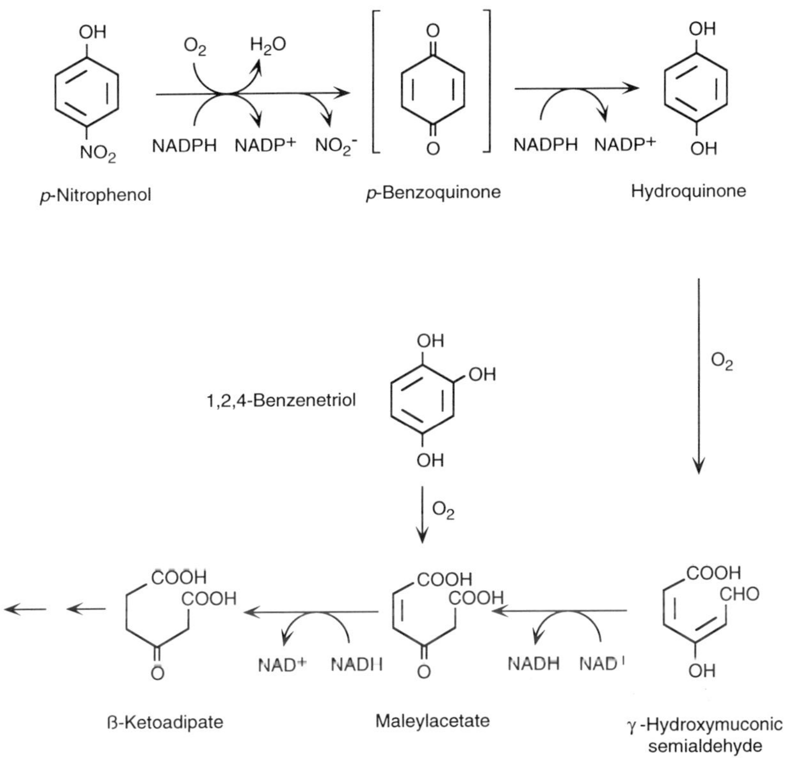

Figure 1. Pathway for degradation of PNP and 1,2,4-benzenetriol (45).

nol degraded by the *Moraxella* is routed through 4-nitrocatechol, but the major route of catabolism is via hydroquinone.

Recently, additional evidence has been obtained for the catabolism of 4-nitrophenol via 4-nitrocatechol. A monooxygenase enzyme that catalyzes the conversion of 4-nitrophenol to 4-nitrocatechol was purified from cells of a *Nocardia* grown on 4-nitrophenol (27). Another strain of *Nocardia* produced 4-nitrocatechol from 4-nitrophenol when the cells were grown in the presence of phenols (18). These studies provided rigorous proof for the existence of enzymes capable of the conversion of 4-nitrophenol to 4-nitrocatechol. Very recently, a complete pathway for the catabolism of 4-nitrophenol via 4-nitrocatechol was characterized in an *Arthrobacter* grown on 4-nitrophenol (21). The first step appears to be a hydroxylation of 4-nitrophenol adjacent to the hydroxyl group to produce 4-nitrocatechol (Fig. 2). The evidence for this step is extensive, but circumstantial; the reaction has not been demonstrated unequivocally. Rigorous proof will require characterization of the enzyme in vitro. The second reaction is a monooxygenase-catalyzed removal of the nitro group which results in the production of 1,2,4-benzenetriol, the substrate for ring fission. The resultant maleylacetate is further metabolized via ß-ketoadipate as described above. In cells of the *Arthrobacter* grown on 4-nitrophenol hydroquinone is not metabolized. Thus, there seems to be only a single route for 4-nitrophenol metabolism in this strain.

Degradation of 2-nitrophenol is initiated by a monooxygenase-catalyzed removal of the nitro group in a *Pseudomonas putida* studied by Zeyer and his coworkers (55, 56). The authors purified the soluble nitrophenol oxygenase and suggested that the reaction produces 1,2-benzoquinone which is subsequently reduced to catechol. However, they were not able

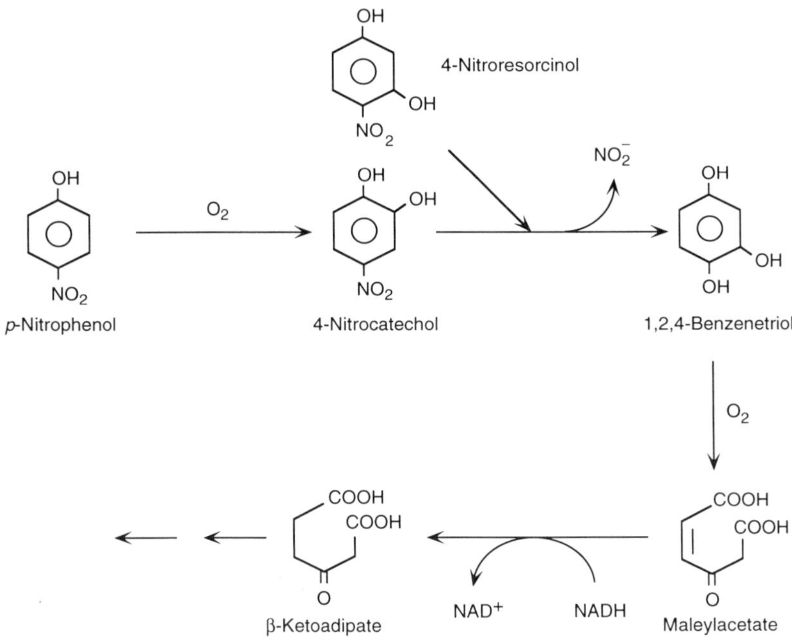

Figure 2. Biodegradation of 4-nitrophenol by an *Arthrobacter* (21).

to demonstrate rigorously the accumulation of 1,2-benzoquinone because it would have been nonenzymatically reduced to catechol by the NADPH required for the initial reaction. The same strain grew on 3-nitrophenol, but the pathway could not be determined (55). It seems to involve reduction of the nitro group because growth on 3-nitrophenol was accompanied by accumulation of ammonium ions.

Dioxygenase Catalyzed Removal of the Nitro Group

Dioxygenase enzymes are known to catalyze the insertion of two atoms of molecular oxygen into the ring of a wide variety of aromatic compounds (12). When the substrate is an aromatic hydrocarbon the product of the reaction is a dihydrodiol which can be oxidized by a dehydrogenase enzyme to produce the corresponding catechol (Fig. 3). When the substrate is a substituted aromatic compound the dioxygenase reaction can lead to the elimination of certain functional groups. Thus, sulfonyl groups and hydroxyl groups can be eliminated from substituted naphthalenes by the action of naphthalene dioxygenase (4). Toluene dioxygenase can catalyze the elimination of hydroxyl groups from substituted benzenes (47). Chloro substituents (11, 25) and amino substituents (2) can be eliminated by similar mechanisms. Dioxygenation of these substituted aromatic compounds forms a nonaromatic dihydroxy derivative of the parent molecule which undergoes a spontaneous anionic elimination of the functional group to produce a 1,2-dihydroxy aromatic compound (Fig. 3). The participation of a dehydrogenase, as required for the rearomatization of the ring

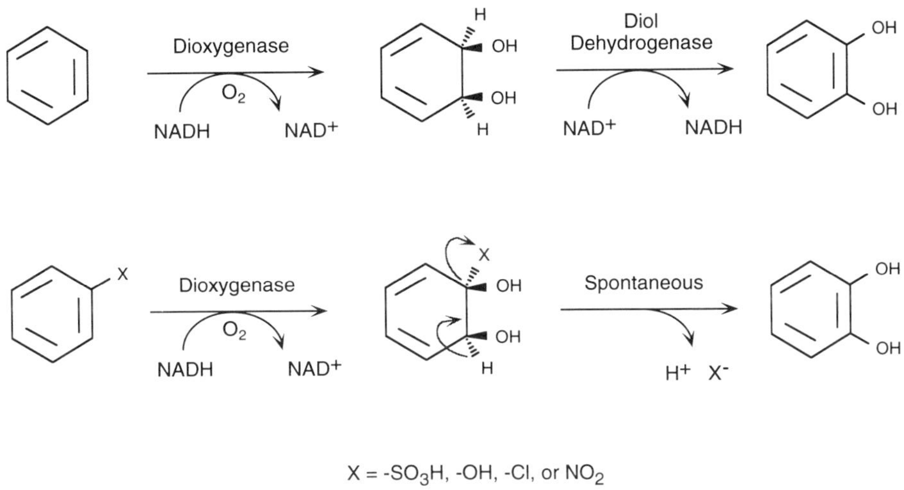

Figure 3. Dioxygenase catalyzed insertion of molecular oxygen and elimination of functional groups.

and release of carboxyl substituents (37), is not necessary for the elimination of anionic leaving groups.

Based on studies involving cooxidation of 2,6-dinitrophenol by *Alcaligenes eutrophus* JMP222 Ecker et al. (9, 10) suggested that dioxygenase enzymes could also catalyze the anionic elimination of the nitro group from an aromatic ring. Subsequently, Sander et al. (39) provided evidence that a dioxygenase enzyme in a strain of *Pseudomonas* grown on chlorobenzenes could eliminate the nitro group from 2,4,5-trichloronitrobenzene. The reactions in both of these systems were proposed to be analogous to the spontaneous eliminations described above.

The first evidence that bacterial dioxygenase enzymes could catalyze the replacement of the aromatic nitro group in a productive pathway was obtained from studies of a *Pseudomonas* isolated for its ability to grow on 2,4-dinitrotoluene (2,4-DNT) (48). *Pseudomonas* sp. strain DNT released stoichiometric amounts of nitrite during growth on 2,4-DNT. Oxygen was required for the initial reaction and 4-methyl-5-nitrocatechol accumulated transiently during induction of the enzymes for growth on 2,4-DNT. Experiments conducted in the presence of $^{18}O_2$ provided rigorous proof that the initial reaction involves a dioxygenase mechanism.

A 4-methyl-5-nitrocatechol monooxygenase, partially purified from 2,4-DNT-grown cells catalyzes the oxidation of 4-methyl-5-nitrocatechol to a quinone and releases nitrite (15). The reaction requires one mole of oxygen and one mole of NADPH for the conversion of 4-methyl-5-nitrocatechol to 2-hydroxy-5-methylquinone. These observations provided the first strong evidence of the participation of a quinone intermediate in the monooxygenase-catalyzed replacement of an aromatic nitro group by a hydroxyl group. The quinone was further reduced to 2,4,5-trihydroxytoluene by an inducible quinone reductase (Fig. 4). Thus, in this system and in the nitrophenol monooxygenases described previously (46, 56) the replacement of a nitro group by a hydroxyl group requires two moles of NADPH and one mole of molecular oxygen. The mechanism of the reactions catalyzed by these flavoprotein monooxygenases has not been studied in detail.

2,4,5-Trihydroxytoluene (THT) produced by the above sequence of reactions serves as the ring fission substrate in *Pseudomonas* sp. strain DNT. It is clear that the opening of the ring is catalyzed by a dioxygenase, but the position of the attack has not been rigorously

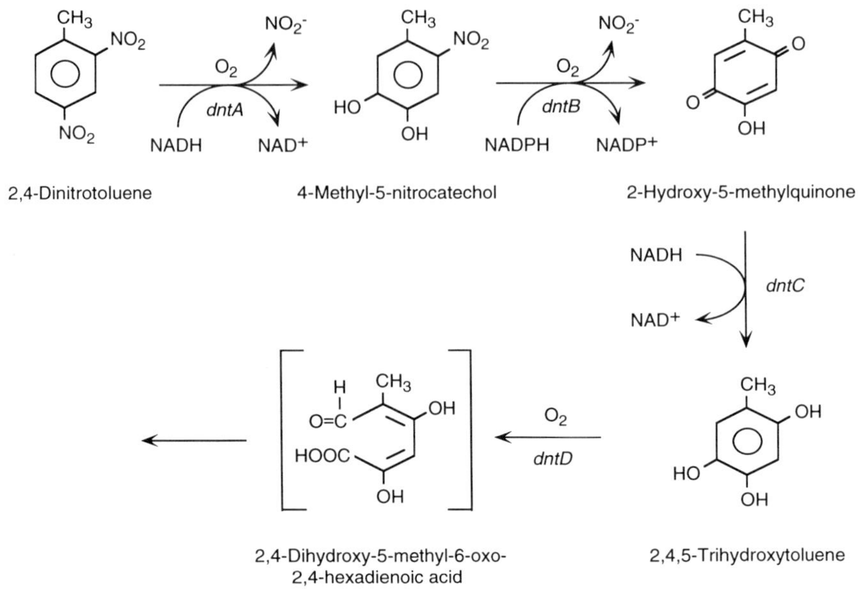

Figure 4. Degradation of 2,4-dinitrotoluene by *Pseudomonas* sp. strain DNT (52).

established (15). Based on analogy with other systems three alternative mechanisms can be postulated for the opening of the ring of THT (Fig. 5). An *ortho* mechanism would produce 3-hydroxy-4-methylmuconic acid whereas *meta* fission would yield either 2,4-dihydroxy-5-methyl-6-oxo-2,4-hexadienoic acid or 2,5-dihydroxy-4-methyl-6-oxo-2,4-hexadienoic acid. Purified THT dioxygenase catalyzes the oxygen-dependent conversion of THT to a transient intermediate with a maximum A_{272}. All attempts to isolate the intermediate have resulted in the isolation of a compound with a molecular weight of 154 which could be explained by a dehydration of any of the three potential ring fission products shown in Figure 5. The isolated compound is not further metabolized by cells or cell extracts which indicates

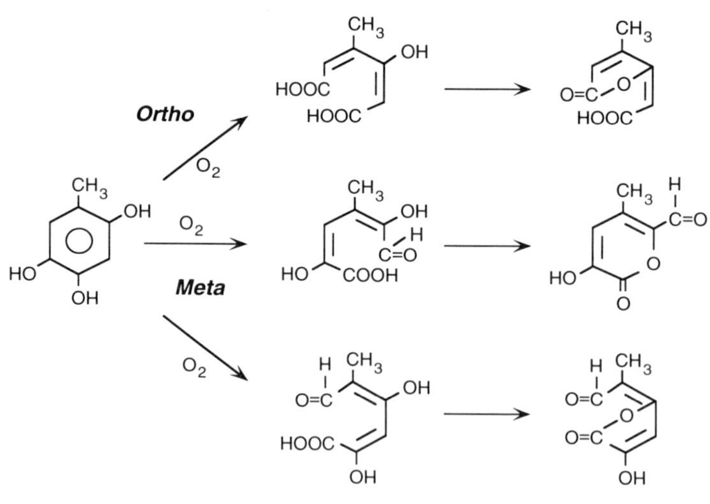

Figure 5. Potential pathways for ring cleavage of 2,4,5-trihydroxytoluene (15).

that it is an artifact of the isolation procedure. The purified enzyme catalyzes what appears to be a *meta* ring fission of 1,2,4-trihydroxybenzene and clearly catalyzes a *meta* ring fission of 4-methylcatechol. The available evidence strongly suggests that THT dioxygenase catalyzes an attack at the 5,6 position of THT. However, the evidence is circumstantial and the other possible mechanisms cannot be eliminated.

The genes that encode the enzymes involved in the DNT degradative pathway are located on a 180 kb plasmid (pJS1) in *Pseudomonas* sp. strain DNT (52). The genes for the DNT dioxygenase enzyme system (*dntA*) have been cloned on a 6.8 kb fragment of DNA. The cloned sequence contains four open reading frames that code for a three component enzyme complex closely related to naphthalene dioxygenase (51). The three components include a flavo-iron-sulfur reductase, a Rieske iron-sulfur ferredoxin, and a Rieske iron-sulfur oxygenase. The cloned DNT dioxygenase catalyzes the conversion of naphthalene to the corresponding *cis*-dihydrodiol and exhibits a substrate range, monooxygenase activity and desaturase activity similar to those of naphthalene dioxygenase (54). The only obvious difference in the substrate range is the ability of DNT dioxygenase to attack 2,4-DNT.

The gene for 4-methyl-5-nitrocatechol monooxygenase (*dntB*) was expressed from a recombinant plasmid containing a 2.2 kb insert. The gene for the 2-hydroxy-5-methylquinone reductase was not isolated because significant constitutive levels of quinone reductase were present in both the wild type *Pseudomonas* sp. strain DNT and in the *E. coli* used as the host. Therefore, the detection of expression of *dntC* was not possible under the conditions used for selection. It is clear that 2-hydroxy-5-methylquinone reductase is induced during growth on DNT (15). However, the location of the gene and its regulation remain to be elucidated. The gene that encodes THT oxygenase (*dntD*) was expressed by a recombinant clone containing a 1.2 kb fragment of DNA from strain DNT. Sequence analysis should reveal whether the gene is more closely related to *ortho* or *meta* ring fission enzymes in other systems. Such information will provide additional evidence to answer questions raised above concerning the mechanism of oxidation of THT.

The genes that encode the enzymes of the DNT pathway are not contiguous on the wild type plasmid pJS1 and they seem to be regulated independently. The enzymes are present at low constitutive levels and seem to be induced sequentially when cells are transferred to media containing 2,4-DNT. The practical result of this mode of regulation is variable and unpredictable lag periods, slow growth, and transient accumulation of 4-methyl-5-nitrocatechol or 2-hydroxy-5-methylquinone. In contrast, a more recently isolated *Pseudomonas* sp. strain R34 grows more rapidly and predictably on 2,4-DNT (Jain and Spain, unpublished). It is also able to tolerate higher concentrations of the growth substrate. Strain R34 contains a 165 kb plasmid that carries the genes for 2,4-DNT degradation. The genes are clustered and are regulated in two operons that are strictly inducible. One operon contains the genes for DNT dioxygenase and 4-methyl-5-nitrocatechol monooxygenase whereas the other contains the gene for THT oxygenase and possibly the genes for the subsequent metabolism of the ring fission product. It seems that strain R34 is better adapted than strain DNT for growth on 2,4-DNT.

Strains able to grow on 2,4-DNT have been isolated from a variety of soils contaminated with 2,4-DNT (Nishino and Spain, unpublished results). To date, they all seem to use the same catabolic pathway as the two strains mentioned above. It will be interesting to compare the genes of such strains to assess whether they share a common origin or arose independently.

Several bacterial isolates with the ability to grow on 1,3-dinitrobenzene have been reported (8, 30). The evidence presented to date indicates that the initial steps in the catabolic pathway are analogous to those described above for the degradation of 2,4-DNT. The first reaction is a dioxygenase catalyzed insertion of molecular oxygen at the 3,4-position. The details of the degradation of the resultant 4-nitrocatechol have not been reported.

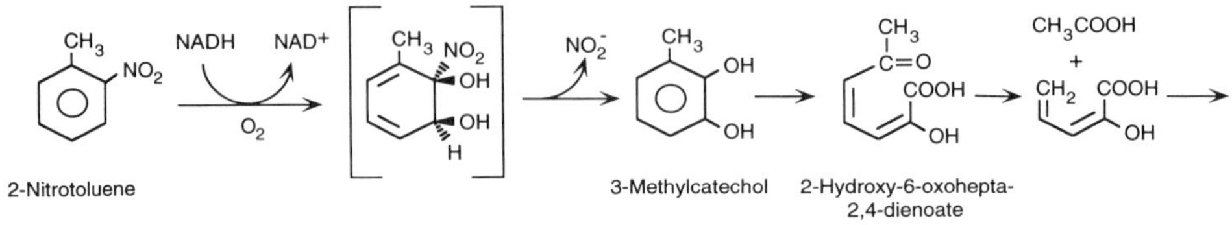

Figure 6. Degradation of 2-nitrotoluene by *Pseudomonas* sp. strain JS42 (17).

To date, no bacteria with the ability to grow on 2,6-DNT have been isolated. This is difficult to explain because most of the field sites contaminated with 2,4-DNT also contain 2,6-DNT. Both isomers are produced by the nitration process used for production of TNT. Similarly, no bacteria able to degrade 2,4,6-trinitrotoluene aerobically have been isolated. This is thought to be because the electron withdrawing properties of the nitro group make the aromatic ring less susceptible to electrophilic attack by oxygenase enzymes. However, strains able to carry out dioxygenase reactions on nitrobenzene (32), 2-nitrotoluene (17), and 2,4-DNT(48) have been isolated. Thus, steric hindrance may play an additional role in the failure of microbial oxygenase enzymes to attack 2,6-dinitro- and 2,4,6–trinitrotoluene.

Recently a strain of *Pseudomonas* able to use 2-nitrotoluene as the sole source of carbon and nitrogen was isolated from contaminated soil (17). The catabolic pathway involves an initial dioxygenation that results in the release of the nitro group as nitrite. The resultant 3-methylcatechol is subsequently degraded by a ring fission reaction catalyzed by an extradiol type of dioxygenase. 2-Hydroxy-6-oxohepta-2,4-dienoate formed by the ring fission reaction is further degraded by a hydrolytic pathway (Fig. 6).

Several bacteria able to grow on 3-nitrobenzoate as the sole source of carbon and nitrogen have recently been isolated (29). One strain was a spontaneous mutant derived from *Pseudomonas* sp. strain JS150 which was previously isolated for its ability to grow on a variety of chloroaromatic compounds. In addition, a *Comamonas* strain able to grown on 3-nitrobenzoate was isolated from an industrial waste treatment system. Both isolates released nitrite during growth on 3-nitrobenzoate. Earlier reports (5) suggested that 3-nitrobenzoate was degraded via sequential monooxygenations that replaced both the nitro group and a proton in the 4 position with hydroxyl groups to produce protocatechuate. Recent studies (29) have shown by measuring the stoichiometry of oxygen utilization and by $^{18}O_2$ incorporation experiments that the initial reaction in both *Pseudomonas* and *Comamonas* is a dioxygenase catalyzed oxygenolytic elimination of nitrite. The protocatechuate produced

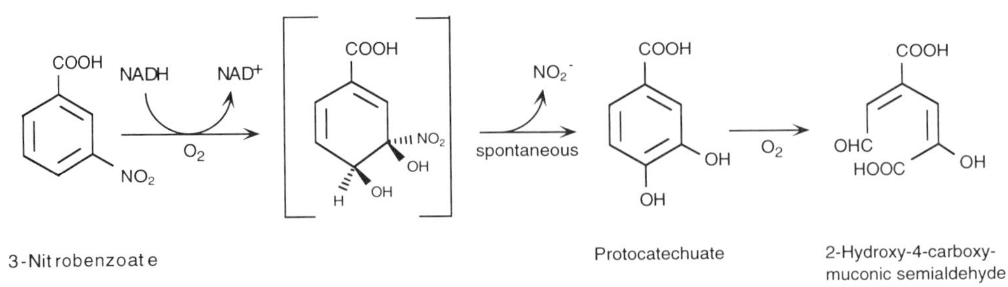

Figure 7. Oxidation of 3-nitrobenzoic acid (29).

by the initial reaction is oxidized by a protocatechuate-4,5-dioxygenase to yield a yellow ring fission product which is further degraded (Fig. 7).

To date all of the bacteria found to synthesize dioxygenase enzymes for elimination of aromatic nitro groups have been gram negative. Very recently, a *Comamonas* strain was isolated from activated sludge based on its ability to grow with nitrobenzene as the sole carbon and nitrogen source (32). The isolate, designated strain JS765 releases nitrite from nitrobenzene which suggests that the initial reactions are oxidative. Cells grown on nitrobenzene rapidly degrade nitrobenzene and catechol, but not nitrophenols or nitrocatechols. Nitrobenzene grown cells contain high levels of an inducible, heat stable catechol-2,3-dioxygenase which indicates that catechol is on the catabolic pathway. Experiments with $^{18}O_2$ demonstrate conclusively that the initial reaction is catalyzed by a dioxygenase. Richard Bartha and his colleagues have also carried out preliminary investigations with an independently isolated strain that seems to use a similar pathway (Richard Bartha ASM 1994).

PARTIAL REDUCTION OF THE NITRO GROUP

Reduction Catalyzed by Nitroreductases

Previously, reduction of aromatic nitro groups has been thought to increase the resistance of the molecule to biodegradation. Recent work (3, 34) has shown that reduction of the nitro group to the amine level can result in productive metabolism under anaerobic conditions. Complete reduction of nitro groups has been widely reported to occur in aerobic bacteria (20, 40). However, the amines produced by such reactions seem to be dead end metabolites under aerobic conditions.

The enzymatic reduction of the nitro group involves the stepwise addition of six reducing equivalents potentially derived from reduced pyridine nucleotides (Fig. 8). The first reaction yields a nitroso derivative which is subsequently reduced to a hydroxylamine; the hydroxylamino compound is then reduced to the amine. In most systems studied to date (Cerniglia and Somerville, this volume) a single nitroreductase enzyme is responsible for all three reactions and there is little or no accumulation of the intermediates. However, reduction of nitro compounds does not seem to be the physiological function of the enzymes that have been reported to carry out these reactions. Diaphorases (23), ferredoxin-NADPH reductase (33), and a variety of other enzymes from procaryotes and eucaryotes have been shown to catalyze the fortuitous reduction of aromatic nitro groups.

Degradation of 4-Nitrobenzoate and 4-Nitrotoluene

A *Comamonas acidovorans* isolated for its ability to use 4-nitrobenzoate as the sole source of carbon and nitrogen contains an inducible nitroreductase that seems to be specific for 4-nitrobenzoate (13, 14). The purified enzyme reduces 4-nitrobenzoate to 4-hydroxylaminobenzoate but not to the corresponding amine. Instead, a hydroxylami-

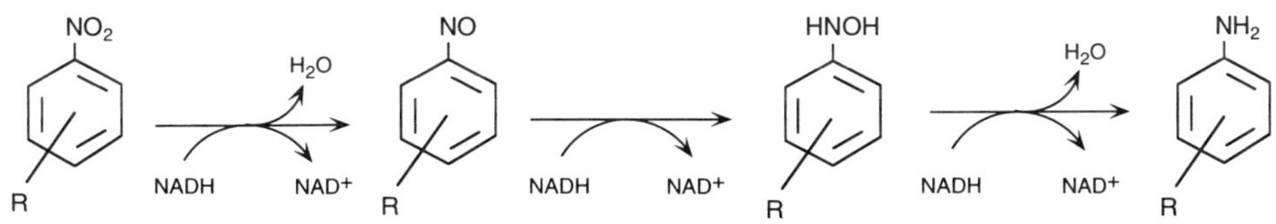

Figure 8. Reduction of the aromatic nitro group.

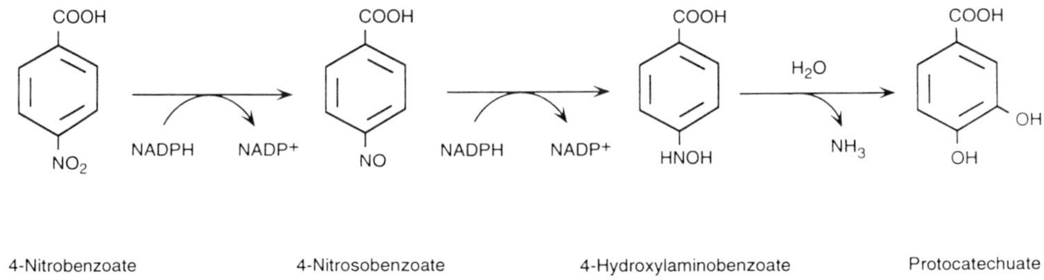

Figure 9. Partial reduction of 4-nitrobenzoate (14).

nolyase converts 4-hydroxylaminobenzoate to protocatechuate which serves as a substrate for ring fission and further metabolism (Fig. 9). The nitrogen is released as ammonia during the reaction, but the details of the mechanism are not clear (Meulenberg and de Bont, this volume).

Two *Pseudomonas* strains isolated for the ability to grow at the expense of 4-nitrotoluene release ammonia during growth which suggests that the pathway involves reduction of the nitro group (16, 38). Pseudomonads harboring the TOL plasmid and grown on toluene are known to oxidize the methyl group of 4-nitrotoluene to a carboxyl group (7). However, these strains cannot grow on 4-nitrotoluene because it does not serve as an effector for induction of the necessary enzymes. The two strains able to grow on 4-nitrotoluene also oxidize the methyl group by reactions analogous to those of the pathway encoded on the TOL plasmid. However, the genes for the initial steps in the pathway do not seem to be closely related to those encoded by the TOL pathway (38). Oxidation of the methyl group of 4-nitrotoluene produces 4-nitrobenzoate which is then mineralized by the pathway described above involving partial reduction of the nitro group and subsequent formation of protocatechuate (Fig. 10). The key enzymes in this sequence are the nitroreductases and the hydroxylaminolyases. These enzymes are discussed in detail by Meulenberg and de Bont (this volume).

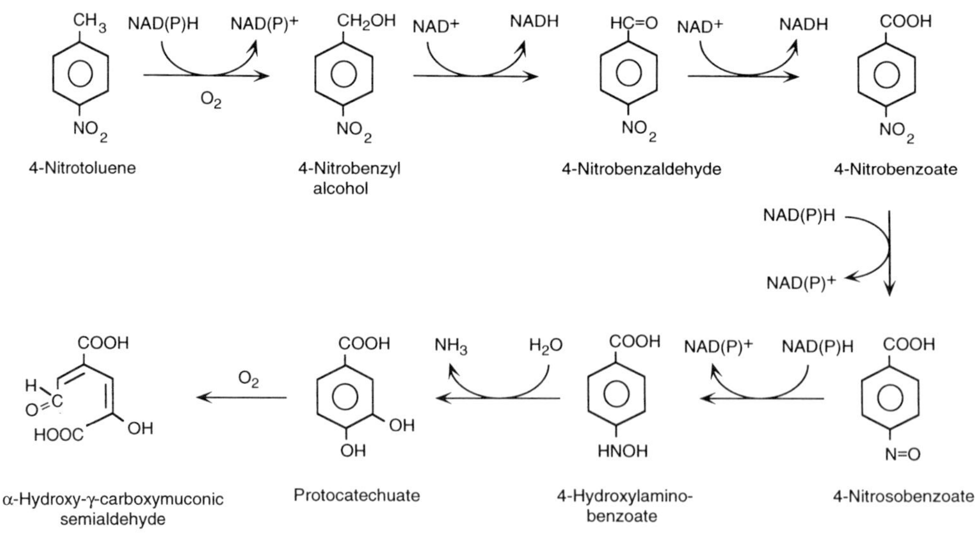

Figure 10. Degradation of 4-nitrotoluene by a *Pseudomonas* (16).

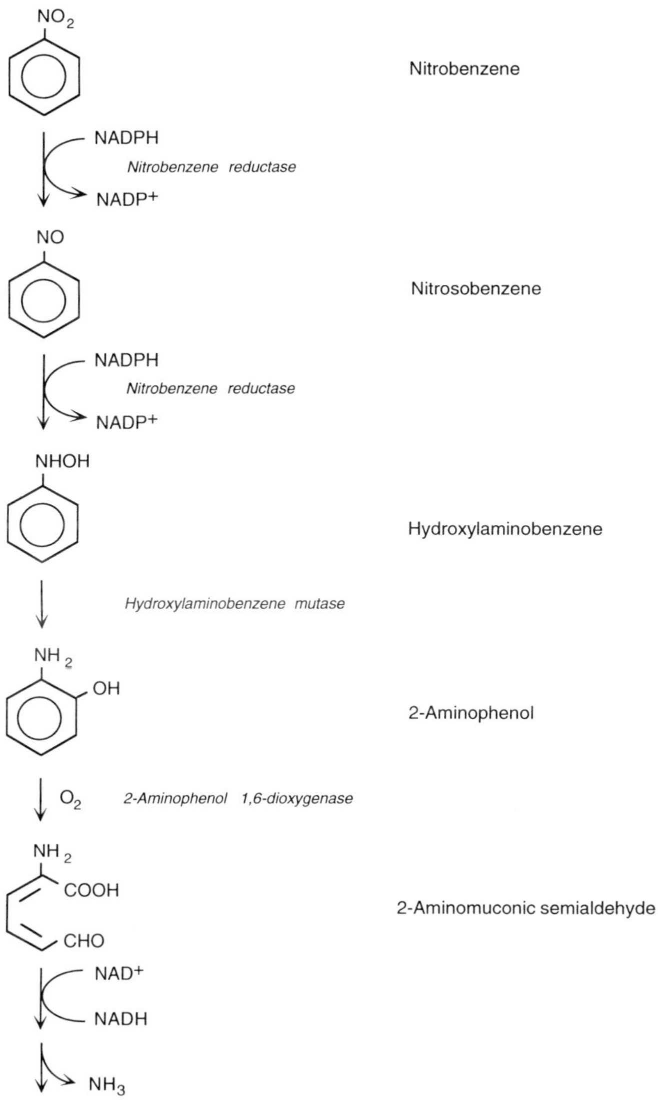

Figure 11. Biodegradation of nitrobenzene by *Pseudomonas pseudoalcaligenes* (31).

Reductive Degradation of Nitrobenzene

Nitrobenzene is the simplest nitroaromatic compound, yet the pathways for its degradation by aerobic bacteria have only recently been elucidated. The aerobic pathway initiated by a dioxygenase described above seems to be relatively rare. Of nine isolates from six contaminated sites, only one releases nitrite from nitrobenzene (32). The rest release ammonia, which strongly suggests the operation of a reductive pathway. One of these strains, a *Pseudomonas pseudoalcaligenes* designated strain JS45, has been studied in some detail and the catabolic pathway shown in Figure 11 has been proposed based on the following evidence (31).

Preliminary experiments revealed that cultures of strain JS45 grown on nitrobenzene do not degrade aniline as would be expected if the pathway involved complete reduction of

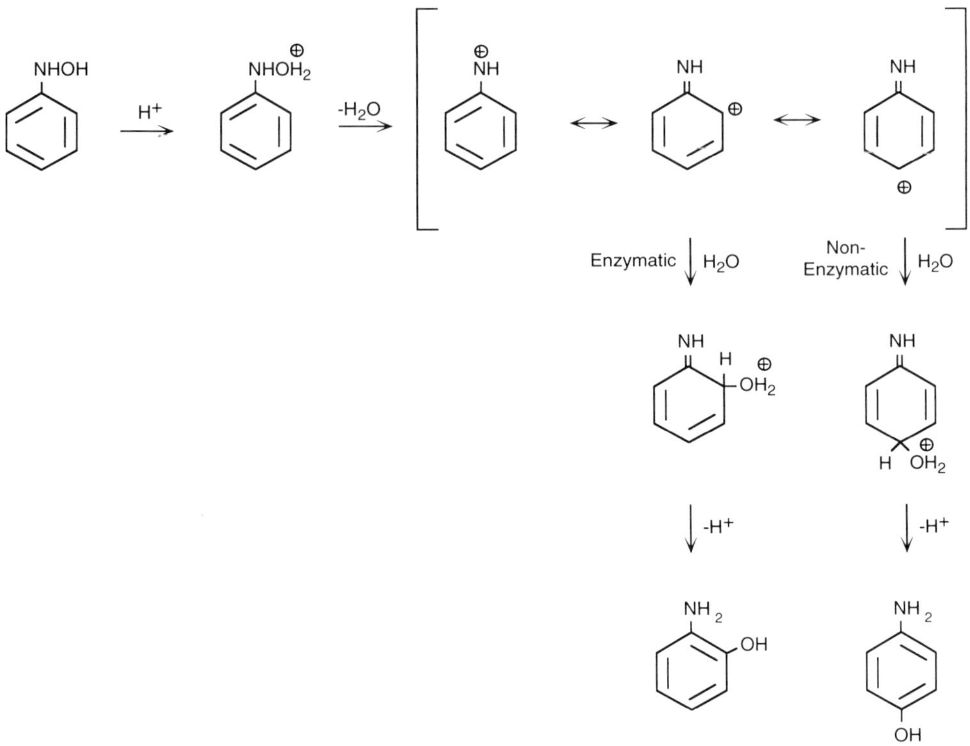

Figure 12. Proposed mechanism for the Bamberger-like rearrangement of hydroxylaminobenzene to 2-aminophenol catalyzed by a mutase enzyme in *Pseudomonas pseudoalcaligenes*.

the nitro group. Catechol would be an intermediate if nitrobenzene were degraded via partial reduction followed by the hydroxylaminolyase catalyzed reaction described above in the pathway for degradation of 4-nitrobenzoate. Surprisingly, catechol is only slowly degraded by nitrobenzene grown cells, whereas hydroxylaminobenzene and 2-aminophenol are degraded rapidly. Extracts prepared from strain JS45 catalyze the reduction of nitrobenzene to hydroxylaminobenzene via nitrosobenzene. The reaction sequence requires the oxidation of two moles of NADPH. The nitrobenzene reductase has been purified and identified as a flavoprotein with a molecular weight of 30 kDa (43). It is active as a monomer and is relatively specific for nitrobenzene. Neither the purified enzyme nor crude cell extracts catalyze the further reduction of hydroxylaminobenzene.

Under aerobic conditions, enzymes in crude extracts convert nitrobenzene to products with no detectable absorbance in the UV range. In the absence of oxygen both nitrobenzene and hydroxylaminobenzene are converted stoichiometrically to 2-aminophenol. An unusual enzyme designated hydroxylaminobenzene mutase catalyzes a rearrangement of hydroxylaminobenzene to 2-aminophenol in the absence of added cofactors (31). The reaction is analogous to the Bamberger rearrangement (Fig. 12) in which, under mild acidic conditions, hydroxylaminoaromatic compounds undergo a rearrangement to aminophenols (41, 44). The nonenzymatic reaction yields predominantly (99%) 4-aminophenol whereas the enzyme catalyzed reaction yields almost exclusively 2-aminophenol. To date, this is the only reported example of such an enzyme in a bacterial catabolic pathway. It has not yet been purified or characterized. Enzymes that catalyze similar reactions have been reported in eucaryotic systems including rats (49), and a *Rhodosporidium* sp. that converts chlorohydroxylaminobenzene to the the corresponding chloroaminophenols (6, Corbett, this volume).

Another unusual enzyme is responsible for the further metabolism of 2-aminophenol. A ferrous iron dependent dioxygenase catalyzes the opening of the ring at the 1,6 position to form 2-aminomuconic semialdehyde. The reaction is unusual because dioxygenase-catalyzed opening of the aromatic ring typically requires the presence of two hydroxyl groups on the ring. Recently, an enzyme capable of opening the ring of 5-aminosalicylate was discovered in a strain of *Pseudomonas* grown on 5-aminosalicylate (50). A similar reaction is catalyzed by catechol 1,2-dioxygenase from *Pseudomonas arvilla* (35) and *Pseudomonas aeruginosa* (22). However, 2-aminophenol is not the physiological substrate for either of these enzymes and the rate of the reaction is up to 1000-fold lower than the rate with catechol. In contrast, the enzyme from *Pseudomonas pseudoalcaligenes* oxidizes 2-aminophenol 50 times faster than catechol. The requirement for ferrous iron and the sensitivity of the enzyme to heat suggest that it is related to the gentisate dioxygenases (19) and hydroquinone dioxygenase (45), but the enzyme has not been purified and characterized.

In crude cell extracts containing added NAD^+, 2-aminomuconic semialdehyde is degraded with concomitant release of ammonia. The reactions have not been studied in detail so it is not known whether the NAD^+ is required for oxidation of the aldehyde before replacement of the amino group. The mechanism involved in the release of ammonia is also unknown (31).

Comparison of Nitrobenzene Catabolic Pathways

Based on known reactions, it is possible to postulate several alternative pathways for the catabolism of nitrobenzene (Fig.13). The simplest and most direct pathway would involve an initial dioxygenase reaction and subsequent metabolism of catechol by a common pathway as described above for the *Comamonas* strain (32). A second option would be complete reduction of the nitro group and degradation of the resultant aniline by the well known route of dioxygenase attack and formation of catechol (2). The third pathway demonstrated in *Pseudomonas pseudoalcaligenes* (31) seems to be the most complex alternative. It not only involves more steps, but also requires the participation of three unusual enzymes. The nitroreductase is unusual because it does not reduce hydroxylaminobenzene. Two other bacterial nitroreductases have been recently reported to partially reduce the nitro group (1, 13). The mutase that catalyzes the Bamberger-like rearrangement of hydroxylaminobenzene has not been reported previously in procaryotes. Finally, the enzyme responsible for the ring-fission reaction seems to exhibit a very unusual substrate specificity. Yet, the pathway involving partial reduction of the nitro group seems to be the most common among the isolates we have obtained from contaminated soil. Examination of the cofactor and oxygen requirements of the three alternative pathways reveals a likely explanation. In order to produce the hypothetical common intermediate, oxalocrotonate, the pathway involving an initial dioxygenase attack requires two moles of molecular oxygen. If nitrobenzene serves as the nitrogen source for growth, the reduction of nitrite to ammonia would require three additional moles of NAD(P)H. The hypothetical sequence including aniline would require two moles of oxygen and three moles of reduced pyridine nucleotide. In contrast, the more complex pathway in *Pseudomonas pseudoalcaligenes* requires only one mole of oxygen and one mole of reducing equivalents to convert nitrobenzene to oxalocrotonate. In an ecosystem where nitrobenzene is the sole source of carbon, nitrogen, and energy and where oxygen is limiting as it is in contaminated aquifers the advantages of the partial reductive pathway are obvious. This is consistent with the fact that all of the strains that use the partial reductive pathway were isolated from contaminated groundwater. In contrast, the strain that degrades nitrobenzene via an initial dioxygenase attack was isolated

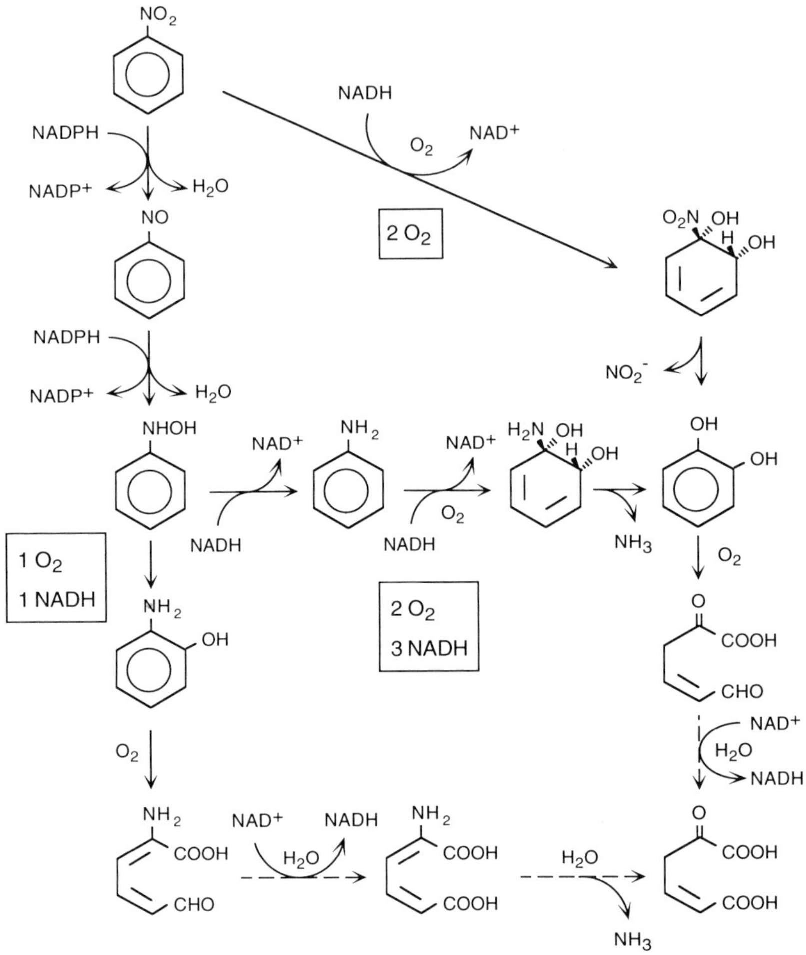

Figure 13. Alternative pathways for bacterial degradation of nitrobenzene.

from an activated sludge waste treatment system where oxygen is less likely to be limiting. The origin and evolutionary history of the reductive pathway will make an exciting topic for future research.

CONCLUSION

The catabolic pathways described in this chapter illustrate the wide range of enzyme catalyzed reactions available for conversion of nitroaromatic compounds into intermediates that can serve as growth substrates for bacteria. This survey of what is currently known is only a small portion of what remains to be discovered about these pathways. The mechanisms of several of the more unusual reactions are unknown, and the reasons for the unusually high specificities of the enzymes involved are not understood. Insight into the mechanisms and specificities of the reactions might allow the use of the novel enzymes for useful biotransformations not related to biodegradation. The molecular biology of these systems is almost completely unexplored. Increased understanding about the regulation of enzyme synthesis and the molecular basis of structural characteristics should allow researchers to extend the substrate ranges of existing pathways and construct novel pathways for degradation of recalcitrant molecules. Finally, very little is known about the origin and evolution of these pathways, and nothing is known about their interactions or microbial ecology.

It is clear that the number of nitroaromatic compounds known to serve as growth substrates for bacteria has risen dramatically over the past four years. It is not clear whether this is because more effort has been expended to find such strains or whether they are evolving rapidly in contaminated ecosystems. Discovery of microbes that mineralize nitroaromatic compounds provides assurance that such compounds can be removed from the environment by bioremediation under appropriate conditions. Compounds that cannot serve as growth substrates are much less likely to be susceptible to bioremediation because of the daunting challenges of providing active biomass and alternate substrates. Therefore, high priority should be given to discovering or constructing strains able to grow on recalcitrant contaminants such as 2,4,6-trinitrotoluene and 2,6-dinitrotoluene.

REFERENCES

1. Anlezark, G. M., R. G. Melton, R. F. Sherwood, B. Coles, F. Friedlos and R. Knox. 1992. The bioactivation of 5-(aziridin-1-YL)-2,4-dinitrobenzamide (CB1954)-1. Purification and properties of a nitroreductase enzyme from *E. coli* - a potential enzyme for antibody-directed enzyme prodrug therapy (ADEPT). Biochem. Pharmacol. 44:2289-2295.
2. Bachofer, R., F. Lingens and W. Schafer. 1975. Conversion of aniline into pyrocatechol by a *Nocardia* sp.: Incorporation of oxygen-18. FEBS Lett. 50:288-290.
3. Boopathy, R. and C. F. Kulpa. 1993. Nitroaromatic compounds serve as nitrogen source for *Desulfovibrio* sp. (B strain). Can. J. Microbiol. 39:430-433.
4. Brilon, C., W. Beckmann and H. Knackmuss. 1981. Catabolism of naphthalenesulfonic acids by *Pseudomonas* sp. A3 and *Pseudomonas* sp. C22. Appl. Environ. Microbiol. 42:44-55.
5. Cartwright, N. J. and R. B. Cain. 1959. Bacterial degradation of nitrobenzoic acids. Biochem. J. 71:248-261.
6. Corbett, M. D. and B. R. Corbett. 1981. Metabolism of 4-chloronitrobenzene by the yeast *Rhodosporidium* sp. Appl. Environ. Microbiol. 41:942-949.
7. Delgado, A., M. G. Wubbolts, M.-A. Abril and J. L. Ramos. 1992. Nitroaromatics are substrates for the TOL plasmid upper-pathway enzymes. Appl. Environ. Microbiol. 58:415-417.
8. Dickel, O. and H.-J. Knackmuss. 1991. Catabolism of 1,3-dinitrobenzene by *Rhodococcus* sp. QT-1. Arch. Microbiol. 157:76-79.
9. Ecker, S., H.-J. Knackmuss and C. Bruhn. 1989. Catabolism of 2,6-dinitrophenol by *Alcaligenes eutrophus* JMP222, abstr. Q-198, p. 168. *In* Abstr. Annu. Meet. Am. Soc. Microbiol. 1989.
10. Ecker, S., T. Widmann, H. Lenke, O. Dickel, P. Fischer, C. Bruhn and H.-J. Knackmuss. 1992. Catabolism of 2,6-dinitrophenol by *Alcaligenes eutrophus* JMP134 and JMP222. Arch. Microbiol. 158:149-154.
11. Fetzner, S., R. Muller and F. Lingens. 1989. A novel metabolite in the microbial degradation of 2-chlorobenzoate. Biochemical and Biophysical Research Communications. 161:700-705.
12. Gibson, D. T. and V. Subramanian. 1984. Microbial degradation of aromatic hydrocarbons, p. 181-252. *In* Gibson, D. T. (ed.), Microbial degradation of organic compounds. Marcel Dekker, Inc. New York.
13. Groenewegen, P. E. J., P. Breeuwer, J. M. L. M. van Helvoort, A. A. M. Langenhoff, F. P. de Vries and J. A. M. de Bont. 1992. Novel degradative pathway of 4-nitrobenzoate in *Comamonas acidovorans* NBA-10. J. Gen. Microbiol. 138:1599-1605.
14. Groenewegen, P. E. J. and J. A. M. de Bont. 1992. Degradation of 4-nitrobenzoate via 4-hydroxylaminobenzoate and 3,4-dihydroxybenzoate in *Comamonas acidovorans* NBA-10. Arch. Microbiol. 158:381-386.
15. Haigler, B. E., S. F. Nishino and J. C. Spain. 1994. Biodegradation of 4-methyl-5-nitrocatechol by *Pseudomonas* sp. strain DNT. J. Bacteriol. 176:3433-3437.
16. Haigler, B. E. and J. C. Spain. 1993. Biodegradation of 4-nitrotoluene by *Pseudomonas* sp. strain 4NT. Appl. Environ. Microbiol. 59:2239-2243.
17. Haigler, B. E., W. H. Wallace and J. C. Spain. Biodegradation of 2-nitrotoluene by *Pseudomonas* sp. strain JS42. Appl. Environ. Microbiol. 60:3466-3469.

18. Hanne, L. F., L. L. Kirk, S. M. Appel, A. D. Narayan and K. K. Bains. 1993. Degradation and induction specificity in actinomycetes that degrade *p*-nitrophenol. Appl. Environ. Microbiol. 59:3505-3508.
19. Harpel, M. R. and J. D. Lipscomb. 1990. Gentisate-1,2-dioxygenase from *Pseudomonas*. Purification, characterization, and comparison of the enzymes from *Pseudomonas testosteroni* and *Pseudomonas acidovorans*. J. Biol. Chem. 265:6301-6311.
20. Higson, F. K. 1992. Microbial degradation of nitroaromatic compounds. Adv. Appl. Microbiol. 37:1-19.
21. Jain, R. K., J. H. Dreisbach and J. C. Spain. Biodegradation of *p*-nitrophenol via 1,2,4-benzenetriol by an *Arthrobacter*. Appl. Environ. Microbiol. 60:3030-3032.
22. Kataeva, I. A. and L. A. Golovleva. 1990. Catechol 2,3-dioxygenases from *Pseudomonas aeruginosa* 2x. Meth. Enzymol. 188:115-121.
23. Kato, R. ., T. Oshima and A. Takanaka. 1969. Studies on the mechanism of nitroreduction by rat liver. Mol. Pharmacol. 5:487-498.
24. Lenke, H. and H.-J. Knackmuss. 1992. Initial hydrogenation during catabolism of picric acid by *Rhodococcus erythropolis* HL 24-2. Appl. Environ. Microbiol. 58:2933-2937.
25. Marcus, A., U. Klages, S. Krauss and F. Lingens. 1984. Oxidation and dehalogenation of 4-chlorophenylacetate by a two component enzyme system from *Pseudomonas* sp. strain CBS3. J. Bacteriol. 160:618-621.
26. McCormick, N. G., F. F. Feeherry and H. S. Levinson. 1976. Microbial transformation of 2,4,6-trinitrotoluene and other nitroaromatic compounds. Appl. Environ. Microbiol. 31:949-958.
27. Mitra, D. and Vaidyanathan. 1984. A new 4-nitrophenol 2-hydroxylase from a *Nocardia* sp. Biochem. Internat. 8:609-615.
28. Munnecke, D. M. and D. P. H. Hsieh. 1974. Microbial decontamination of parathion and *p*-nitrophenol in aqueous media. Appl. Microbiol. 28:212-217.
29. Nadeau, L. J. and J. C. Spain. 1994. The bacterial degradation of *m*-nitrobenzoic acid, abstr. Q-114, p. 408. *In* Abstr. Annu. Meet. Am. Soc. Microbiol. 1994.
30. Nishino, S. F. and J. C. Spain. 1992. Initial steps in the bacterial degradation of 1,3-dinitrobenzene, abstr. Q-135, p. 358. *In* Abstr. 92nd Annu. Meet. Am. Soc. Microbiol. 1992.
31. Nishino, S. F. and J. C. Spain. 1993. Degradation of nitrobenzene by a *Pseudomonas pseudoalcaligenes*. Appl. Environ. Microbiol. 59:2520-2525.
32. Nishino, S. F. and J. C. Spain. Oxidative pathway for the biodegradation of nitrobenzene by a *Comamonas* sp. strain JS765. Appl. Environ. Microbiol. In press.
33. Orna, M. V. and R. P. Mason. 1989. Correlation of kinetic parameters of nitroreductase enzymes with redox properties of nitroaromatic compounds. J. Biol. Chem. 264:12379-12384.
34. Preuss, A., J. Fimpel and G. Diekert. 1993. Anaerobic transformation of 2,4,6-trinitrotoluene (TNT). Arch. Microbiol. 159:345-353.
35. Que, L. 1978. Extradiol cleavage of *o*-aminophenol by pyrocatechase. Biochem. Biophys. Res. Commun. 84:123-129.
36. Raymond, D. G. M. and M. Alexander. 1971. Microbial metabolism and cometabolism of nitrophenols. Pest. Biochem. Physiol. 1:123-130.
37. Reiner, A. M. 1972. Metabolism of aromatic compounds in bacteria. Purification and properties of the catechol-forming enzyme, 3,5-cyclohexadiene-1,2-diol-1-carboxylic acid (NAD^+) oxidoreductase (decarboxylating). J. Biol. Chem. 247:4960-4965.
38. Rhys-Williams, W., S. C. Taylor and P. A. Williams. 1993. A novel pathway for the catabolism of 4-nitrotoluene by *Pseudomonas*. J. Gen. Microbiol. 139:1967-1972.
39. Sander, P., R.-M. Wittaich, P. Fortnagel, H. Wilkes and W. Francke. 1991. Degradation of 1,2,4-trichloro- and 1,2,4,5-tetrachlorobenzene by *Pseudomonas* strains. Appl. Environ. Microbiol. 57:1430-1440.
40. Schackmann, A. and R. Müller. 1991. Reduction of nitroaromatic compounds by different *Pseudomonas* species under aerobic conditions. Appl. Microbiol. Biotechnol. 34:809-813.
41. Shine, H. J. 1967. The rearrangement of phenylhydroxylamines, p. 182-190. *In* Aromatic Rearrangements. Elsevier Publishing Company. Amsterdam, London, New York.
42. Simpson, J. R. and W. C. Evans. 1953. The metabolism of nitrophenols by certain bacteria. Biochem. J. 55:XXIV.
43. Somerville, C. C., S. F. Nishino and J. C. Spain. Purification and characterization of nitrobenzene nitroreductase from *Pseudomonas pseudoalcaligenes* JS45. J. Bacteriol. In press.
44. Sone, T., K. Hamamoto, Y. Seiji, S. Shinkai and O. Manabe. 1981. Kinetics and mechanisms of the Bamberger rearrangement. Part 4. Rearrangement of sterically hindered phenylhydroxyamines to 4-aminophenols in aqueous sulphuric acid solution. J. Chem. Soc. Perkin Trans. 2. 1981:1596-1598.
45. Spain, J. C. and D. T. Gibson. 1991. Pathway for biodegradation of *p*-nitrophenol in a *Moraxella* sp. Appl. Environ. Microbiol. 57:812-819.

46. Spain, J. C., O. Wyss and D. T. Gibson. 1979. Enzymatic oxidation of *p*-nitrophenol. Biochem. Biophys. Res. Commun. 88:634-641.
47. Spain, J. C., G. J. Zylstra, C. K. Blake and D. T. Gibson. 1989. Monohydroxylation of phenol and 2,5-dichlorophenol by toluene dioxygenase in *Pseudomonas putida* F1. Appl. Environ. Microbiol. 55:2648-2652.
48. Spanggord, R. J., J. C. Spain, S. F. Nishino and K. E. Mortelmans. 1991. Biodegradation of 2,4-dinitrotoluene by a *Pseudomonas* sp. Appl. Environ. Microbiol. 57:3200-3205.
49. Sternson, L. A. and R. E. Gammans. 1975. A mechanistic study of aromatic hydroxylamine rearrangement in the rat. Bioorg. Chem. 4:58-63.
50. Stoltz, A., B. Nörtemann and H.-J. Knackmuss. 1992. Bacterial metabolism of 5-aminosalicylic acid. Initial ring cleavage. Biochem. J. 282:675-680.
51. Suen, W.-C., B. E. Haigler and J. C. Spain. 1994. 2,4-Dinitrotoluene dioxygenase genes from *Pseudomonas* sp. strain DNT: homology to naphthalene dioxygenase, abstr. Q-397, p. 458. *In* Abstr. Annu. Meet. Am. Soc. Microbiol. 1994.
52. Suen, W.-C. and J. C. Spain. 1993. Cloning and characterization of *Pseudomonas* sp. strain DNT genes for 2,4-dinitrotoluene degradation. J. Bacteriol. 175:1831-1837.
53. Vorbeck, C., H. Lenke, P. Fischer and H.-J. Knackmuss. 1994. Identification of a hydride-Meisenheimer complex as a metabolite of 2,4,6-trinitrotoluene by a *Mycobacterium* strain. J. Bacteriol. 176:932-934.
54. Wackett, L. P., L. D. Kwart and D. T. Gibson. 1988. Benzylic monooxygenation catalyzed by toluene dioxygenase from *Pseudomonas putida*. Biochem. 27:1360-1367.
55. Zeyer, J. and P. C. Kearney. 1984. Degradation of *o*-nitrophenol and *m*-nitrophenol by a *Pseudomonas putida*. J. Agric. Food Chem. 32:238-242.
56. Zeyer, J. and H. P. Kocher. 1988. Purification and characterization of a bacterial nitrophenol oxygenase which converts *ortho*-nitrophenol to catechol and nitrite. J. Bacteriol. 170:1789-1794.

3

MICROBIAL PRODUCTION OF CATECHOLS FROM NITROAROMATIC COMPOUNDS

Rogier Meulenberg and Jan A. M. de Bont

Division of Industrial Microbiology
Department of Food Science
Wageningen Agricultural University
P.O. Box 8129
6700 EV Wageningen
The Netherlands

INTRODUCTION

Nitroaromatic compounds are partly or fully degraded by a wide range of microorganisms. For this reason, a growing interest in the degradative characteristics of bacteria and fungi towards nitroaromatic compounds has emerged. This interest is aimed at exploiting these capacities to remove nitroaromatic compounds from the environment, or better, to prevent their release. Another potentially beneficial aspect of the great diversity in metabolic pathways involved in the degradation of nitroaromatic compounds is found in possible biotechnological applications. Nitroaromatic compounds may be used as cheap feedstocks in the biocatalytic production of valuable products by enzymes or intact cells involved in their conversion. This paper concentrates on these biocatalytic aspects and deals with research aimed at the production of catechols from nitroaromatic compounds.

DEGRADATION OF NITROAROMATIC COMPOUNDS

Microbial mineralization of nitroaromatic compounds can be initiated by different reaction mechanisms (for reviews see 28 and 37).

Oxygenolytic Removal of the Nitro Group

In the presence of oxygen, several microorganisms start the degradation of nitroaromatic compounds with the oxidative removal of the nitro group, yielding nitrite and a hydroxylated aromatic compound. Oxygenolytic attack on nitroaromatic compounds can be accomplished by monooxygenase and dioxygenase enzyme systems. Oxidation of 2-nitrophenol in *Pseudomonas putida* B2 was catalysed by a monooxygenase (70, 72). The

Biodegradation of Nitroaromatic Compounds
Edited by J.C. Spain, Plenum Press, New York, 1995

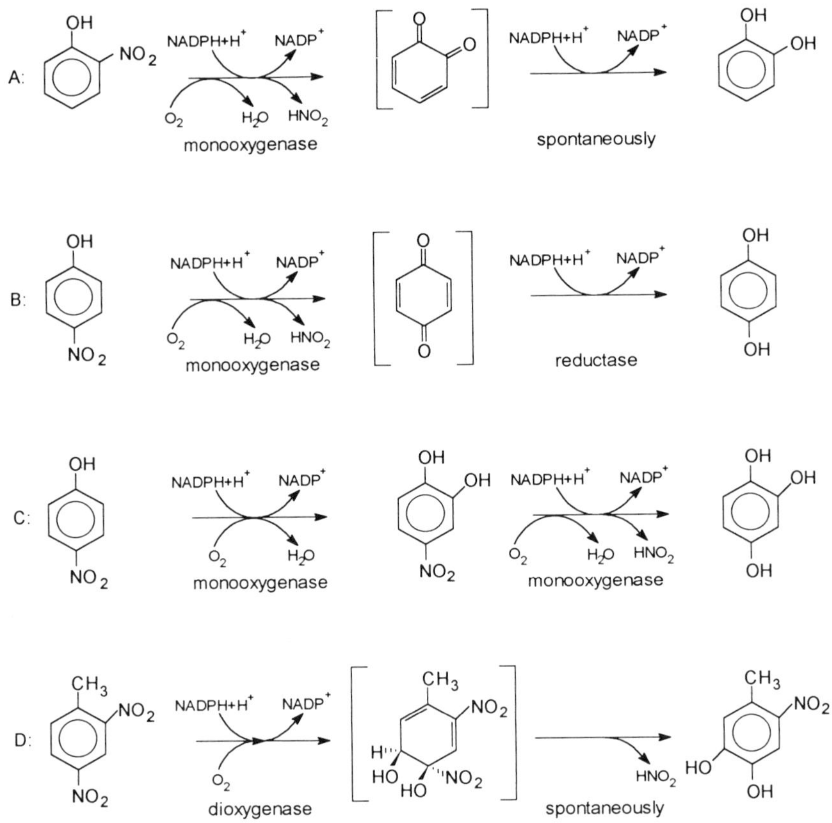

Figure 1. Proposed pathways for monooxygenolytic degradation of 2-nitrophenol (A) and 4-nitrophenol (B and C), and dioxygenolytic breakdown of 2,4-dinitrotoluene (D) (31, 58, 61, 71).

enzyme converted 2-nitrophenol to catechol and nitrite via the intermediary formation of 1,2-benzoquinone (71, Fig. 1A).

For the monooxygenase-mediated degradation of 4-nitrophenol different pathways have been described. *Arthrobacter aurescens* TW17 and a strain of *Moraxella* converted 4-nitrophenol to hydroquinone and nitrite, and 1,4-benzoquinone was suggested as an intermediate (27, 58, 60, Figure 1B). Monooxygenase activity on 4-nitrophenol by *Flavobacterium* sp., *Pseudomonas* sp., *Arthrobacter* sp. JS443 and *Nocardia* sp. TW2 revealed the intermediary formation of 4-nitrocatechol (27, 31, 48, 63). In *Arthrobacter* sp. JS443 this intermediate was subsequently further oxygenated to 1,2,4-benzenetriol (31, Fig. 1C).

The activity of a dioxygenase on mono- or dinitrosubstituted phenols, toluenes and benzoic acids in all cases yielded substituted catechols and nitrite (14, 15, 61). The nitro group was spontaneously eliminated from the dihydroxy cyclohexadiene intermediate which subsequently rearranged to the catechol (Fig. 1D).

Initial Hydrogenation of the Aromatic Ring

For a strain of *Rhodococcus erythropolis*, Lenke and Knackmuss (35) described the initial reduction of the aromatic nucleus of picric acid (2,4,6-trinitrophenol) by a hydride ion (H^-), yielding a hydride-Meisenheimer complex (Fig. 2). With the addition of a proton, the nitro group was subsequently released as nitrite and 2,4-dinitrophenol was produced.

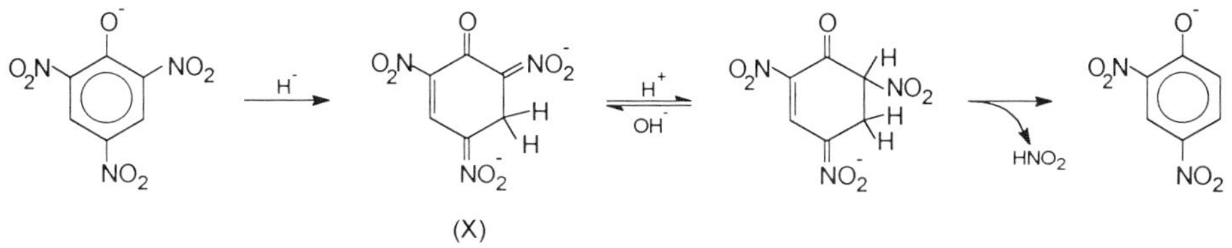

Figure 2. Hydride (H⁻) reduction of picric acid resulting in the intermediary formation of a hydride-Meisenheimer complex (X). Upon addition of a proton, nitrite is eliminated from this complex, yielding 2,4-dinitrophenol (35).

Hydrogenation of picric acid only occurred in the presence of oxygen and succinate as a carbon and energy source (52). Under these conditions, picric acid was used as a nitrogen source. The product of hydrogenation and subsequent protonation of picric acid, 2,4-dinitrophenol, could be used as the sole carbon, nitrogen and energy source by *R. erythropolis* HLPM-1. Very recently, initial hydride reduction has also been described for the degradation of 2,4,6-trinitrotoluene by a *Mycobacterium* strain (65).

Initial Reduction of the Nitro Group to an Amine

Nitroaromatic compounds can be initially reduced to aniline derivatives via the intermediary formation of nitroso- and hydroxylamino derivatives (Fig. 3). Subsequently, the amines can be reductively deaminated, yielding benzene derivatives and ammonium (5, 6). Several bacteria are only capable of partial degradation of nitroaromatic compounds and either use the released ammonium as a nitrogen source for growth without further degradation of the aromatic ring (5, 6, 45), or use the nitroaromatic compound only as an electron acceptor under anaerobic conditions (2, 3, 12, 34, 38, 47). In the latter case, the amine produced by the reduction of the nitroaromatic compound is not further converted.

Partial Reduction of the Nitro Group to a Hydroxylamino Group

Nitroaromatic compounds can be partially reduced to the corresponding hydroxylamines, which are then hydrolysed to a catechol and ammonium (Fig. 4A). This degradative pathway, which will be discussed in detail below, has been described for the mineralization of 4-nitrobenzoate by *Comamonas acidovorans* NBA-10 (20, 21) and for degradation of 4-nitrotoluene by two *Pseudomonas* species (26, 51). In the latter case, the substrate is first oxidized to 4-nitrobenzoate and further degraded as depicted in Figure 4A.

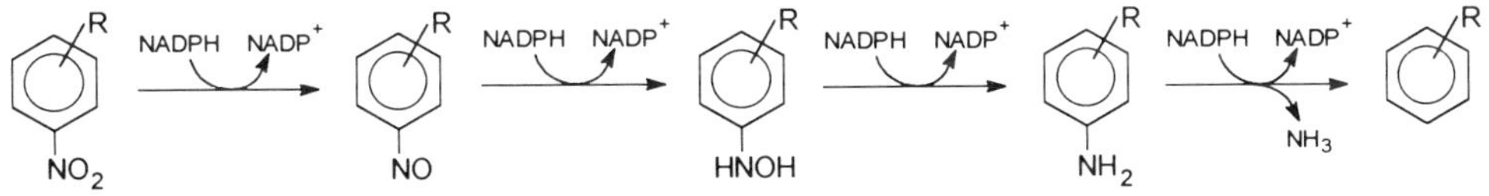

Figure 3. Initial reduction of nitroaromatic compounds to an aniline derivative. Several bacteria are able to reductively deaminate the amine to a benzene derivative and ammonium (5, 6).

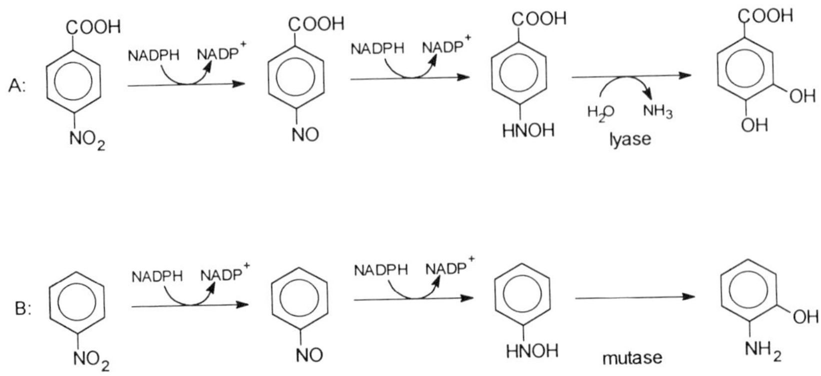

Figure 4. Degradation of nitroaromatic compounds via partial reduction of the nitro group to a hydroxylaminoaromatic group. *Comamonas acidovorans* NBA-10 converted 4-hydroxylaminobenzoate to 3,4-dihydroxybenzoate and ammonium (A) (20, 21) and a *Pseudomonas pseudoalcaligenes* converted hydroxylaminobenzene to 2-aminophenol (B) (41).

BIOTECHNOLOGICAL PRODUCTION OF CATECHOLS BY DIRECT OXIDATION

Catechols are important intermediates in agrochemical, pharmaceutical, petrochemical and flavor industries. However, organic chemical synthesis of these compounds by direct hydroxylation of the aromatic ring is difficult, as the reaction products are usually not stable under the reaction conditions employed (24, 36, 44). Biotechnological production processes for catechols are, therefore, of great interest. As outlined below, several processes based on the direct biological oxidation of aromatic structures have been studied.

Monooxygenase-Mediated Catechol Formation

Monooxygenases introduce one atom of molecular oxygen into the organic substrate, while the other oxygen atom is reduced to water with NAD(P)H or ascorbic acid as the electron donor (23, 43).

The formation of catechol from 2-nitrophenol (Fig. 1A) and the conversion of 4-nitrophenol to 4-nitrocatechol (Fig. 1C) are examples of monooxygenase-catalysed catechol formation. Industrial applications of these types of reactions are the production of L-3,4-dihydroxyphenylalanine (L-DOPA) by hydroxylation of L-tyrosine by *Vibrio tyrosinaticus* and several *Pseudomonas* species (46, 57, 64, 68, 69, Fig. 5A) and the conversion of 4-hydroxyphenylacetic acid to 3,4-dihydroxyphenylacetic acid by a flavin-dependent monooxygenase (1, 30, 39, 64, 67, Fig. 5B). Production of both catechols is of industrial interest because L-DOPA is an important drug for treatment of Parkinson's disease and 3,4-dihydroxyphenylacetic acid is a widely used precursor for the production of synthetic antibiotics.

Dioxygenase-Mediated Catechol Formation

Dioxygenases catalyse the introduction of two atoms of molecular oxygen into the substrate. There are two different types of dioxygenases with a different class of substrates. One type which is NAD(P)H-dependent (43) accomplishes the double hydroxylation of the aromatic ring, yielding a 1,2-*cis*-dihydrodiol (17, 18, 19, 29). Subsequent dehydrogenation

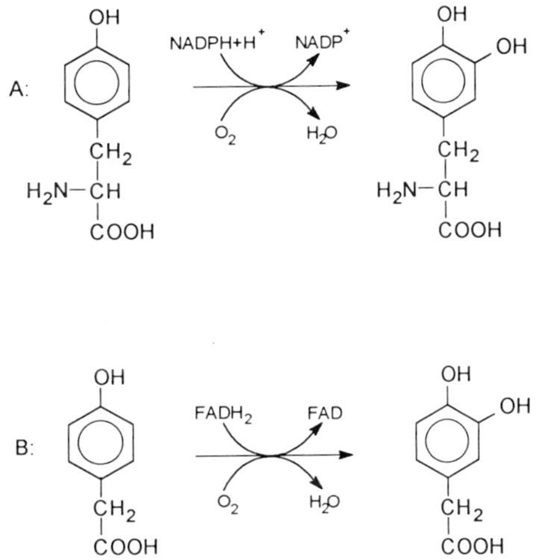

Figure 5. Hydroxylation of L-tyrosine to L-3,4-dihydroxyphenylalanine (A) and conversion of 4-hydroxyphenylacetic acid to 3,4-dihydroxyphenylacetic acid by *Klebsiella oxytoca* (B) (64).

yields a catechol (32, 50). The other type of dioxygenase cleaves the aromatic ring of a catechol by the introduction of molecular oxygen into the ring at the 1,2-position (*intradiol*-cleavage) or at the 2,3-position (*extradiol*-cleavage), yielding muconic acid or hydroxymuconic semialdehyde. It is obvious that for the production of catechols the activity of the first type of dioxygenase is of interest and that the activity of any ring-opening dioxygenase must be prohibited.

There are many microorganisms capable of producing catechols from substituted aromatic compounds. For instance, benzene, toluene, halogenated benzenes, *o*-xylene, benzoic acid and benzoic acid derivatives, phthalic acid and nitrobenzene are dioxygenated by members of the genera *Pseudomonas*, *Alcaligenes*, *Corynebacterium* and *Moraxella*, without modification of the functional group on the aromatic ring (4, 17, 18, 25, 29, 49, 53, 54, 59, 66).

Dioxygenolytic replacement of substituents on the aromatic ring, leading to formation of catechols has been described for a variety of bacteria. Dioxygenation of 2-fluorobenzoate by the benzoate 1,2-dioxygenase in a *Pseudomonas* species yielded 2- and 6-fluoro-1,2-dihydro-1,2-dihydroxybenzoate. The intermediates spontaneously rearranged to catechol, fluoride and CO_2; and 3-fluorocatechol and CO_2 (40, 64), respectively. A *Moraxella* species was able to convert 4-chloroaniline and 4-chloroaniline derivatives by an aniline dioxygenase to 4-chlorocatechol (73), and a *Pseudomonas* species yielded the same product from monooxygenation of 4-chloro-2-nitrophenol (42). Dioxygenolytic replacement of nitro groups, leading to catechol production, has already been described above (Fig. 1D).

BIOTECHNOLOGICAL PRODUCTION OF CATECHOLS FROM NITROAROMATIC COMPOUNDS

For practical application in catechol production the oxygenase-catalysed reactions mentioned above suffer from several disadvantages. The need for expensive cofactor regeneration makes these conversions unfavourable for application on an industrial scale, particularly if cell-free systems are to be used. Furthermore, the toxicity of the oxygen-labile catechols will lead to severe problems in large scale production processes. Catechols easily autooxidize or polymerize in the presence of oxygen and the products from these reactions are often very toxic to the microorganisms (56). The latter disadvantage would be overcome if a hydrolytic, rather than an oxygenolytic, removal of the nitro group were possible.

Recently, such a hydrolytic replacement reaction has been described (Fig. 4A). The main advantage of this replacement reaction is that it can take place under anaerobic conditions. Therefore, the produced catechol is not susceptible to further oxidation or polymerization and, hence, a stable product is achieved. In the next section this hydrolytic replacement will be discussed in detail.

MICROBIAL PRODUCTION OF 3,4-DIHYDROXYBENZOATE FROM 4-NITROBENZOATE

Two years ago a new degradation pathway for nitroaromatic compounds, leading to catechol production, was found in a *Comamonas acidovorans* species (20, 21). *C. acidovorans* NBA-10 was isolated on 4-nitrobenzoate as the sole source of carbon and energy. In the presence of oxygen, the organism converted 4-nitrobenzoate to 3,4-dihydroxybenzoate via 4-nitroso- and 4-hydroxylaminobenzoate, but not via 4-aminobenzoate (Fig. 4A). A nitrobenzoate reductase catalysed the reduction of 4-nitrobenzoate to 4-hydroxylaminobenzoate, which was further converted to 3,4-dihydroxybenzoate by a hydroxylaminolyase. Reduction of 4-nitrobenzoate to 4-hydroxylaminobenzoate by intact cells of *C. acidovorans* NBA-10 could also be accomplished in the absence of oxygen, but then ethanol had to be added for cofactor regeneration.

This degradation pathway for nitroaromatic compounds has recently also been suggested for degradation of 4-nitrotoluene in two *Pseudomonas* species (26, 51). These two organisms initially oxidized the methyl group of 4-nitrotoluene to a carboxyl group and further mineralized the resulting 4-nitrobenzoate via 4-hydroxylaminobenzoate and 3,4-dihydroxybenzoate, by a pathway identical to the one in *C. acidovorans* NBA-10 (26, 51).

Nitrobenzoate reductase, which catalyses the reduction of 4-nitrobenzoate to 4-hydroxylaminobenzoate, was purified from cell-free extracts of *C. acidovorans* NBA-10 (21). The enzyme was NAD(P)H-dependent and NADPH was 15 times more effective as a cofactor than NADH. Incubation of 4-nitrobenzoate with purified enzyme and NADPH resulted in a stoichiometric conversion of the substrate to 4-hydroxylaminobenzoate. The amount of $NADP^+$ formed was approximately twice the amount of 4-nitrobenzoate converted. With 4-nitrosobenzoate as a substrate, equimolar amounts of 4-hydroxylaminobenzoate and $NADP^+$ were produced. 4-Nitrosobenzoate was unstable in the presence of NADPH and absence of 4-nitrobenzoate reductase, but upon addition of purified enzyme the conversion rate drastically increased.

Enzyme activity was not affected by the presence or absence of oxygen. The optimum temperature was 30°C and at 45°C enzyme activity was completely absent. The substrate specificity of the 4-nitrobenzoate reductase was very narrow. 4-Nitrophenylacetate and 3-nitrobenzoate, respectively, were reduced at relative enzyme activities of 6.9 and 1.9% (activity with 4-nitrobenzoate = 100%) and other nitroaromatic compounds tested were reduced at even lower relative activities or were not converted at all (Table 1).

Hydroxylaminolyase, catalysing the hydrolysis of 4-hydroxylaminobenzoate to 3,4-dihydroxybenzoate and ammonium, was also purified from cell-free extracts. The native enzyme had a molecular weight of 45 kDa. Optimum temperature and pH were 32°C and 7.0, respectively, and the affinity constant K_m of hydroxylaminolyase for 4-hydroxylaminobenzoate was 1.3 mM.

The specific activities of nitrobenzoate reductase and hydroxylaminolyase in crude cell-free extracts were approximately 0.6 and 0.15 U•(mg protein)$^{-1}$, respectively (20,

Table 1. Substrate specificity of purified 4-nitrobenzoate reductase from *Comamonas acidovorans* NBA-10 (data from 21, 100% corresponds with a specific enzyme activity of 11.6 U•(mg protein)$^{-1}$).

Substrate	Relative specific activity (%)
2-Nitrobenzoate	0
3-Nitrobenzoate	1.9
4-Nitrobenzoate	100
4-Nitrophenylalanine	0.4
4-Nitrophenylacetate	6.9
4-Nitrophenylchloride	0.5

21). The high affinity constant of hydroxylaminolyase for its substrate suggests that 4-hydroxylaminobenzoate accumulates during conversion of 4-nitrobenzoate. However, when intact cells were incubated with 4-nitrobenzoate under aerobic or oxygen-limited conditions, only traces (< 0.01 mM) of 4-hydroxylaminobenzoate could be detected in the incubation mixture (21). It is therefore questionable, also considering the specific activities of both enzymes in crude cell-free extracts mentioned above, whether this high affinity constant for 4-hydroxylaminobenzoate is also found *in vivo*.

Under anaerobic conditions, without any external cofactor added, crude cell-free extracts of *C. acidovorans* NBA-10 stoichiometrically converted 4-hydroxylaminobenzoate to 3,4-dihydroxybenzoate and ammonium. Dialysis of the crude extract, both under aerobic and anaerobic conditions, resulted in a complete loss of hydroxylaminolyase activity. Surprisingly, addition of NADH or NADPH to the incubation mixture restored enzyme activity, but only under anaerobic conditions. In the presence of oxygen, these cofactors did not restore enzyme activity.

Because the hydrolysis of 4-hydroxylaminobenzoate to 3,4-dihydroxybenzoate is a redox neutral conversion, it was difficult to interpret the role of NAD(P)H in this reaction. Determination of the NAD(P)H concentrations with HPLC during the conversion revealed that no net consumption of the cofactors occurred. However, this did not exclude the possibility that the reduced nucleotides were oxidized and regenerated by the same enzyme. Therefore, 4-hydroxylaminobenzoate was incubated with purified enzyme, NADH and $NADP^+$. Use and regeneration of cofactors would then yield NAD^+ and NADPH. Under these conditions, however, no formation of NAD^+ and NADPH was found, while the conversion rate of 4-hydroxylaminobenzoate was not affected (Fig. 6).

To investigate if NAD(P)H was only used to lower the redox potential, other reducing compounds were tested for this property. Indeed, ascorbate, cysteine and dithiothreitol also restored enzyme activity to 50, 69 and 100% of the activity obtained with NAD(P)H. Addition of dithiothreitol even revealed enzyme activity under aerobic, but unstirred, conditions. From these results it was concluded that NAD(P)H was not a substrate in the reaction, but was only used to lower the redox potential. Whether this low redox potential is needed for stabilization of the enzyme or for other reasons is still unclear.

Although Groenewegen and De Bont (20) and Groenewegen et al. (21) were the first to describe this novel degradation pathway for nitroaromatic compounds, doubts had already been raised in the older literature as to whether aminoaromatic compounds were true intermediates in the reductive breakdown of nitroaromatic compounds. Ke et al. (33) reported that a *Flavobacterium* species, grown on 2-nitrobenzoate, could immediately degrade 2-nitrosobenzoate and 2-hydroxylaminobenzoate, while conversion of 2-aminobenzoate showed a lag phase of approximately one hour. Inhibition of *de novo* protein synthesis

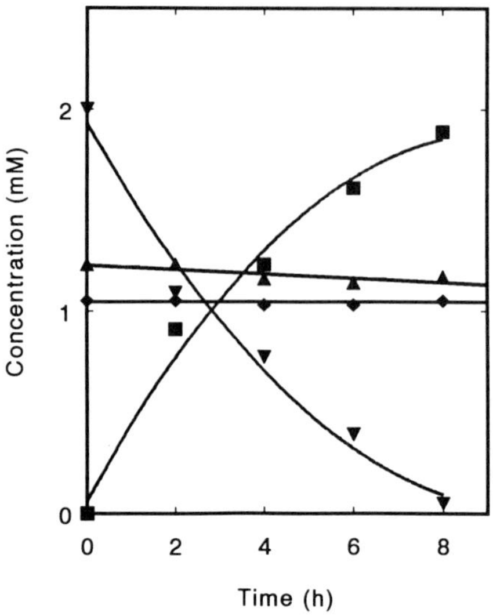

Figure 6. Conversion of 4-hydroxylaminobenzoate (▼) to 3,4-dihydroxybenzoate (■) in the presence of NADH (▲) and NADP$^+$ (◆) by purified hydroxylaminolyase under anaerobic conditions (data from 20).

by UV-irradiation prohibited degradation of 2-aminobenzoate, but conversion of 2-nitrosobenzoate and 2-hydroxylaminobenzoate was not affected (33).

Nocardia opaca transiently accumulated 2-aminobenzoate from 2-nitrobenzoate under oxygen-limited conditions (8). Upon appearance of 2-aminobenzoate, a degradative enzyme system for this compound was induced. This induction was very effectively inhibited by D-chloramphenicol, while growth on 2-nitrobenzoate was not affected in the presence of this antibiotic (8). When thorough aeration yielded fully aerobic conditions during growth on 2-nitrobenzoate, 2-aminobenzoate could not be detected and the organism was not adapted to this compound. Cain (8, 9) concluded that 2-aminobenzoate was not an intermediate in the degradation of 2-nitrobenzoate in this organism.

Nocardia erythropolis, grown on 4-nitrobenzoate, was able to oxidize 4-nitroso-, 4-hydroxylamino- and 3,4-dihydroxybenzoate, but was not induced for the oxidation of 4-aminobenzoate (11, 12). However, small quantities of 4-aminobenzoate were transiently formed in cultures growing on 4-nitrobenzoate. The authors suggested that formation of 4-aminobenzoate was a non-specific reaction of the nitroreductase, because many other non-degradable nitroaromatic compounds were reduced to their amino analogues by these cultures (12). When grown on 4-nitrobenzoate, but not when grown on 3,4-dihydroxybenzoate, *N. erythropolis* was able to oxidize 4-hydroxybenzoate. The latter compound could even be isolated from the supernatant of cultures growing on 4-nitrobenzoate (11). Therefore, Cartwright and Cain (11) suggested 4-nitroso-, 4-hydroxylamino-, 4-hydroxy- and 3,4-dihydroxybenzoate as intermediates in the degradation of 4-nitrobenzoate by *N. erythropolis*.

In *Pseudomonas putida* and some other members of this genus, no evidence was found for the intermediary formation of 3-aminophenol in the conversion of 3-nitrophenol (16, 70). Also no adaptation to 2- or 4-aminobenzoate was found during growth of these organisms on 2- or 4-nitrobenzoate, respectively (16).

Nishino and Spain (41) isolated a *Pseudomonas pseudoalcaligenes* that reduced nitrobenzene to hydroxylaminobenzene. Hydroxylaminobenzene was further converted to 2-aminophenol (Fig. 4B) and not to catechol and ammonium, as depicted in Figure 4A. This enzyme-mediated reaction parallels the chemical Bamberger rearrangement (55).

Sternson and Gammans (62) studied the metabolism of aromatic hydroxylamines in rat liver homogenates and found a hydroxylamino isomerase, catalysing the conversion of hydroxylaminoaromatic compounds to 2-aminophenols. In 1960, Cain and Cartwright (10) described the formation of 3-hydroxy-4-aminobenzoate in cultures of *N. erythropolis* growing on 4-nitrobenzoate and production of 3-hydroxy-2-aminobenzoate by *N. opaca* growing on 2-nitrobenzoate. These products may have been formed via a chemical or enzymatic isomerization of the transiently accumulating 2- and 4-hydroxylaminobenzoate. Conversion of a 4-chloro-hydroxylaminobenzene via a Bamberger rearrangement to 4-chloro-2-hydroxyaniline has also been observed in a yeast, *Rhodosporium* sp. (13) and rabbit-liver microsomes were also able to convert substituted hydroxylaminoaromatic compounds to the corresponding aminophenols (7).

HYPOTHETICAL PATHWAY FOR CONVERSION OF 4-HYDROXYLAMINOBENZOATE TO 3,4-DIHYDROXYBENZOATE

Two different hypothetical pathways can be suggested for the conversion of 4-hydroxylaminobenzoate to 3,4-dihydroxybenzoate (22). Both reaction schemes start with the protonation and rearrangement of 4-hydroxylaminobenzoate to a positively charged intermediate (X) (Fig. 7). In pathway A, this compound undergoes a nucleophilic attack of water on the positively charged carbon in the aromatic ring. Transfer of a proton from the incoming water molecule to the amino group results in the release of ammonia and a proton and 3,4-dihydroxybenzoate (Z) is produced.

In the other reaction scheme (pathway B, Fig. 7), deprotonation of the intermediary compound (X) results in the formation of the stable intermediate 3-hydroxy-4-aminobenzoate (Y). In the further conversion of the aminophenol, the carboxyl function on the aromatic ring is very important as it stabilizes the different resonance structures. These resonance structures enable the nucleophilic attack of a water molecule on the positively charged carbon atom in the ring. After a few rearrangement reactions, ammonia leaves the ring and the catechol (Z) is formed (Fig. 7).

The main difference between the pathways depicted in Figure 7 is the involvement of the carboxyl function in the formation of the resonance-stabilized structures susceptible to further reaction. If the carboxyl function at the 1-position is indeed important in the conversion of hydroxylaminoaromatic compounds to catechols, this might explain why catechol formation has only been found from 4-hydroxylaminobenzoate (20, 26, 51) and not from hydroxylaminobenzene (41), 1-hydroxy-1-phenyl-3-methyl urea (62), 1-chloro-4-hydroxylaminobenzene (13) and hydroxylaminobiphenyls (7). These hydroxylaminoaromatic compounds, lacking a carboxyl function at the 1-position, were converted to the corresponding aminophenols.

Conversion of hydroxylaminoaromatic compounds via scheme A (Fig. 7) also implies catechol formation from substrates without a carboxyl group at the 1-position. The results in the literature, therefore, suggest that hydroxylaminoaromatic compounds are converted via the hypothetical scheme B. Substrates with a resonance structure-stabilizing function at the 1-position which are able to create a positively charged aromatic nucleus by withdrawal of a pair of electrons, like a carboxyl or a cyanide group, are then expected to lead to catechol formation and other substrates are expected to yield aminophenols. However, thorough characterization of the intermediates formed during conversion of different substituted hydroxylaminoaromatic compounds, together with the purification and characterization of the enzymes involved, must yield the experimental evidence for the correct reaction pathway.

To investigate the possibility of producing catechols from a variety of substituted hydroxylaminoaromatic compounds, we are currently investigating the substrate specificity of the hydroxylaminolyase from *C. acidovorans* NBA-10 and more recently isolated organisms. These studies should also yield additional information about the pathway of the degradation of hydroxylaminoaromatic compounds by these types of enzymes.

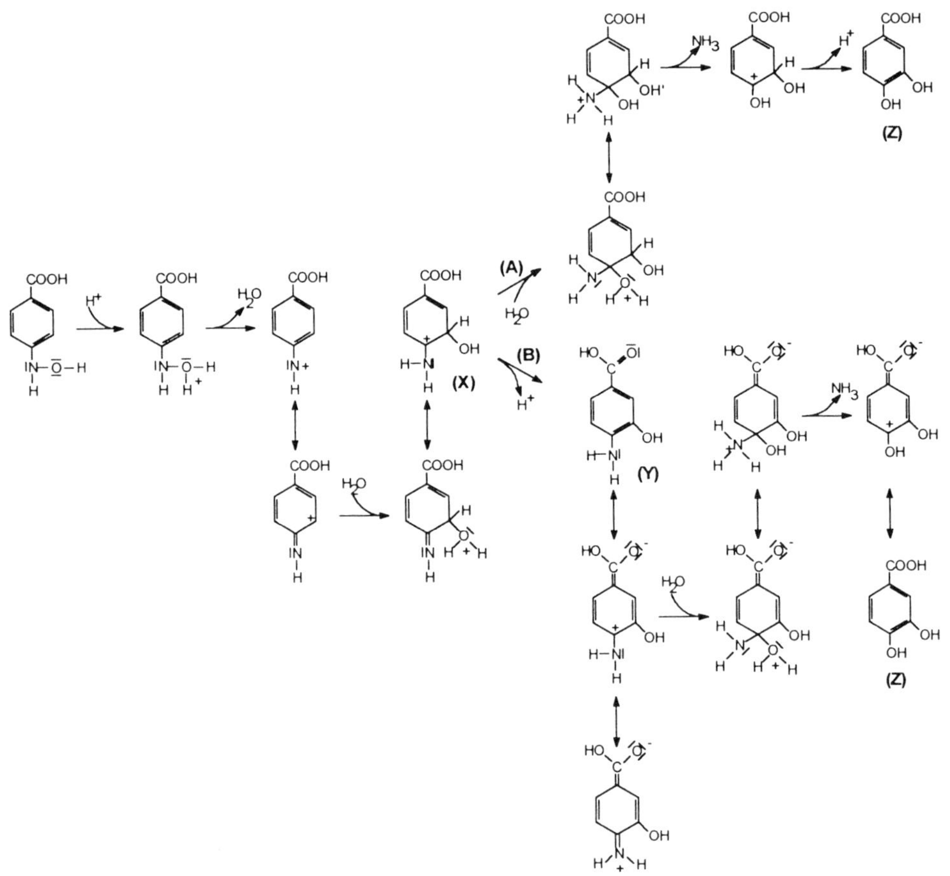

Figure 7. Hypothetical pathways for conversion of 4-hydroxylaminobenzoate to 3,4-dihydroxybenzoate and ammonium.

PRODUCTION OF SUBSTITUTED CATECHOLS FROM HYDROXYLAMINOAROMATIC COMPOUNDS

In the absence of oxygen, washed cells of *Comamonas acidovorans* NBA-10 stoichiometrically converted 4-nitrobenzoate to 3,4-dihydroxybenzoate if ethanol was added for cofactor regeneration. Although growing cells were totally inhibited by 4-nitrobenzoate concentrations above 5 mM (21), resting cell suspensions of *C. acidovorans* NBA-10 converted 50 mM concentrations of 4-nitrobenzoate at full activity (Fig. 8). The range of catechols which potentially can be produced by *C. acidovorans* NBA-10 and new isolates from a variety of nitroaromatic compounds is now being investigated.

Eight microorganisms were isolated on 4-nitrobenzoate. Intact cells and cell-free extracts of five of these strains (F1, F3, F4, F5 and L4) converted 4-nitrobenzoate to 3,4-dihydroxybenzoate and ammonium under anaerobic conditions and, hence, contained a hydroxylaminolyase. Seven microorganisms were isolated on 2-nitrobenzoate and four on 4-nitrophenylalanine, but none of these strains contained a hydroxylaminolyase.

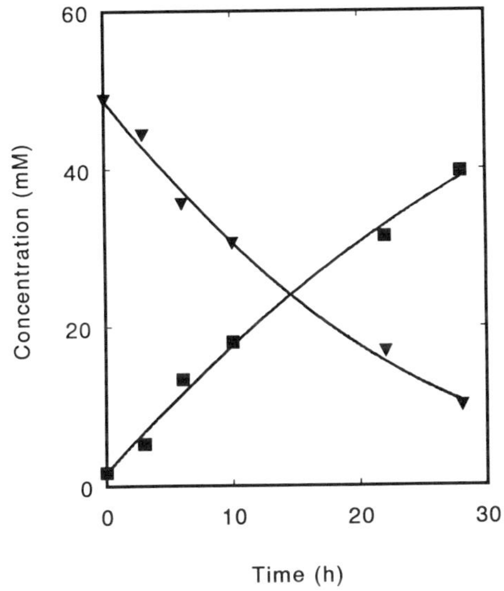

Figure 8. Anaerobic conversion of 4-nitrobenzoate (▼) to 3,4-dihydroxybenzoate (■) by washed cells of *Comamonas acidovorans* NBA-10.

In contrast to *C. acidovorans* NBA-10 and *Pseudomonas* 4NT (26), the new isolates F1, F3, F4, F5 and L4 were all able to grow on 2- and 4-aminobenzoate. Although 4-aminobenzoate is not an intermediate in the degradation pathway of 4-nitrobenzoate, growth on this compound after a short lag by 4-nitrobenzoate-grown cells has been reported previously for *Nocardia erythropolis* (11, 12) and *Pseudomonas putida* (16). The isolates F4 and F5 could also grow on 4-aminotoluene and L4 used 4-aminoanisole as a growth substrate. None of the lyase-containing microorganisms were able to grow on nitrobenzene, aniline, 2-nitrotoluene, 4-nitrotoluene, 2-aminotoluene, 2-nitrobenzoate, 3-nitrobenzoate, 3-aminobenzoate, 4-nitrobenzonitrile, 4-aminobenzonitrile, 4-nitrophenylacetic acid, 4-aminophenylacetic acid, 4-nitroanisole and 1-chloro-4-nitrobenzene and its amino derivative.

To study the substrate specificity of the hydroxylaminolyases, a set of substituted hydroxylaminoaromatic compounds was synthesized (Fig. 9). In initial experiments, the substrate-dependent oxygen uptake by 4-nitrobenzoate-grown cell suspensions was investigated (Table 2). The synthesized hydroxylaminoaromatic compounds and the corresponding nitroaromatic compounds were used as substrates. It is clear from Table 2 that 4-nitrobenzoate and 4-hydroxylaminobenzoate showed the highest oxygen uptake rate. However, 1-chloro-4-nitrobenzene, 4-nitrobenzonitrile, 4-nitroanisole and, to a

Hydroxylaminoaromatic	R1	R2	R3
Hydroxylaminobenzene	H	H	H
2-Hydroxylaminotoluene	CH3	H	H
4-Hydroxylaminotoluene	H	H	CH3
2-Hydroxylaminobenzoic acid	COOH	H	H
3-Hydroxylaminobenzoic acid	H	COOH	H
4-Hydroxylaminobenzoic acid	H	H	COOH
4-Hydroxylaminobenzonitrile	H	H	CN
4-Hydroxylaminoanisol	H	H	OCH3
4-Hydroxylaminophenylacetic acid	H	H	CH2COOH
Methyl-4-hydroxylaminobenzoate	H	H	COOCH3
1-Chloro-4-hydroxylaminobenzene	H	H	Cl

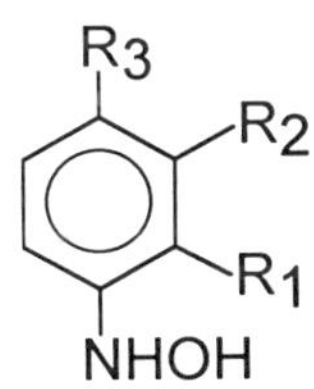

Figure 9. Substituted hydroxylaminoaromatic compounds, synthesized for the investigation of the substrate specificity of the hydroxylaminolyases.

Table 2. Substrate-dependent oxygen uptake rates by 4-nitrobenzoate-grown cell suspensions. The synthetic hydroxylaminoaromatic compounds and the corresponding nitroaromatic compounds were used as substrates

	Relative oxygen uptake rate (%)						
Substrate	F1	F3	F4	F5	L4	NBA-10[1]	4NT[2]
Nitrobenzene	0	8	20	18	0	12	0
Hydroxylaminobenzene	5	21	33	34	0	23	16
2-Nitrotoluene	ND	ND	ND	23	0	28	0
2-Hydroxylaminotoluene	5	8	22	18	0	12	11
4-Nitrotoluene	0	10	15	31	0	31	23
4-Hydroxylaminotoluene	0	9	16	18	0	9	6
2-Nitrobenzoate	0	7	0	8	0	5	0
2-Hydroxylaminobenzoate	0	0	0	6	9	0	0
3-Nitrobenzoate	0	0	7	11	0	0	0
3-Hydroxylaminobenzoate	0	0	20	11	0	0	5
4-Nitrobenzoate	100	100	100	100	100	100	100
4-Hydroxylaminobenzoate	79	116	133	37	149	88	99
4-Nitrobenzonitrile	13	27	46	47	0	12	0
4-Hydroxylaminobenzonitrile	8	12	15	16	0	0	7
4-Nitroanisole	18	14	29	27	14	19	0
4-Hydroxylaminoanisole	11	13	21	18	0	0	0
4-Nitrophenylacetic acid	0	0	0	11	0	8	0
4-Hydroxylaminophenylacetic acid	13	21	12	20	0	0	0
Methyl-4-nitrobenzoate	ND	ND	ND	56	0	23	0
Methyl-4-hydroxylaminobenzoate	12	16	20	20	0	0	10
1-Chloro-4-nitrobenzene	40	62	61	48	0	27	0
1-Chloro-4-hydroxylaminobenzene	14	13	22	26	0	6	11

[1] *Comamonas acidovorans* NBA-10.
[2] *Pseudomonas* 4NT.
ND: not determined.

lesser extent, hydroxylaminobenzene, 4-nitrotoluene and 1-chloro-4-hydroxylaminobenzene also revealed a significant increase in oxygen uptake rate with several isolates. The isolates F4 and F5 showed an increased oxygen uptake rate with many different nitro- and hydroxylaminoaromatic compounds, while L4 and *Pseudomonas* 4NT hardly oxidized any substrate except 4-hydroxylamino- and 4-nitrobenzoate (Table 2). A reason for this apparent limited oxygen uptake with substrates other than the growth substrate may be the narrow substrate specificity of an involved transport system. If hydroxylamino- and nitroaromatic compounds have to be actively transported into the cytoplasm, the 4-nitrobenzoate transport protein may not be able to transport other substrates, hence preventing their oxidation. Another possibility is that the substrates are converted into the corresponding aminophenols whose oxidation capacity is not induced, or that the substrates are indeed converted into catechols which are not further oxidized by the 3,4-dihydroxybenzoate dioxygenase. In view of the production of substituted catechols from nitroaromatic compounds the latter possibility would be very interesting. Therefore, to determine the substrate spectrum specifically of the hydroxylaminolyases and to characterize the reaction products, anaerobic incubations of 4-nitrobenzoate-grown cell suspensions and cell-free extracts with hydroxylaminoaromatic compounds must be performed.

DIRECTIONS FOR FUTURE RESEARCH

In view of the environmental problems associated with the presence of nitroaromatic compounds in waste and surface waters, soils and air, the need for remediation processes and microorganisms able to degrade these compounds will increase within the next few years. In practice, polluted sites will often contain different nitroaromatic compounds, therefore, microorganisms containing degradative enzyme systems with broad substrate specificities will have advantages over the specialists. Isolation of versatile organisms and investigation of their growth and degradation of nitroaromatic compounds is, therefore, not only a challenge for the potential production of substituted catechols for the chemical and pharmaceutical industry, as described in this chapter, but may also become of great environmental importance in the bioremediation of polluted sites.

ACKNOWLEDGEMENTS

The authors are grateful to Dr. J. C. Spain (Tyndall Air Force Base, Florida, USA) and Dr. J. A. Jongejan (Delft University of Technology, The Netherlands) for kindly providing *Pseudomonas* 4NT and preparation of 4-hydroxylaminophenylacetic acid, respectively.

REFERENCES

1. Adachi, K., Y. Takeda, S. Senoh, and H. Kita. 1964. Metabolism of *p*-hydroxyphenylacetic acid in *Pseudomonas ovalis*. Biochim. Biophys. Acta 93:483-493.
2. Angermeier, L., and H. Simon. 1983. On the reduction of aliphatic and aromatic nitro compounds by clostridia, the role of ferredoxin and its stabilization. Hoppe-Seyler's Z. Physiol. Chem. 364:961-975.
3. Angermeier, L., and H. Simon. 1983. On nitroaryl reductase activities in several clostridia. Hoppe-Seyler's Z. Physiol. Chem. 364:1653-1664.
4. Bont, J. A. M. de, M. J. A. W. Vorage, S. Hartmans, and W. J. J. van den Tweel. 1986. Microbial degradation of 1,3-dichlorobenzene. Appl. Environ. Microbiol. 52:677-680.
5. Boopathy, R., and C. F. Kulpa. 1993. Nitroaromatic compounds serve as nitrogen source for *Desulfovibrio* sp. (B strain). Can. J. Microbiol. 39:430-433.
6. Boopathy, R., C. F. Kulpa, and M. Wilson. 1993. Metabolism of 2,4,6-trinitrotoluene (TNT) by *Desulfovibrio* sp. (B strain). Appl. Microbiol. Biotechnol. 39:270-275.
7. Booth, J., and E. Boyland. 1964. The biochemistry of aromatic amines. 10. Enzymic N-hydroxylation of arylamines and conversion of arylhydroxylamines into *o*-aminophenols. Biochem. J. 91:362-369.
8. Cain, R. B. 1966. Induction of an anthranilate oxidation system during the metabolism of *ortho*-nitrobenzoate by certain bacteria. J. Gen. Microbiol. 42:197-217.
9. Cain, R. B. 1966. Utilization of anthranilic and nitrobenzoic acids by *Nocardia opaca* and a *Flavobacterium*. J. Gen. Microbiol. 42:219-235.
10. Cain, R. B., and N. J. Cartwright. 1960. Intermediary metabolism of nitrobenzoic acids by bacteria. Nature 185:868-869.
11. Cartwright, N. J., and R. B. Cain. 1959. Bacterial degradation of the nitrobenzoic acids. Biochem. J. 71:248-260.
12. Cartwright, N. J., and R. B. Cain. 1959. Bacterial degradation of the nitrobenzoic acids. 2. Reduction of the nitro group. Biochem. J. 73:305-314.
13. Corbett, M. D., and B. R. Corbett. 1981. Metabolism of 4-chloronitrobenzene by the yeast *Rhodosporidium* sp. Appl. Environ. Microbiol. 41:942-949.
14. Dickel, O., and H.-J. Knackmuss. 1991. Catabolism of 1,3-dinitrobenzene by *Rhodococcus* sp. QT-1. Arch. Microbiol. 157:76-79.
15. Ecker, S., T. Widmann, H. Lenke, O. Dickel, P. Fischer, C. Bruhn, and H.-J. Knackmuss. 1992. Catabolism of 2,6-dinitrophenol by *Alcaligenes eutrophus* JMP 134 and JMP 222. Arch. Microbiol. 158:149-154.

16. Germanier, R., and K. Wuhrmann. 1963. Über den aeroben mikrobiellen Abbau aromatischer Nitroverbindungen. Path. Microbiol. 26:569-578.
17. Gibson, D. T., G. E. Cardini, F. C. Maseles, and R. E. Kallio. 1970. Incorporation of oxygen-18 into benzene by *Pseudomonas putida*. Biochemistry 9:1631-1635.
18. Gibson, D. T., M. Hensley, H. Yoshioka, and T. J. Mabry. 1970. Formation of (+)-*cis*-2,3-dihydroxy-1-methylcyclohexa-4,6-diene from toluene by *Pseudomonas putida*. Biochemistry 9:1626-1630.
19. Gibson, D. T., J. R. Koch, C. L. Schuld, and R. E. Kallio. 1968. Oxidative degradation of aromatic hydrocarbons by microorganisms. II. Metabolism of halogenated aromatic hydrocarbons. Biochemistry 7:3795-3802.
20. Groenewegen, P. E. J., and J. A. M. de Bont. 1992. Degradation of 4-nitrobenzoate via 4-hydroxylaminobenzoate and 3,4-dihydroxybenzoate in *Comamonas acidovorans* NBA-10. Arch. Microbiol. 158:381-386.
21. Groenewegen, P. E. J., P. J. Breeuwer, J. M. L. M. van Helvoort, A. M. M. Langenhoff, F. P. de Vries, and J. A. M. de Bont. 1992. Novel degradative pathway of 4-nitrobenzoate in *Comamonas acidovorans* NBA-10. J. Gen. Microbiol. 138:1599-1605.
22. Groot, A. E. de (Wageningen Agricultural University, The Netherlands). 1994. Personal communication.
23. Gunsalus, I. C., T. C. Pederson, and S. G. Sligar. 1975. Oxygenase-catalyzed biological hydroxylations. Ann. Rev. Biochem. 44:377-407.
24. Gunstone, F. D. 1960. Hydroxylation methods. Adv. Org. Chem. 1:103-147.
25. Haigler, B. E., and J. C. Spain. 1991. Biotransformation of nitrobenzene by bacteria containing toluene degradative pathways. Appl. Environ. Microbiol. 57:3156-3162.
26. Haigler, B. E., and J. C. Spain. 1993. Biodegradation of 4-nitrotoluene by *Pseudomonas* sp. strain 4NT. Appl. Environ. Microbiol. 59:2239-2243.
27. Hanne, L. F., L. L. Kirk, S. M. Appel, A. D. Narayan, and K. K. Bains. 1993. Degradation and induction specificity in *Actinomycetes* that degrade *p*-nitrophenol. Appl. Environ. Microbiol. 59:3505-3508.
28. Higson, F. K. 1992. Microbial degradation of nitroaromatic compounds. Adv. Appl. Microbiol. 37:1-19.
29. Högn, T., and L. Jaenicke. 1972. Benzene metabolism of *Moraxella* sp. Eur. J. Biochem. 30:369-375.
30. Hosokawa, K., and R. Y. Stanier. 1966. Crystallization and properties of *p*-hydroxybenzoate hydroxylase from *Pseudomonas putida*. J. Biol. Chem. 241:2453-2460.
31. Jain, R. K., J. H. Dreisbach, and J. C. Spain. 1994. Biodegradation of *p*-nitrophenol through 1,2,4-benzenetriol by an *Arthrobacter* sp. Appl. Environ. Microbiol. 60:3030-3032.
32. Johnston, J. B., and V. Renganathan. 1987. Production of substituted catechols from substituted benzenes by a *Pseudomonas* sp. Enz. Microb. Technol. 9:706-708.
33. Ke, Y.-H., L. L. Gee, and N. N. Durham. 1959. Mechanism involved in the metabolism of nitrophenylcarboxylic acids by microorganisms. J. Bacteriol. 77:593-598.
34. Kinouchi, T., Y. Manabe, K. Wakisaki, and Y. Ohnishi. 1982. Biotransformation of 1-nitropyrene in intestinal anaerobic bacteria. Microbiol. Immunol. 26:993-1005.
35. Lenke, H., and H.-J. Knackmuss. 1992. Initial hydrogenation during catabolism of picric acid by *Rhodococcus erythropolis* HL 24-2. J. Bacteriol. 58:2933-2937.
36. March, J. 1985. Advanced organic chemistry. Reactions, mechanisms, and structure, p. 498. Wiley & Sons, New York.
37. Marvin-Sikkema, F. D., and J. A. M. de Bont. 1994. Degradation of nitroaromatic compounds by microorganisms. Appl. Microbiol. Biotechnol.42:499-507.
38. McCormick, N. G., F. E. Feeherry, and H. S. Levinson. 1976. Microbial transformation of 2,4,6-trinitrotoluene and other nitroaromatic compounds in sewage effluent. Appl. Environ. Microbiol. 31:949-958.
39. Michalover, J. L., and D. W. Ribbons. 1973. 3-Hydroxybenzoate 4-hydroxylase from *Pseudomonas testosteroni*. Biochim. Biophys. Res. Comm. 55:888-896.
40. Milne, G. W. A., P. Goldman, and J. L. Holzman. 1968. The metabolism of 2-fluorobenzoic acid. II. Studies with $^{18}O_2$. J. Biol. Chem. 243:5347-5376.
41. Nishino, S. F., and J. C. Spain. 1993. Degradation of nitrobenzene by a *Pseudomonas pseudoalcaligenes*. Appl. Environ. Microbiol. 59:2520-2525.
42. Nörtemann, B., C. Bruhn, and H.-J. Knackmuss. 1986. Recruitment of complementary catabolic activities for mineralization of aminonaphthalene sulfonates and chloronitrophenols. EMBO Workshop: Genetic Manipulation of Pseudomonads. Applications in Biotechnology and Medicine.
43. Nozaki, M., and O. Hayaishi. 1984. Dioxygenases and monooxygenases, p. 68-104. *In* J. V. Bannister and W. H. Bannister (ed.) The biology and chemistry of active oxygen. Elsevier, Oxford.
44. Olah, G. A., A. P. Fung, and T. Keumi. 1981. Oxyfunctionalization of hydrocarbons. Hydroxylation of benzene and alkylbenzenes with hydrogen peroxide in hydrogen fluoride/boron trifluoride. J. Org. Chem. 46:4306-4307.

45. Preuss, A., J. Fimpel, and G. Diekert. 1993. Anaerobic transformation 2,4,6-trinitrotoluene (TNT). Arch. Microbiol. 159:345-353.
46. Pshirkov, S. Y., O. I. Boiko, E. A. Kiprianova, and I. I. Starovoitov. 1982. Transformation of L-tyrosine into L-dihydroxyphenylalanine by *Pseudomonas* cultures. Mikrobiologiya 51:272-274.
47. Rafii, F., W. Franklin, R. H. Reflich, and C. E. Cerniglia. 1991. Reduction of nitroaromatic compounds by anaerobic bacteria isolated from the human gastrointestinal tract. Appl. Environ. Microbiol. 57:962-968.
48. Raymond, D. G. M., and M. Alexander. 1971. Microbial metabolism and cometabolism of nitrophenols. Pest. Biochem. Physiol. 1:123-130.
49. Reiner, A. M., and G. D. Hegeman. 1971. Metabolism of benzoic acid by bacteria. Accumulation of (-)-3,5-cyclohexadiene-1,2-diol-1-carboxylic acid by a mutant strain of *Alcaligenes eutrophus*. Biochemistry 13:2530-2536.
50. Renganathan, V., and J. B. Johnston. 1989. Catechols of novel substrates produced using the toluene ring oxidation pathway of *Pseudomonas* sp. strain T-12. Appl. Microbiol. Biotechnol. 31:419-424.
51. Rhys-Williams, W., S. C. Taylor, and P. A. Williams. 1993. A novel pathway for the catabolism of 4-nitrotoluene by *Pseudomonas*. J. Gen. Microbiol. 139:1967-1972.
52. Rieger, P.-G, A. Preuss, H. Lenke, H.-J Knackmuss. 1994. H^--additions as initial steps of aerobic bacterial degradation of 2,4,6-trinitrophenol (picric acid), abstr. Q-120, p. 409. Abstr. 94th Annu. Meet. Am. Soc. Microbiol. 1994.
53. Schraa, G., M. L. Boone, M. S. M. Jetten, A. R. W. van Neerven, P. J. Colberg, A. J. B. Zehnder. 1986. Degradation of 1,4-dichlorobenzene by *Alcaligenes* sp. strain A175. Appl. Environ. Microbiol. 52:1374-1381.
54. Schraa, G., B. M. Bethe, A. R. W. van Neerven, W. J. J. van den Tweel, E. van der Ende, and A. J. B. Zehnder. 1987. Degradation of 1,2-dimethylbenzene by *Corynebacterium* strain C125. Antonie van Leeuwenhoek 53:159-170.
55. Shine, H. J. 1967. Aromatic rearrangements, p. 182-190. *In* C. Eaborn and N. B. Chapman (ed.) Reaction mechanisms in organic chemistry. Elsevier Publishing Company, Amsterdam.
56. Shirai, K. 1987. Catechol production from benzene through reaction with resting and immobilized cells of a mutant strain of *Pseudomonas*. Agric. Biol. Chem. 51:121-128.
57. Singh, D. V., P. P. Mukherjee, S. P. Pal, and P. K. Bhattacharyya. 1973. Microbial synthesis of L-DOPA (L-3,4-dihydroxy L-phenylalanine) by a *Pseudomonas* mutant D101. J. Ferment. Technol. 51:713-718.
58. Spain, J. C., and D. T. Gibson. 1991. Pathway for biodegradation of *p*-nitrophenol in a *Moraxella* sp. Appl. Environ. Microbiol. 57:812-819.
59. Spain, J. C., and S. F. Nishino. 1987. Degradation of 1,4-dichlorobenzene by a *Pseudomonas* sp. Appl. Environ. Microbiol. 53:1010-1019.
60. Spain, J. C., O. Wyss, and D. T. Gibson. 1979. Enzymatic oxidation of *p*-nitrophenol. Biochem. Biophys. Res. Comm. 88:634-641.
61. Spanggord, R. J., J. C. Spain, S. F. Nishino, and K. E. Mortelmans. 1991. Biodegradation of 2,4-dinitrotoluene by a *Pseudomonas* sp. Appl. Environ. Microbiol. 57:3200-3205.
62. Sternson, L. A., and R. E. Gammans. 1975. A mechanistic study of aromatic hydroxylamine rearrangement in the rat. Bioorg. Chem. 4:58-63.
63. Sudhakar-Barik, R. Siddaramappa, P. A. Wahid, and N. Sethunathan. 1978. Conversion of *p*-nitrophenol to 4-nitrocatechol by a *Pseudomonas* sp. Antonie van Leeuwenhoek 44:171-176.
64. Tweel, W. J. J. van den. 1988. Thesis. Wageningen Agricultural University, The Netherlands.
65. Vorbeck, C., H. Lenke, P. Fischer, and H.-J Knackmuss. 1994. Identification of a hydride-Meisenheimer complex as a metabolite of 2,4,6-trinitrotoluene by a *Mycobacterium* strain. J. Bacteriol. 176:932-934.
66. Whited, G. M., W. R. McCombie, L. D. Kwart, and D. T. Gibson. 1986. Identification of *cis*-diols as intermediates in the oxidation of aromatic acids by a strain of *Pseudomonas putida* that contains a TOL plasmid. J. Bacteriol. 166:1028-1039.
67. Yamamoto, S., M. Katagiri, H. Maeno, and O. Hayaishi. 1965. Salicylate hydroxylase, a monooxygenase requiring flavin adenine dinucleotide. Purification and properties. J. Biol. Chem. 240:3408-3413.
68. Yoshida, H., Y. Tanaka, and K. Nakayama. 1973. Production of 3,4-dihydroxyphenyl-L-alanine (L-DOPA) and its derivatives by *Vibrio tyrosinaticus*. Agric. Biol. Chem. 37:2121-2126.
69. Yoshida, H., Y. Tanaka, and K. Nakayama. 1974. Production of 3,4-dihydroxyphenyl-L-alanine (L-DOPA) and its derivatives by *Pseudomonas melanogenum*. Agric. Biol. Chem. 38:455-462.
70. Zeyer, J., and P. C. Kearney. 1984. Degradation of *o*-nitrophenol and *m*-nitrophenol by a *Pseudomonas putida*. J. Agric. Food Chem. 32:238-242.
71. Zeyer, J., and H. P. Kocher. 1988. Purification and characterization of a bacterial nitrophenol oxygenase which converts *ortho*-nitrophenol to catechol and nitrite. J. Bacteriol. 170:1789-1794.

72. Zeyer, J., H. P. Kocher, and K. N. Timmis. 1986. Influence of *para*-substituents on the oxidative metabolism of *o*-nitrophenols by *Pseudomonas putida* B2. Appl. Environ. Microbiol. 52:334-339.
73. Zeyer, J., A. Wasserfallen, and K. N. Timmis. 1985. Microbial mineralization of ring-substituted anilines through an *ortho*-cleavage pathway. Appl. Environ. Microbiol. 50:447-453.

4

POTENTIAL OF TOLUENE-DEGRADING SYSTEMS FOR THE CONSTRUCTION OF HYBRID PATHWAYS FOR NITROTOLUENE METABOLISM

Juan L. Ramos, Alí Haïdour, Asunción Delgado, Estrella Duque, María-Dolores Fandila, Matilde Gil, and Guadalupe Piñar

Consejo Superior de Investigaciones Científicas
Estación Experimental del Zaidín
Department of Biochemistry and Molecular and Cellular Plant Biology
Apdo 419
18008 Granada
Spain

INTRODUCTION

Wastes from manufacturing processes, together with the widespread use of some nitroaromatic compounds, have polluted waters and soils. Because of their toxicity and mutagenicity, some of these compounds have been designated pollutants whose elimination is a high priority (25). A variety of alternative treatments including physical, chemical and biological (21), have been developed to reduce or eliminate contamination due to hazardous nitroorganic compounds. Physical treatments comprise absorption to activated carbon, filtration and incineration. Chemical approaches involve solvent extraction or surfactant precipitation. These treatments are expensive, and may in some cases generate unwanted products (22). The biological approach, including continuous and batch treatments of liquid wastes, and composting and soil treatments, is usually cheaper and involves less risk to human health or the environment (12, 23, 54). However, nitroaromatic compounds are relatively refractory to mineralization by microbes.

It has been suggested that aromatic hydrocarbons with nitro substituents are resistant to electrophilic attack by dioxygenases (2). This is attributed to the electron-withdrawing properties of nitro groups on the aromatic ring. Not all nitroaromatic compounds are equally recalcitrant; for example, polar nitroaromatic compounds, i.e. mononitrobenzoates and the mononitrophenols, are mineralized by a wide variety of bacteria and yeasts (6, 7, 14, 24, 49, 58). In contrast, nonpolar nitroaromatic compounds are more resistant to biodegradation. However, it was recently reported that some nitrotoluenes, nitrobenzenes and other compounds are degraded under laboratory conditions, in activated sludges, and in several municipal and industrial plant waste water facilities (10, 50). Several studies have shown that bacteria can transform nitrotoluenes, via nitroso and hydroxylamino intermediates, to aromatic amines, which usually accumulate in the culture medium as byproducts (5, 15, 21, 23, 30, 31, 35, 36, 37). In the aerobic biotransformation of dinitro- and trinitrotoluenes, azoxy compounds arise from the condensation of partially reduced intermediates (15, 30, 31). None of these reduction products seem to be further degraded.

Biodegradation of hazardous nitroaromatic compounds can be considered complete only when nitro groups have been removed and the carbon skeleton has been broken down to intermediates, which will eventually be oxidized to CO_2 (mineralization). The biotreatment of waste streams containing nitroaromatic compounds may require concerted action by a series of microorganisms to achieve full degradation. In this context, the metabolism of chemicals by the action of a consortium has been observed. For example the degradation of the insecticide DDT and of herbicides such as dalapon and parathion, requires several microorganisms (46, 48, 51). Alternatively, a single microorganism able to eliminate nitroaromatic compounds may be produced by appropriate recruitment of catabolic functions, as described for the metabolism of haloaromatic compounds (39, 42, 44).

Strategies for biodegradation of the isomeric nitrotoluenes can be developed through: i) an intensive search for microbes able to use these compounds as the sole carbon and nitrogen sources; ii) through the establishment of consortia, so that partial metabolism by two or more microbes leads to mineralization of the target compound; or iii) through the assemblage of several pathway segments in a single microorganism to achieve the conversion of nonpolar nitroaromatic compounds into biodegradable polar intermediates, which will eventually be mineralized through the host pathways.

The evolution of new metabolic activities in the laboratory may be helpful because the frequency and type of genetic events needed (mutation, alteration of gene expression, gene transfer, etc.) can be carefully controlled, and selective conditions can be optimized (40). There are basically three experimental approaches to the laboratory evolution of metabolic pathways (40). Chemostat selection provides long-term selective pressure for the evolution of organisms able to eliminate novel substrates. *In vivo* gene transfer involves the recruitment of critical genes into a pathway of another organism through the exploitation of natural processes such as conjugation, transformation and transduction. This phenomenon occurs in nature because catabolic pathways are frequently located on transmissible plasmids. *In vitro* evolution involves the selective introduction of cloned and well-characterized genes in a host organism in order to create a new pathway (see 40 for a review). Although all three approaches constitute powerful experimental tools to bring about the desired phenotypic changes in microbes, *in vitro* evolution has the added advantage of being highly controlled, of producing predictable changes, and of making multiple changes possible in a single step, without the need to directly select the final phenotype desired. However, it requires detailed information on the genetics and biochemistry of the key elements of the pathways and the enzymatic steps to be manipulated. In contrast, chemostat selection and *in vivo* evolution can often be done successfully without extensive prior characterization of the biochemistry and the genetics of the catabolic pathway.

MINERALIZATION AND BIOTRANSFORMATION OF MONONITROTOLUENES

Mineralization of 4-Nitrotoluene by *Pseudomonas* sp.

4-Nitrotoluene and 2-nitrotoluene (17) are currently known to be mineralized by microbes, whereas 3-nitrotoluene is more recalcitrant. Independent reports from the groups of Peter Williams (45) and Jim Spain (16) documented mineralization of 4-nitrotoluene by bacteria belonging to the genus *Pseudomonas*.

Spain's group isolated a strain of *Pseudomonas* from nitrobenzene-contaminated soil by selective enrichment on 4-nitrotoluene as the sole source of carbon, nitrogen, and energy. Williams' group in the U.K. used selective enrichment to isolate eleven different pseudomonads able to grow on 4-nitrotoluene as the sole carbon source, although only one isolate, TW3, was characterized in detail. Both *Pseudomonas* sp. 4NT and TW3 also grew on 4-nitrobenzyl alcohol, 4-nitrobenzaldehyde, and 4-nitrobenzoate. Haigler and Spain (16) detected 4-nitrobenzoate and ammonia in the culture fluid of glucose-grown cells during induction with 4-nitrotoluene. Rhys-Williams et al. (45) found 4-nitrobenzaldehyde and 4-nitrobenzoate when 4-nitrotoluene-grown cells were incubated with 4-nitrobenzyl alcohol. Washed suspensions of 4-nitrotoluene- or 4-nitrobenzoate-grown cells oxidized 4-nitrotoluene, 4-nitrobenzyl alcohol, 4-nitrobenzaldehyde and protocatechuate. Both groups found that extracts prepared from induced cells contained 4-nitrobenzaldehyde dehydrogenase and 4-nitrobenzyl alcohol dehydrogenase, and both groups showed that the oxidation of nitrobenzyl alcohol was dependent on NAD^+, however; differences in the electron acceptor were found for benzaldehyde dehydrogenases. Metabolism of 4-nitrobenzoate occurred via its conversion to 4-hydroxylaminobenzoate in NADPH-dependent reactions. 4-Hydroxylaminobenzoate was then converted to protocatechuate (16, 45). Haigler and Spain (16) found that the ring fission of protocatechuate was catalyzed by protocatechuate 4,5-dioxygenase, whereas Rhys-Williams et al. (45) suggested that the strain they isolated used protocatechuate 3,4-dioxygenase. These results indicate that 4-nitrotoluene was degraded by an initial oxidation of the methyl group to form 4-nitrobenzyl alcohol, which was in turn converted to 4-nitrobenzoate via 4-nitrobenzaldehyde (Fig. 1). The 4-nitrobenzoate was reduced to 4-hydroxylaminobenzoate, which was then converted to protocatechuate (Fig. 1, see also Spain, this volume).

The detection of 4-nitrobenzaldehyde and 4-nitrobenzyl alcohol dehydrogenase and 4-nitrotoluene oxygenase activities in 4-nitrobenzoate-grown cells suggests that 4-nitrobenzoate is an inducer of the 4-nitrotoluene degradative pathway. The 4-nitrotoluene oxidation pathway is similar to the pathway for the oxidation of toluene to benzoate as catalyzed by the TOL plasmid "upper" pathway, which has been shown to be able to deal with 4-nitro substituted substrates (9), however; the TOL plasmid pathway is induced by the pathway

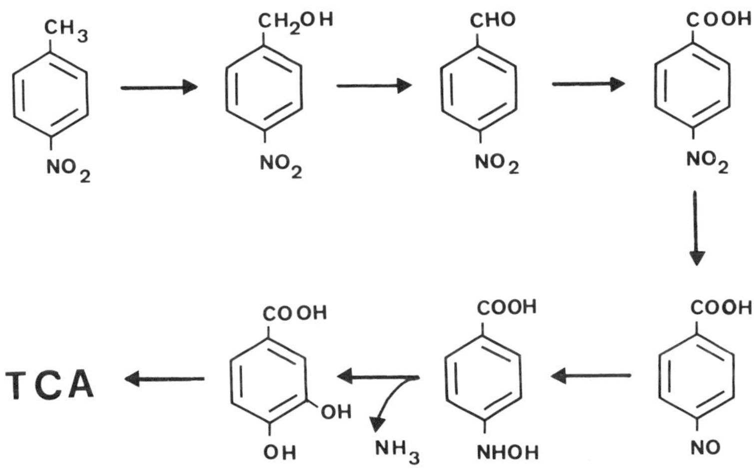

Figure 1. Proposed catabolic pathway for the metabolism of 4-nitrotoluene by *Pseudomonas* sp.

substrates (toluene and benzyl alcohol) but not by the end product. Furthermore, Rhys-Williams et al. (45) found no DNA homology between the TW3 pathway for metabolism of 4-nitrotoluene and the TOL "upper" pathway, as revealed by the lack of hybridization of the TOL genes used as a probe against the whole DNA of the 4-nitrotoluene isolate (45).

Potential of Toluene-Oxidizing Enzymes for Biotransformation of Nitrotoluenes

Microbial metabolism of 2- and 3-nitrotoluene seems to be more difficult than metabolism of 4-nitrotoluene. However, a variety of mechanisms are known for the initial attack on unsubstituted toluene by bacteria (Fig. 2). Strains that contain the TOL plasmid oxidized the methyl group to the corresponding alcohol, aldehyde and carboxylic acid derivatives (55). *Pseudomonas putida* F1 and similar strains produce a toluene dioxygenase that initially attacks at the 2,3- position on the ring. This yields a *cis*-dihydrodiol that is then

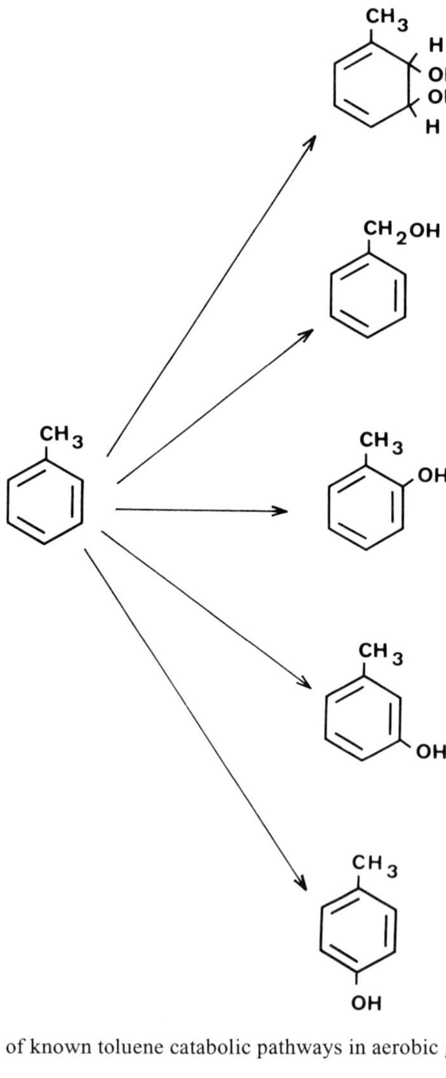

Figure 2. Initial steps of known toluene catabolic pathways in aerobic gram-negative bacteria.

converted to 3-methylcatechol (13). Strains have recently been described that catalyze an initial monooxygenase attack at the 2-, 3- or 4- position on the ring (47, 53, 56). As an approach to evaluating the potential use of these pathways for the vertical evolution of the catabolism of nonpolar nitroaromatic compounds, the ability of some of these enzymes to biotransform nitrotoluenes was tested and the nature of the resulting products was determined.

Delgado et al. (9) examined the ability of nitrosubstituted compounds to serve as substrates for the catabolic enzymes of the TOL pathway, namely, toluene monooxygenase, benzyl alcohol dehydrogenase and benzaldehyde dehydrogenase. The *xylMA* genes from the TOL plasmid encoding for toluene monooxygenase were cloned under the control of the P*lac* promoter in pWA21 and induced in *E. coli* in the presence of IPTG. Toluene monooxygenase oxidized 3- and 4-nitrotoluene to the corresponding benzyl alcohol derivatives. The specific activity of the enzyme with 3- or 4-nitrotoluene as substrate was around 70 U/mg protein, which was about 40% of the rate of oxidation recorded with *m*-xylene (57). The nitrobenzyl alcohols were subsequently oxidized to the corresponding benzaldehydes. This is due to a benzyl alcohol dehydrogenase activity associated with toluene monooxygenase, which functions with unsubstituted and methyl-substituted benzyl alcohols (18). Delgado et al. (9) found that 2-nitrotoluene and 3,4-dinitrotoluene were not substrates of the toluene monooxygenase of TOL.

The activities of the TOL plasmid-encoded benzyl alcohol dehydrogenase and benzaldehyde dehydrogenase were assayed in cell-free extracts of *Pseudomonas* (pWW0) grown on *m*-xylene. These enzymes were able to oxidize 3- and 4-nitro-substituted substrates (9). For benzyl alcohol dehydrogenase, the rate of oxidation of nitro substituted substrates was about 30-50% of the rate recorded with unsubstituted benzyl alcohol. Benzaldehyde dehydrogenase was also active against 3- and 4-nitrobenzaldehyde. The rate of transformation of these nitrobenzaldehydes was 1.7 to 2.2-fold higher than with the unsubstituted substrate. The results of these preliminary studies clearly revealed that enzymes of the TOL catabolic pathways can metabolize 3- and 4-nitrotoluene to their corresponding 3- and 4-nitrobenzoates (Fig. 3).

It is well known that in *P. putida*(pWW0) the genes for the metabolism of toluene to benzoate are organized in a single operon, whose expression is under transcriptional control (29). Positive control from the Pu promoter ("upper" pathway operon promoter) is mediated by the *xylR* gene product. The *xylR* gene on TOL is not linked to the "upper" pathway operon. Stimulation of transcription from the Pu promoter is relatively complex in the sense that in addition to the activation of XylR by an effector, at least two chromosomal functions are also required; these include a sigma factor, called RpoN, and a protein called IHF, involved in DNA bending. Abril et al. (1) assayed the effector specificity of XylR by using a fusion of Pu to a promoterless *lacZ* gene with the *xylR* gene in *trans*, and tests were done in the absence and in the presence of a wide variety of aromatic compounds. The XylR regulator was very relaxed in effector specificity; it recognized a wide variety of aromatic hydrocarbons, benzyl alcohols and 4-chlorobenzaldehyde as effectors. Delgado and Ramos (8) later found that XylR recognized 2- and 4-nitrotoluene, but not 3-nitrotoluene, as effectors. So far, the XylR protein is the only regulator described that recognizes nitrotoluenes as effectors.

In summary, it can be said that the potential of the TOL "upper" pathway seems to be limited to degradation of 4-nitrotoluene, because the compound by itself is an effector of the pathway, and it and its metabolites are substrates of the catabolic enzymes. In contrast, 2-nitrotoluene, an effector for the regulator, is not a substrate for the catabolic enzymes, and 3-nitrotoluene and its metabolites, which are substrates for the catabolic enzymes, are not effectors for the regulator.

Expansion of the substrate range of the enzymes by mutagenizing all the "upper" pathway genes of the TOL plasmid could potentially allow the complete metabolism of

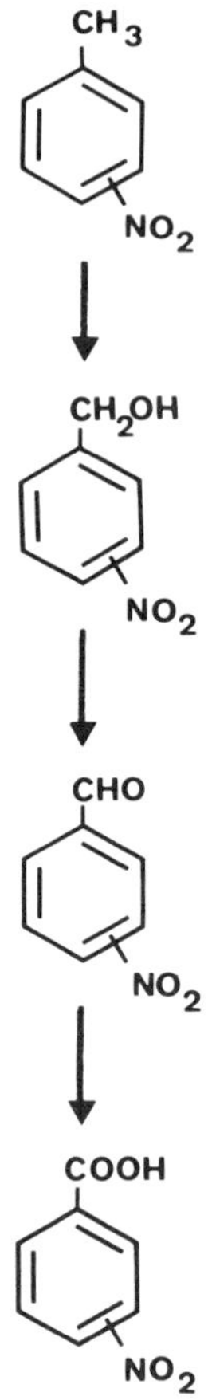

Figure 3. Biotransformation of nitrotoluene by toluene monooxygenase, produced by the TOL plasmid pWW0 of *Pseudomonas putida*.

2-nitrotoluene, but this approach is technically problematic since the substrate specificity of three enzymes must be altered. We postulated that the isolation of XylR mutants able to recognize 3-nitrotoluene would increase the potential usefulness of the TOL "upper" pathway for the *in vitro* evolution of nitroaromatic pathways. To this end we designed a strategy to isolate mutants of the XylR regulator able to recognize 3-nitrotoluene as an effector (8). The strategy was similar to that used by Ramos et al. (39) to isolate mutants of other catabolic pathway regulators and to use them to engineer the catabolic potential of microbes. Briefly, the promoter of a catabolic pathway is fused to a promoterless reporter gene such as *tet, cam, lux,* or *lacZ*, so that basal expression of the reporter gene from the fusion can be distinguished from expression of the reporter protein in the presence of an active regulator. The regulator cloned in a plasmid is mutagenized chemically with nitrosoguanidine (NTG) or ethylmethane sulfonate, and the mutagenized plasmids from the population surviving mutagenesis are isolated and transferred to the host bearing the target fusion. Clones expressing the reporter gene under selective conditions are chosen on the basis of the phenotype conferred by the reporter, for example, resistance to antibiotics, light emission, or color changes in the presence of chromogenic substrates.

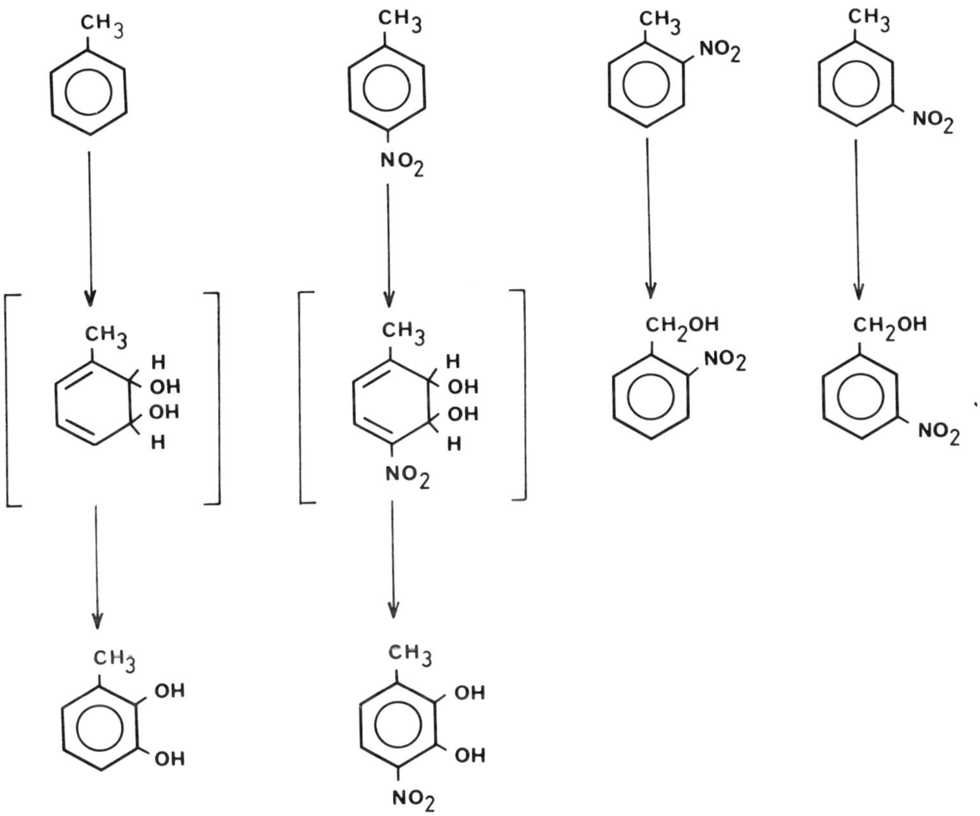

Figure 4. Biotransformation of toluene and nitrotoluenes by the chromosomally encoded toluene dioxygenase of *Pseudomonas* sp. F1.

Colonies of *E. coli* MC4100 bearing the plasmid pRD579 (Pu::'*lacZ*) and pTS174 (*xylR*), are colorless on MacConkey agar, but they turn red if a XylR effector such as *m*-xylene is supplied in the vapor phase. We mutagenized the *xylR* gene on pTS174 *in vivo* with NTG, then isolated the plasmid and transformed it into *E. coli* MC4100 (pRD579). Colonies that turned red in the presence of 3-nitrotoluene were then isolated. One mutant that recognized 3-nitrotoluene as an effector was analyzed in detail. The mutant exhibited a single point mutation, which, when translated into the protein sequence, resulted in an amino acid change at position 172: a glutamic acid residue was replaced by a lysine residue. The mutant XylR, called XylRE172K, gained the ability to recognize 3-nitrotoluene, and also retained the ability of the wild type to recognize 2- and 4-nitrotoluene and other effectors (8). This mutant regulator, together with the TOL wild type catabolic pathway, now provides the potential for the biotransformation of 3-nitrotoluene into 3-nitrobenzoate.

The initial reaction in the degradation of toluene catalyzed by toluene dioxygenase from *Pseudomonas putida* F1 and *Pseudomonas* sp. JS150 yields *cis*-toluene dihydrodiol as the first metabolite. Subsequent oxidation of the dihydrodiol leads to the formation of 3-methylcatechol (Fig. 4), which undergoes ring fission. Analogous reactions were expected when toluene-grown cells of *P. putida* F1 and JS150 were incubated with 2-, 3- and 4-nitrotoluene. However, the major products (43) were 2- and 3-nitrobenzyl alcohol from 2- and 3-nitrotoluene respectively, and 3-methyl-6-nitrocatechol from 4-nitrotoluene

(Fig. 4). The same products were found when the *todC1C2BA* genes encoding toluene dioxygenase were expressed in *E. coli* under the IPTG-inducible *tac* promoter. These results clearly suggest that the toluene dioxygenase of *P. putida* strain F1 exhibits broad substrate specificity. It functions as a monooxygenase when it oxidizes the alkyl side chain of 2- and 3-nitrotoluene, this is in contrast with its more common function as a dioxygenase with other substrates such as toluene and 4-nitrotoluene (Fig. 4). The rate of transformation of 3-nitrotoluene into 3-nitrobenzyl alcohol was about 34 μmol/mg cell protein per hour. Toluene monooxygenase from TOL and the enzyme from F1 or JS150 (acting as a monooxygenase) apparently open a pathway for the metabolism of 3-nitrotoluene to 3-nitrobenzoate. The full exploitation of the toluene monooxygenase potential will require further biotransformations of the 2- and 3-nitrobenzyl alcohol into the corresponding 2- and 3-nitrobenzoates via benzyl alcohol dehydrogenase and benzaldehyde dehydrogenase enzymes. In the case of 3-nitro substituted substrates, the enzymes of the TOL pathway can already be used for this purpose, whereas these enzymes are not able to act on 2-nitro substituted substrates.

VERTICAL EXPANSION OF THE PATHWAYS FOR MONONITROBENZOATES AND PUTATIVE HOSTS FOR EXPLOITATION OF THE CATABOLIC POTENTIAL OF ENZYMES ABLE TO BIOTRANSFORM NITROTOLUENES

The vertical expansion of a catabolic pathway involves the addition of a series of catabolic cassettes to an existing pathway. The strategy for introduction of the catabolic capability into a specific host will depend on the nature of the host and its catabolic pathway(s).

Construction of a hybrid pathway is simplified when the genes for the catabolic pathways are borne by conjugative plasmids, as is the case for the toluene catabolic pathway on the TOL plasmid. If the catabolic genes are borne on non-conjugative plasmids or on the chromosome, the genes encoding the catabolic functions can be cloned by using a wide battery of genetic tools, and the target cloned genes can be delivered to the host either to the chromosome (via mini-transposons, for example) or via a broad-host range plasmid (19, 26, 27, 28). The success of the construction of the hybrid route will require the stable inheritance of the recombinant genes. In addition, a functional hybrid pathway will require the effective expression of the genes in the host.

The host range in which catabolic plasmids are replicated depends on the plasmid. In the TOL plasmid, the catabolic pathway of which has been used for the evolution of pathways for biodegradation of aromatic compounds (20, 42), the host range seems to be limited to some pseudomonads belonging to rRNA group I, and to some enterobacteria (34, 41). This imposes natural barriers to catabolic expansion. Although the TOL plasmid pWW0 is generally stable, examples of instability of TOL are known in which a segment of 40 kb including the catabolic genes is deleted (32). Therefore, selective pressure may be needed to maintain the catabolic trait. The expression of TOL promoters and regulators has been faithfully reproduced in *Escherichia coli* and several pseudomonads by several research groups (reviewed by Marqués and Ramos, 29), and therefore should not represent a major obstacle for the construction of hybrid pathways.

If catabolic genes must be inserted onto the host chromosome to improve their stability, many genetic devices working in a wide range of gram-negative bacteria are available. A battery of mini-Tn*5* transposons, developed in the laboratory of K.N. Timmis (19, 28), provides not only an easy way to transfer genes into target hosts, but also a battery of cognate promoters that can perform efficiently *in vivo* to control expression under different promoters or cascade systems (27). Broad host range plasmids are another potential way to deliver genes to other hosts. This strategy was used by Lehrbach et al. (26) to expand the range of chlorosalicylates metabolized by *Pseudomonas* sp. B13, through a modified *ortho*-cleavage pathway.

The metabolism of 4-nitrobenzoate by bacteria was first reported by Cartwright and Cain (6, 7). Their later studies (3, 4) suggested the progressive reduction of the nitro group on the aromatic ring via nitrosobenzoate to hydroxylaminobenzoate, which was converted to protocatechuate. This pathway has been analyzed in some detail in *Comamonas* (14) and recently Rhys-Williams et al. (45) and Haigler and Spain (16) suggested that 4-nitrobenzoate resulting from the metabolism of 4-nitrotoluene by *Pseudomonas* sp. could be metabolized through the same set of reactions (Fig. 1). Delgado, Haïdour and Ramos (unpublished) have extended the observation of this pathway to some enterobacteria which used 4-nitrobenzoate as the sole source of carbon and nitrogen. They have identified 4-hydroxylaminobenzoate as an intermediate in the pathway, but not the 4-amino derivative. Cells grown on 4-nitrobenzoate accumulated protocatechuate from 4-nitrobenzoate when incubated under anaerobiosis, and assimilated this after transfer to aerobiosis. Therefore, *Comamonas*, *Pseudomonas* and some enterobacteria such as *Klebsiella* and *Citrobacter*, which grow on 4-nitrobenzoate as the sole source of carbon and nitrogen, are good candidates for the vertical expansion of 4-nitrobenzoate pathways to include 4-nitrotoluene. Because the different 4-nitrobenzoate degraders colonize different ecosystems, the vertical expansion of this catabolic pathway in different microbes will allow not only the expansion of the pathway but also its potential use in different polluted niches.

Metabolism of 2-nitrobenzoate by different bacteria has been reported, although details of the catabolic pathways are not well defined (3, 4). We have found several *Pseudomonas*, *Klebsiella* and other microbes able to grow on 2-nitrobenzoate. These organisms can be useful hosts for the expansion of 2-nitrobenzoate pathways to include 2-nitrotoluene. The work of Robertson et al. (43) provided evidence for the existence of enzymes able to oxidize 2-nitrotoluene to 2-nitrobenzyl alcohol; however, the exploitation of this bioconversion awaits the isolation of benzyl alcohol dehydrogenases and benzaldehyde dehydrogenases able to use the corresponding 2-nitrosubstituted substrates.

Metabolism of 3-nitrobenzoate seems to be rather refractory, although a recent report suggested that some *Pseudomonas* strains can mineralize this compound via protocatechuate (33). In our laboratory we have established a community formed by a consortium of at least four microbes (a *Candida*, two different *Pseudomonas* strains, and a *Klebsiella*), which uses 3-nitrobenzoate as the sole nitrogen source (Fandila and Ramos, unpublished). We are currently trying to expand the catabolic potential of the community to include the metabolism of 3-nitrotoluene by introducing a *Pseudomonas* strain bearing the wild type pWW0 plasmid and the *xylR* allele encoding for the regulator XylRE172K.

A HYBRID PATHWAY FOR THE METABOLISM OF 2,4,6-TRINITROTOLUENE BY *PSEUDOMONAS* SP. CLONE A

Isolation of Bacteria Able to Use TNT as the Sole Nitrogen Source

Many reports have described the metabolism of TNT by soil microbes, but in most cases metabolism involved the reduction of one or more of the nitro groups on the aromatic ring, and the accumulation of aminonitrotoluenes. In addition, under aerobic conditions azoxy compounds resulting from the condensation of partially reduced intermediates have been found (5, 11, 30, 31, 35, 36, 37).

Duque et al. (11) isolated a *Pseudomonas* sp. strain from soils of an explosives factory that used TNT as the sole nitrogen source but required an exogeneous carbon source for growth. The isolate, belonging to *Pseudomonadaceae* group I as determined by rRNA analysis, was called *Pseudomonas* sp. clone C1S1. Growth of *Pseudomonas* sp. C1S1 with fructose as the carbon source and with TNT at saturation (100 mg/liter) or supplied in excess as crystals (1 to 10 g/liter) was optimal at 30 to 32°C within a pH range of 5.5 to 9. The generation time with excess crystals of TNT as a nitrogen source and with 0.5% (wt/vol) fructose as the carbon source was about 22 h. The clone was not able to use TNT as a nitrogen source at concentrations below 20 mg/liter. In addition to TNT, *Pseudomonas* sp. C1S1 used 2,4- and 2,6-dinitrotoluene, 2-nitrotoluene, nitrate, nitrite and ammonium as nitrogen sources. Growth on TNT, 2,4- and 2,6-dinitrotoluene and 2-nitrotoluene led to the accumulation of up to 0.5 mM nitrite in the culture medium (11). Because bacteria growing on nitrate did not accumulate nitrite, we measured nitrite reductase in nitrate- and TNT-grown cells (11). The level of nitrite reductase in nitrate-grown cells was 28 U/mg protein, whereas in TNT grown-cells it was only about one-sixth as high. The evidence suggests that this strain is limited in NO^{-}_{2} assimilation.

The accumulation of nitrite from TNT in the culture medium, and the fact that this nitroarene was not used as a carbon source, led us to analyze culture supernatants to look for metabolites lacking one or more of the nitro groups on the aromatic ring (11). Mass spectra revealed the presence of 2,4-and 2,6-dinitrotoluene, 2-nitrotoluene and toluene in culture supernatants, suggesting sequential removal of all three nitro groups. No hydroxylated derivatives bearing one or two nitro groups on the aromatic ring were found. In addition to removing nitro groups from the aromatic ring, strain C1S1 was able to reduce the nitro groups to amino groups via hydroxylamine, and azoxy compounds were also found (see below). Some products resulting from the oxidation of the alkyl side group were also detected. Duque et al. (11) proposed that elimination of the first two nitro groups on the TNT ring takes place via the production of Meisenheimer complexes. ^{1}H-NMR analysis of trace compounds in the culture medium of this *Pseudomonas* strain revealed the presence of the hydride-Meisenheimer complex for TNT (15). The following spectroscopic characteristics were found for this compound: H3, 3.92 ppm; H5, 8.31 ppm; 1-CH_3 2.55. Figure 5 summarizes the sequence of events believed to be involved in the removal of the first nitro group from the TNT aromatic ring. The TNT hydride-Meisenheimer complex was also detected by Vorbeck et al. (52) when a *Mycobacterium* sp. was exposed to TNT.

We believe that the second nitro group is eliminated through a similar mechanism, but have not yet obtained chemical evidence for this process. The mechanism of removal of the nitro group from the 2-nitrotoluene ring also remains to be elucidated. The major problem in analyzing the metabolism of 2-nitrotoluene by this isolate is that although it grows on 2-nitrotoluene, when it is supplied exogenously the bacterium is unstable and loses the

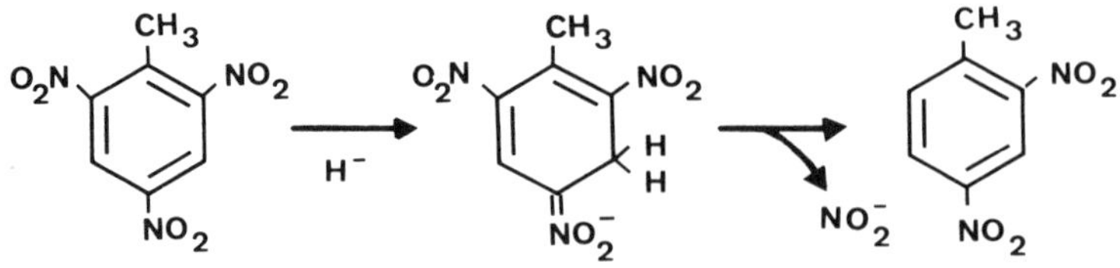

Figure 5. The hydride-Meisenheimer complex in the metabolism of trinitrotoluene by *Pseudomonas* sp. clone A.

character. Nonetheless we have recently observed that culturing the bacteria with trace amounts of yeast extract and 2-nitrotoluene allows the 2-nitrotoluene-phenotype to be maintained (Duque and Ramos, unpublished).

Selection of Derivatives of *Pseudomonas* sp. Strain C1S1 That Use TNT Most Efficiently

Because of the relatively slow growth of *Pseudomonas* sp. C1S1, the accumulation of nitrite in the culture medium, the clone's inability to use TNT when supplied at concentrations below 20 mg/liter, and the misrouting of part of the TNT through reduction of nitro groups, we designed experiments to derive more efficient TNT utilizing-strains from C1S1. Derivatives of *Pseudomonas* sp. C1S1 that grew more efficiently on TNT as a nitrogen source were sought in two steps (11). The first involved batch subculture of the strain in minimal medium with TNT supplied in excess (1 g/liter) and with fructose as the carbon source. Cells were transferred sequentially, and successive subculturing resulted in the isolation of a strain designated *Pseudomonas* sp. clone A, that grew faster than its parental strain. The generation time in minimal medium with TNT as the nitrogen source supplied in excess as crystals and 0.5% (wt/vol) fructose was around 15 h. The derivative bacterium did not accumulate nitrite in the culture medium and was able to grow in a continuous culture with TNT as a growth-limiting factor. The strain growing with TNT as the sole nitrogen source exhibited high nitrite reductase levels (37 U/mg protein) and retained the parental characteristic of reducing the nitro groups of TNT. The strain otherwise conserved other parental characteristics regarding strict requirements of oxygen for growth and the spectrum of carbon sources utilized, as well as the temperature and pH ranges in which the bacterium grew well. A second step in the search for more efficient TNT-utilizers involved the selection of *Pseudomonas* clone A derivatives in continuous culture with TNT as the growth-limiting factor (70 mg/liter). After long-term incubation, a derivative strain called *Pseudomonas* sp. A-R30/12/5/31 was selected which accumulated lower amounts of reduced forms of TNT (Duque and Ramos, unpublished).

A Hybrid Pathway for the Mineralization of TNT

Vertical expansion of catabolic pathways requires the addition of a series of metabolic steps to a preexisting pathway in order to generate a hybrid one. *Pseudomonas putida* (pWW0-Km) is able to grow on toluene as a carbon source because of the catabolic pathway encoded by the TOL plasmid, but cannot grow on TNT because it lacks the enzymes for the elimination of nitro groups from the aromatic ring. On the other hand, the *Pseudomonas* strains of rRNA group I that were able to transform TNT into toluene were logical candidates to receive the TOL plasmid and to yield a hybrid strain able to mineralize TNT. We set up matings of the *Pseudomonas* (pWW0-Km) and the Rif^r derivatives of three possible recipients, namely *Pseudomonas* C1S1, clone A, and clone A-R30/12/5/31. Transconjugants able to grow on toluene were found at a frequency of 10^{-6} per recipient. All toluene utilizers were Km^r, and twenty-four transconjugants of each mating were tested for growth on TNT as the sole carbon and nitrogen sources. All transconjugants could grow on TNT as the sole carbon and nitrogen sources, although the most efficient ones were those derived from clone A-R30/12/5/31. Figure 6 summarizes the hybrid pathway for TNT mineralization. However, as the transconjugants grew on TNT, they accumulated nitrite, suggesting that growth was unbalanced. The addition to the culture of small amounts of fructose (0.01% wt/vol) allowed the bacteria to grow without accumulation of nitrite.

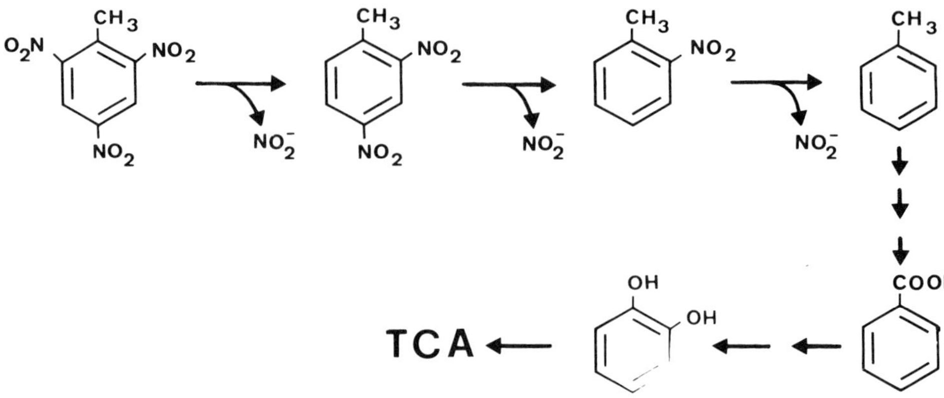

Figure 6. Hybrid pathway for the metabolism of trinitrotoluene by *Pseudomonas* sp. clone A (pWW0-Km)

LIMITATIONS OF THE HYBRID BACTERIA FOR TNT TREATMENTS

The *Pseudomonas* sp. strains provided with the TOL plasmid, although able to grow with TNT as the sole source of carbon and nitrogen, mineralized only about 50% of the total TNT supplied. The rest appeared as reduced metabolites of TNT. Figure 7 shows the products found identified in the majority of the culture supernatants.

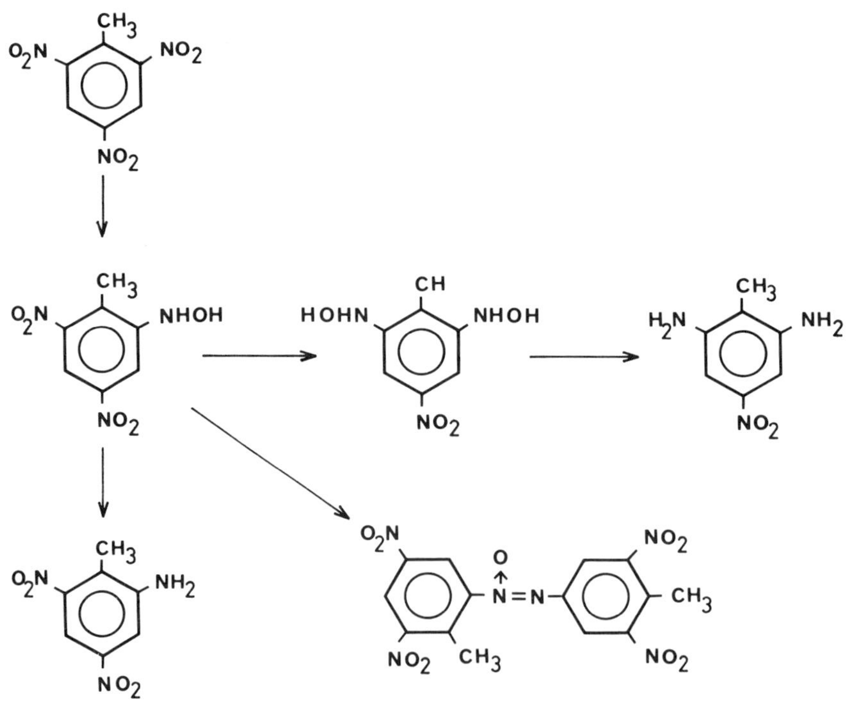

Figure 7. Byproducts and pathway of TNT metabolism by *Pseudomonas* sp. clone A.

We have found (15) that the production of azoxy compounds derived from TNT metabolism is a spontaneous process which occurs under aerobic conditions. Under strict anaerobic conditions *in vitro*, the reduction of TNT with less than stoichiometric amounts of $S(NH_4)_2$ led to the production of monohydroxylaminodinitrotoluenes, without the appearance of azoxy compounds. When the same reaction took place under aerobic conditions, azoxy compounds were formed. Furthermore, monohydroxylaminodinitrotoluenes chemically produced under anaerobiosis, upon transfer to aerobic conditions, formed azoxy compounds. The proposed mechanism of condensation (15) is summarized in the equations below:

$$2\ \text{R-NHOH} + O_2 \rightarrow 2\ \text{R-N=O} + 2\ H_2O \qquad (1)$$

$$2\ \text{R-NHOH} + 2\ \text{R-N=O} \rightarrow 2\ \text{R-}\overset{\overset{O}{\uparrow}}{\text{N}}\text{=N-R} + 2\ H_2O + O_2 \qquad (2)$$

$$4\ \text{R-NHOH} + O_2 \rightarrow 2\ \text{R-}\overset{\overset{O}{\uparrow}}{\text{N}}\text{=N-R} + 4\ H_2O + O_2 \qquad \text{(global)}$$

This finding clearly limits TNT mineralization by microbes that produce enzymes that reduce TNT. Two alternative approaches can be envisaged to solve this problem. One is to search for microbes that do not reduce TNT at all under aerobiosis. These organisms would be the hosts for cloned TNT pathways, which would make the biotransformation of TNT into toluene possible. In addition, the hosts should be able to replicate the TOL plasmid to allow full mineralization of toluene produced from TNT. A number of soil pseudomonads and enterobacteria that do not reduce TNT at all have been isolated by our group and we are using them as hosts for cloning the *Pseudomonas* clone A TNT catabolic pathway. The second alternative is to use microbes that mineralize TNT under anaerobic conditions. Such microoorganisms have been reported (12, 38) in the literature, and represent an important element for the biological treatment of polluted sites.

ACKNOWLEDGMENTS

Work done at the authors' laboratory was supported by grants from the Unión Española de Explosivos, Madrid (Spain) and The Comisión Interministerial de Ciencia y Tecnología (BIO-091-0651 and AMB94-1038-C02-01).

REFERENCES

1. Abril, M. A., C. Michán, K. N. Timmis, and J. L. Ramos. 1989. Regulator and enzyme specificities of the TOL plasmid-encoded upper pathway for degradation of aromatic hydrocarbons and expansion of the substrate range of the pathway. J. Bacteriol. 171:6782-6790.

2. Bruhn, C., H. Lenke, and H.-J. Knackmuss. 1987. Nitrosubstituted aromatic compounds as nitrogen source for bacteria. Appl. Environ. Microbiol. 53:208-210.
3. Cain, R. B. 1966. Induction of an anthranilate oxidation system during the metabolism of *ortho*-nitrobenzoate by certain bacteria. J. Gen. Microbiol. 42: 197-217.
4. Cain, R. B. 1966. Utilization of anthranilic and nitrobenzoic acids by *Nocardia opaca* and *Flavobacterium*. J. Gen. Microbiol. 42:219-235.
5. Carpenter, D. F., N. G. McCormick, J. H. Cornell, and A. M. Kaplan. 1978. Microbial transformation of ^{14}C-labeled 2,4,6-trinitrotoluene in an activated-sludge system. Appl. Environ. Microbiol. 35:949-954.
6. Cartwright, N. J., and R. B. Cain. 1959. Bacterial degradation of the nitrobenzoic acids. Biochem. J. 71:248-261.
7. Cartwright, N. J., and R. B. Cain. 1959. Bacterial degradation of nitrobenzoic acids. 2. Reduction of the nitro group. Biochem. J. 73:305-314.
8. Delgado, A., and J. L. Ramos. 1994. Genetic evidence for activation of the positive transcriptional regulator XylR, a member of the NtrC family of regulators, by effector binding. J. Biol. Chem. 269:8059-8062.
9. Delgado, A., M. G. Wubbolts, M. A. Abril, and J. L. Ramos. 1992. Nitroaromatics are substrates for the TOL plasmid upper-pathway enzymes. Appl. Environ. Microbiol. 58:415-417.
10. Dickel, O., and H.-J. Knackmuss. 1991. Catabolism of 1,3-dinitrobenzene by *Rhodococcus* sp. QT-1. Arch. Microbiol. 157:76-79.
11. Duque, E., A. Haïdour, F. Godoy, and J. L. Ramos. 1993. Construction of a *Pseudomonas* hybrid strain that mineralizes 2,4,6-trinitrotoluene. J. Bacteriol. 175: 2278-2283.
12. Funk, S. B., D. J. Roberts, D. L. Crawford, and R. L. Crawford. 1993. Initial-phase optimization for bioremediation of munition compound-contaminated soils. Appl. Environ. Microbiol. 59:2171-2177.
13. Gibson, D. T., M. Hensley, H. Yoshioka, and T. J. Mabry. 1970. Formation of (+)-*cis*-2,3-dihydroxy-1-methylcyclohexa-4,6-diene from toluene by *Pseudomonas putida*. Biochemistry 9:1626-1630.
14. Groenewegen, P. E. J., P. Breeuwer, J. M. L. M. van Helvoort, A. A. M. Langenhoff, F. P. de Vries, and J. A. M. de Bont. 1992. Novel degradative pathway of 4-nitrobenzoate in *Comamonas acidovorans* NBA-10. J. Gen. Microbiol. 138:1599-1605.
15. Haïdour, A., and J. L. Ramos. Identification of products resulting from the biological reduction of 2,4,6-trinitrotoluene, and 2,4- and 2,6-dinitrotoluene by *Pseudomonas* sp. Appl. Environ. Microbiol., submitted.
16. Haigler, B.E., and J.C. Spain. 1993. Biodegradation of 4-nitrotoluene by *Pseudomonas* sp. strain 4NT. Appl. Environ. Microbiol. 59:2239-2243.
17. Haigler, B.E., W.H. Wallace, and J.C. Spain. 1994. Biodegradation of 2-nitrotoluene by *Pseudomonas* sp. strain JS42. Appl. Environ. Microbiol. 60:3466-3469.
18. Harayama, S., R. A. Leppik, M. Rekik, N. Mermod, P. R. Lehrbach, W. Reinecke, and K. N. Timmis. 1986. Gene order of the TOL catabolic plasmid upper pathway operon and oxidation of both toluene and benzyl alcohol by the *xylA* product. J. Bacteriol. 187:455-461.
19. Herrero, M., V. de Lorenzo, and K. N. Timmis. 1990. Transposon vectors containing non-antibiotic selection markers for cloning and stable chromosomal insertion of foreign DNA in gram-negative bacteria. J. Bacteriol. 172:6557-6567.
20. Jeenes, D. J., W. Reineke, J.-H. Knackmuss, and P. A. Williams. 1982. TOL plasmid pWW0 in constructed halobenzoate-degrading *Pseudomonas* strains: enzyme regulation and DNA structure. J. Bacteriol. 150:180-187.
21. Kaplan, D.L. 1989. Biotransformation pathways of hazardous energetic organo-nitro compounds, p 155-181. *In* D. Kamely, A. Chakrabarty, and G. Omenn (ed.), Biotechnology and biodegradation. Portfolio Publishing Co, Houston.
22. Kaplan, D.L., and A.M. Kaplan. 1982. Mutagenicity of 2,4,6-trinitrotoluene-surfactant complexes. Bull. Environ. Contam. Toxicol. 28:33-38.
23. Kaplan, D.L., and A.M. Kaplan. 1982. Thermophilic biotransformations of 2,4,6-trinitrotoluene under simulated composting conditions. Appl. Environ. Microbiol. 44:757-760.
24. Ke, Y.-H., L.L. Gee, and N.N. Durham. 1959. Mechanism involved in the metabolism of nitrophenyl-carboxylic acid compounds by microorganisms. J. Bacteriol. 77:593-598.
25. Keith, L. H., and W. A. Telliard. 1979. Priority pollutants. I. A perspective view. Environ. Sci. Technol. 13:416-423.
26. Lehrbach, P. R., J. Zeyer, W. Reineke, H.-J. Knackmuss, and K. N. Timmis. 1984. Enzyme recruitment *in vitro*: use of cloned genes to extend the range of haloaromatics degraded by *Pseudomonas* sp. B13. J. Bacteriol. 158:1025-1032.

27. de Lorenzo, V., S. Fernández, M. Herrero, U. Jakubzik, and K. N. Timmis. 1992. Engineering of alkyl- and haloaromatic-responsive gene expression with mini-transposons containing regulated promoters of biodegradative pathways of *Pseudomonas*. Gene 130:41-46.
28. de Lorenzo, V., M. Herrero, V. Jacubzik, and K. N. Timmis. 1990. Mini-Tn*5* transposon derivatives for insertion mutagenesis, promoter probing and chromosomal insertion of cloned DNA in gram-negative bacteria. J. Bacteriol. 172:6568-6572.
29. Marqués, S., and J. L. Ramos. 1993. Transcriptional control of the *Pseudomonas putida* TOL plasmid catabolic pathways. Mol. Microbiol. 9:923-929.
30. McCormick, N. G., J. H. Cornell, and A. M. Kaplan. 1978. Identification of biotransformation products from 2,4-dinitrotoluene. Appl. Environ. Microbiol. 35:945-948.
31. McCormick, N. G., F. F. Feeherry, and H. S. Levinson. 1976. Microbial transformation of 2,4,6-trinitrotoluene and other nitroaromatic compounds. Appl. Environ. Microbiol. 31:949-958.
32. Meulien, P., R.-G. Dowling, and P. Broda. 1981. Excision of the 40 kb segment from the TOL plasmid from *Pseudomonas putida* mt-2 involves direct repeats. Mol. Gen. Genet. 184:97-101.
33. Nadeau, L. J., and J. Spain. 1994. The bacterial degradation of *m*-nitrobenzoic acid, abstr. Q-114, p. 408. Abstr. 94th Annu. Meet. Amer. Soc. Microbiol. 1994.
34. Nakazawa, T. 1978. TOL plasmid in *Pseudomonas aeruginosa* PAO: thermosensitivity of self-maintenance and inhibition of host cell growth. J. Bacteriol. 133:527-535.
35. Nay, M. W., C. W. Randall, Jr., and P. H. King. 1974. Biological treatability of trinitrotoluene manufacturing waste-water. J. Water Pollut. Control Fed. 46:485-497.
36. Parrish, F. W. 1977. Fungal transformation of 2,4-dinitrotoluene and 2,4,6-trinitrotoluene. Appl. Environ. Microbiol. 34:232-233.
37. Pereira, W. E., D. L. Short, D. B. Manigold, and P. K. Roscio. 1979. Isolation and characterization of TNT and its metabolites in groundwater by gas chromoatograph-mass spectrometer-computer techniques. Bull. Environ. Contam. Toxicol. 21:554-562.
38. Preuss, A., J. Fimpel, and G. Diekert. 1993. Anaerobic transformation of 2,4,6-trinitrotoluene (TNT). Arch. Microbiol. 159:345-353.
39. Ramos, J. L., A. Stolz, W. Reineke, and K. N. Timmis. 1986. Altered effector specificities in regulators of gene expression: TOL plasmid *xylS* mutants and their use to engineer expansion of the range of aromatics degraded by bacteria. Proc. Natl. Acad. Sci. USA. 83:8467-8471.
40. Ramos, J. L., and K. N. Timmis. 1987. Experimental evolution of catabolic pathways of bacteria. Microbiol. Sci. 4:228-237.
41. Ramos-González, M. I., E. Duque, and J. L. Ramos. 1991. Conjugational transfer of recombinant DNA in cultures and in soils: host range of *Pseudomonas putida* TOL plasmids. Appl. Environ. Microbiol. 57:3020-3027.
42. Reineke, W., and H.-J. Knackmuss. 1979. Construction of haloaromatics utilising bacteria. Nature 277:385-386.
43. Robertson, J. B., J. C. Spain, J. D. Haddock, and D. T. Gibson. 1992. Oxidation of nitrotoluenes by toluene dioxygenase: evidence for a monooxygenase reaction. Appl. Environ. Microbiol. 58:2643-2648.
44. Rojo, F., D.H. Pieper, K.H. Engesser, H.-J. Knackmuss, and K.N. Timmis. 1987. Assemblage of *ortho*-cleavage routes for degradation of chloro- and methylaromatics. Science 238:1395-1398.
45. Rhys-Williams, W., S. C. Taylor, and P. A. Williams. 1993. A novel pathway for catabolism of 4-nitrotoluene by *Pseudomonas*. J. Gen. Microbiol. 139:1967-1972.
46. Senior, E., A. T. Bull, and J. M. Slater. 1976. Enzyme evolution in a microbial community growing on the herbicide dalapon. Nature 263:476-479.
47. Shields, M. S., S. O. Montgomery, P. J. Chapman, S. M. Cuskey, and P. H. Pritchard. 1989. Novel pathway of toluene catabolism in the trichloroethylene-degrading bacterium G-4. Appl. Environ. Microbiol. 55:1624-1629.
48. Slater, J. M., and A. T. Bull. 1982. Environmental microbiology: biodegradation. Phil. Trans. R. Soc. London B. 297:575-597.
49. Spain, J. C., and D. T. Gibson. 1991. Pathway for biodegradation of *p*-nitrophenol in a Moraxella sp. Appl. Environ. Microbiol. 57:812-819.
50. Spanggord, R. J., J. C. Spain, S. F. Nishino, and K. E. Mortelmans. 1991. Biodegradation of 2,4-dinitrotoluene by a *Pseudomonas* sp. Appl. Environ. Microbiol. 57:3200-3205.
51. Subba-Rao, R. V., and M. Alexander. 1985. Bacterial and fungal cometabolism of 1,1,1-trichloro-2,2-bis(4-chlorophenyl)ethane (DDT) and its breakdown products. Appl. Environ. Microbiol. 49:509-516.
52. Vorbeck, C., H. Lenke, P. Fischer, and H.-J. Knackmuss. 1994. Identification of a hydride-Meisenheimer complex as a metabolite of 2,4,6-trinitrotoluene by a *Mycobacterium* strain. J. Bacteriol. 176:932-934.

53. Whited, G. M., and D. T. Gibson. 1991. Toluene-4-monoxygenase, a three-component enzyme system that catalyzes the oxidation of toluene to *p*-cresol in *Pseudomonas mendocina* KR1. J. Bacteriol. 173:3010-3016.
54. Williams, R. T., P. S. Ziegenfuss, and W. E. Sisk. 1992. Composting of explosives and propellant contaminated soils under thermophilic and mesophilic conditions. J. Ind. Microbiol. 9:137-144.
55. Worsey, M. I., and P. A. Williams. 1975. Metabolism of toluene and xylenes by *Pseudomonas putida* (arvilla) mt-2: evidence for a new function of the TOL plasmid. J. Bacteriol. 124:7-13.
56. Wright, A., and R. H. Olsen. 1994. Self-mobilization and organization of the genes encoding the toluene metabolic pathway of *Pseudomonas mendocina* KR1. Appl. Environ. Microbiol. 60:235-241.
57. Wubbolts, M. G. 1994. Xylene and alkane mono-oxygenases from *Pseudomonas putida*. Ph.D. dissertation. University of Groningen, Groningen, The Netherlands.
58. Zeyer, J., and P. C. Kearney. 1984. Degradation of *o*-nitrophenol and *m*-nitrophenol by a *Pseudomonas putida*. J. Agric. Food. Chem. 32:238-241.

5

ANAEROBIC TRANSFORMATION OF 2,4,6-TRINITROTOLUENE AND OTHER NITROAROMATIC COMPOUNDS

Andrea Preuß* and Paul-Gerhard Rieger†

* Degussa AG/VT-PB
Postfach 1345
63403 Hanau
Germany

† Institut für Mikrobiologie
Universität Stuttgart
Allmandring 31
70569 Stuttgart
Germany

INTRODUCTION

Considerations about transformation of nitroaromatic compounds in soil must include anaerobic reactions, because "the soil" is a heterogenous medium that consists of microhabitats of different oxygen partial pressures. Anaerobic conditions are created by bacteria that consume oxygen during degradation of soil organic matter and respiration. As a consequence, aerobic and anaerobic habitats coexist in or around soil particles within a very small space (Fig. 1). Therefore, it is important to pay attention to transformation reactions occurring under anaerobic conditions. There have been many publications on the reductive transformation of nitroaromatic compounds by strictly anaerobic bacteria. Table 1 indicates the variety of strictly anaerobic bacteria that carry out nitro reduction and the nitro compounds that can be transformed by these organisms.

The fact that many biological systems - anaerobic as well as aerobic - are capable of catalyzing the reduction of nitro substituents is due to the chemical properties of the nitro group, which will be described in the first section of this chapter. The nitro group can be easily reduced, either chemically in the presence of reducing agents or cometabolically by a variety of metabolic enzymes, which will be discussed in the second section. The following sections will address the reduction of nitro groups during the anaerobic transformation of 2,4,6-trinitrotoluene (TNT) and its derivative 2,4-diamino-6-nitrotoluene. TNT is one of the most widely used explosives in ammunition production. Because of its highly xenobiotic

Biodegradation of Nitroaromatic Compounds
Edited by J.C. Spain, Plenum Press, New York, 1995

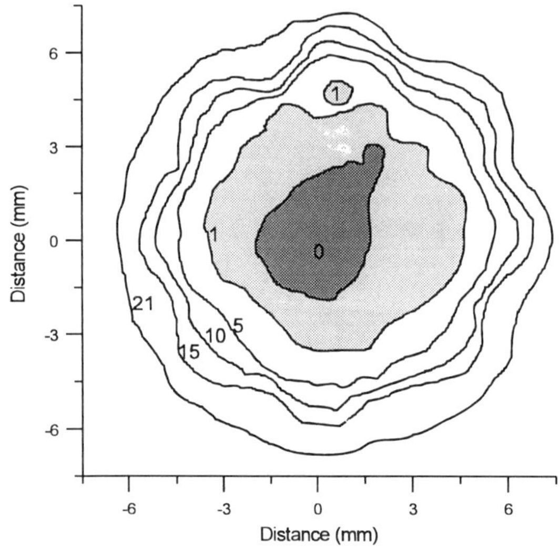

Figure 1. Oxygen concentrations (in %) in a soil particle (air is 21% O_2). Adapted from Brock and Madigan (8).

character as a polynitroaromatic compound, TNT is very persistent in nature. It is therefore of great environmental concern as a pollutant at military sites where ammunition production or handling have resulted in the contamination of large amounts of soil. Transformation of TNT is therefore of special interest with respect to bioremediation of munition-contaminated soils.

Complete anaerobic degradation of aromatic nitro compounds includes the elimination of the resulting amino groups. This will be discussed in the last section of this chapter with respect to the degradation of triaminotoluene, the product of complete reduction of TNT.

1. CHEMICAL PROPERTIES AND REACTIVITY OF NITRO COMPOUNDS

Aromatic nitro groups are easily susceptible to reductive attack due to their chemical structure. The nitro group consists of two different elements that are both highly electronegative and therefore compete for the available electrons. The nitro group is usually described as a resonance hybrid between the structures shown in Figure 2. The electronegativity of the oxygen is higher than that of the nitrogen atom, resulting in a polarization of the N-O-bond. The resulting, partially positive charge on the nitrogen atom, combined with its high electronegativity, makes the nitro group easily reducible.

The reduction rate of nitro compounds is determined by the chemical properties of the whole molecule and is therefore influenced by other ring substituents. McCormick et al. (32) pointed out that the rate of nitro reduction by anaerobic bacteria increased with

Figure 2. Canonical forms of the nitro group.

Table 1. Strictly anaerobic bacteria that have been reported to mediate nitro reduction and the nitro compounds that can be transformed by these organisms

Organism	Nitroaromatic compounds transformed	Reference
Clostridium acetobutylicum	chloramphenicol 2-/3-nitrophenol 2-/3-/4-nitrobenzoate 2-nitrobenzaldehyde	O'Brien and Morris (35)
Veillonella alcalescens *Clostridium pasteurianum*	TNT 40 various nitro compounds, including: nitrophenols nitrobenzoates nitrotoluenes	McCormick et al. (32)
Clostridium kluyveri *Clostridium* Sp.La1 *C. pasteurianum*	4-nitrobenzoate 2-nitroethanol	Angermaier and Simon (3)
C. propionicum *C. butyricum* *C. barkeri* *Proteus mirabilis* *Micrococcus lactilyticus* *Peptostreptococus anaerobius*	4-nitrobenzoate 2-nitroethanol	Angermaier and Simon (2)
Bacteroides fragilis	1-nitropyrene	Kinouchi and Ohnishi (25)
Clostridium leptum *C. paraputrificum* *C. clostridiiforme* *Eubacterium* sp.	1-nitropyrene 1,3-dinitropyrene 1,6-dinitropyrene	Rafii et al. (40)
Haloanaerobium praevalens *Sporohalobacter marismortui*	nitrobenzene 2-/3-/4-nitrophenol 2-/3-/4-nitroaniline 2,4-dinitrophenol 2,4-dinitroaniline	Oren et al. (36)
Desulfovibrio sp. (B strain)	TNT 2,4-/2,6-dinitrotoluene 2,4-dinitrophenol	Boopathy and Kulpa (5,7)
Desulfovibrio sp.	TNT 2,6-/3,4-dinitrotoluene 2-/4-nitrotoluene nitrobenzene	Preuss et al. (39)
Methanobacterium formicicum *M. thermoautotrophicum* *Methanospirillum hungatei* *Methanosarcina* sp.KS2002 *Ms. barkeri* *Ms. frisia* *Methanogenium tationis* *Desulfovibrio desulfuricans* *D.gigas* *Desulfovibrio* sp. AS *Desulfovibrio* sp. HB *Desulfotomaculum* sp. GROL	3-/4-nitrophenol 2,4-dinitrophenol 4-nitrobenzoate 4-nitroaniline	Gorontzy et al. (18)
Desulfotomaculum orientis *Desulfococcus multivorans*	4-nitrophenol	Gorontzy et al. (18)

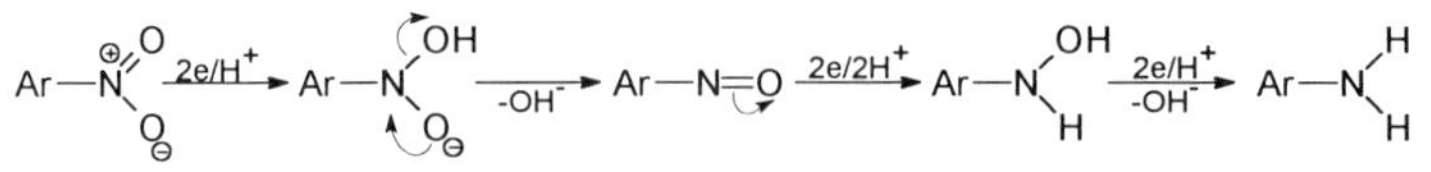

Figure 3. Mechanism of the sequential two-electron reduction of the nitro group (Ar- =Aryl-).

increasing electron-withdrawing power of functional groups *para* to the nitro group. This results in the following order of substituents corresponding to their enhancing effect on the reduction rate: NH_2<-OH<-H<-CH_3<-COOH<-NO_2 (32). As a consequence, in the course of reduction of one nitro group, a substituent with a reduction-rate-accelerating effect (-NO_2) is converted into one with the opposite effect (-NH_2): the amino group increases the electron density of the aromatic ring, thus "deactivating" the molecule and any remaining nitro groups of polynitroaromatic compounds for further reduction.

In addition to the number and kind of substituents at the aromatic ring, the rate of nitro reduction is influenced by the position of the substituents relative to each other. Gorontzy et al. (18) described the preferred abiotic reduction of *ortho*-nitrophenol by sulfide, whereas *meta*- and *para*-nitrophenol were not reduced. The abiotic reduction of 2,4-dinitrophenol yielded 2-amino-4-nitrophenol as the sole isomer detected.

In the same way, biologically mediated reduction mainly occurred at the *ortho*- rather than at the *para*-position in the case of nitroanilines and nitrophenols (32). In contrast, the nitro group at the *para*-position was most readily reduced in nitrobenzoic acids and nitrotoluenes (32). For nitrotoluenes this can be explained by the positive field effect of the methyl group, which mainly affects the *ortho*-position.

Most authors postulate that biologically mediated reduction of nitro groups proceeds via two-electron reduction steps through the corresponding nitroso and hydroxylamino derivatives as the intermediates. This corresponds to controlled electrolytic reduction of nitro substituents, which occurs in exact stages according to the mechanism depicted in Figure 3.

However, there are reports of enzymatic nitro reduction by microsomal enzymes or by microorganisms where the reaction does not proceed in two-electron steps and where the first intermediate was described as the nitro radical anion (10, 31, 38). Peterson et al. (38) described the detection of two nitroreductases in *Escherichia coli*, one was oxygen-insensitive and catalyzed the two-electron-reduction of nitrofurazone, while the other one was oxygen-sensitive and catalyzed the formation of the nitrofurazone radical anion. The "oxygen sensitivity" of the latter reductase is due to the spontaneous reoxidation of the radical anion by molecular oxygen, which of course will play no role under anaerobic conditions.

The occurrence of the nitro radical anion during transformation of nitro compounds with crude extracts or enzyme preparations of strictly anaerobic bacteria has been detected by Angermaier and Simon (2, 3). The authors described the formation of the dianion radical of 4-nitrobenzoate and further reduction reactions that may occur. The nitro radical anion is subject to dismutation with the concomitant formation of the nitroso derivative and the parent nitro compound. This dismutation reaction occurs in competition with further one-electron or two-electron reduction reactions (Fig. 4). According to Angermaier and Simon (3), chemical reduction and dismutation reactions may interfere with the reduction mediated by nitro-reducing enzymes.

The rates at which one-electron reduction of nitro compounds proceeds can be correlated linearly to their one-electron midpoint redox potentials as described by Clark et al. (13), for example. Bryant and DeLuca (10) reported on the purification from *Enterobacter cloacae* of a nitro reductase that was oxygen-insensitive and there-

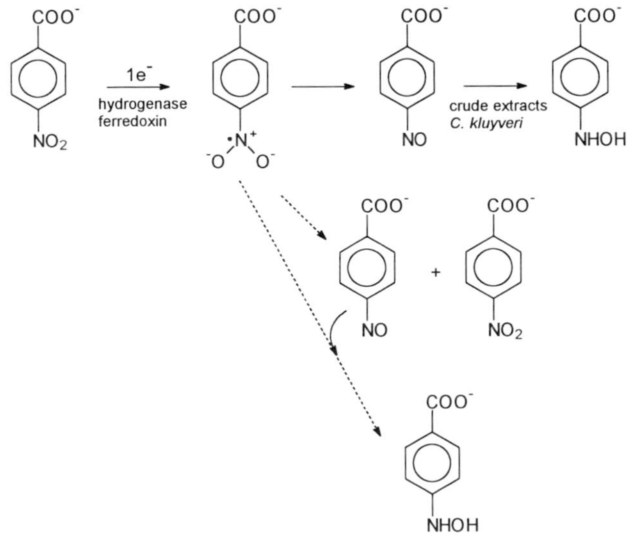

Figure 4. Nitro radical anion formation and possible abiotic dismutations with crude extracts of *Clostridium* sp. according to Angermaier and Simon (3).

fore was believed to catalyze the reduction of nitro compounds in two-electron steps. Nevertheless, the plot of the one-electron midpoint redox potentials of several 4-substituted nitrobenzene derivatives against the $\log_{10}$ of their initial reduction rates showed a linear relationship. Therefore, the authors did not rule out the possibility that a kinetically transient one-electron-reduced species participates in the reaction and determines the transformation rate (10). A linear correlation between the reactivity and the one-electron reduction potential of different nitro compounds was also observed by Glaus et al. (16) in *Streptomyces* exudates. These results suggest that reduction of aromatic nitro groups might generally proceed through the nitro radical anion, though it has not always been detected.

Some particular properties of the TNT molecule are due to its symmetric substitution with three nitro groups. On the one hand, the electron density of the aromatic ring of TNT is considerably lowered as compared to an unsubstituted aromatic system, since the nitro substituents exert a strong electron-withdrawing effect. As a consequence, the common oxygenolytic attack under aerobic conditions is difficult, since oxygen is a molecule with highly electrophilic character. Therefore, a reductive rather than an oxidative primary attack on the aromatic ring or one of the nitro groups seems to take place in certain aerobic organisms. Lenke and Knackmuss (28) proposed the formation of a Meisenheimer complex by addition of a hydride ion to the aromatic ring during growth of *Rhodococcus erythropolis* with picric acid as the nitrogen source. Recently, Rieger et al. (41) confirmed the formation of this Meisenheimer complex as the central intermediate in the productive degradation of picric acid by comparison with a newly synthesized authentic standard. A similar mechanism for the cometabolic reduction of the aromatic ring of trinitrotoluene to the corresponding Meisenheimer complex was proven by Vorbeck et al. (49). On the other hand, the nitro groups of TNT are particularly susceptible to reduction due to the rate-enhancing effect of nitro groups as described by McCormick et al. (32). Under strictly anaerobic conditions all three nitro groups of the TNT molecule will be sequentially reduced to amino groups.

2. ABIOTIC AND COMETABOLIC REDUCTION OF NITRO GROUPS

Abiotic Nitro Reduction by Chemical Agents

The abiotic reduction of nitro groups by reducing agents like cysteine, sulfide, etc., routinely used for the preparation of anaerobic media has been investigated by Gorontzy et al. (18). The authors observed the abiotic reduction of *o*-nitrophenol and of 2,4-dinitrophenol in phosphate buffer with 1.5 mM sulfide, whereas *p*- and *m*-nitrophenol and *p*-nitrobenzoate were not reduced under these conditions. In the presence of reducing agents exhibiting a more negative redox potential, like titanium(III) or dithionite, all these nitro compounds were completely reduced to the corresponding amino compounds.

Other reports on the abiotic reduction of nitro groups indicated the participation of a mediator, a quinone compound or an iron porphyrin, which served as an electron carrier between a bulk electron donor like hydrogen sulfide and different nitrobenzenes (Fig.5) (16, 46). Furthermore, picric acid is abiotically reduced initially to picramic acid under anaerobic conditions by Fe(II) with dithiothreitol as the bulk reductant (Rieger, unpublished). Heijman et al. (20) described the abiotic reduction of 4-chloronitrobenzene with Fe(II), associated with or bound to the surface of magnetite particles in an iron-reducing enrichment culture. The abiotic reduction of TNT by sulfide was observed by Preuss et al. (39) and will be described in section 3.

Microbial Enzymes Mediating Nitro Reduction

"Nitroreductases", which catalyze the reduction of nitro groups to amino groups, are ubiquitous in microorganisms as well as in higher organisms. In eucaryotes, nitro-reducing activity very often is an unspecific side activity of another enzyme. In rat liver, for example, nitro reduction is mediated by various enzymes of normal liver metabolism.

Very early reports described the reduction of nitro substituents by liver xanthine-oxidase (11), an enzyme located in the cytosol. In the microsomal fraction of the rat liver, nitro reduction is catalyzed by a system of cytochrome P450 in combination with NADPH-cytochrome-P450-reductase, as was described, for example, by Harada and Omura (19). Other enzymes that have been shown to reduce nitroaromatic compounds include aldehyde oxidase (51), dihydrolipoic amide dehydrogenase (47), cytochrome b_5 reductase (31), and diaphorases (24).

There are several reports of enzymes involved in reduction of nitro compounds by strictly anaerobic organisms. According to these reports, nitro reduction can be carried out by different anaerobic systems: NAD(P)H-dependent reductases, the physiological function of which is still unclear (2, 25); extracellular nitroreductases from human gastrointestinal

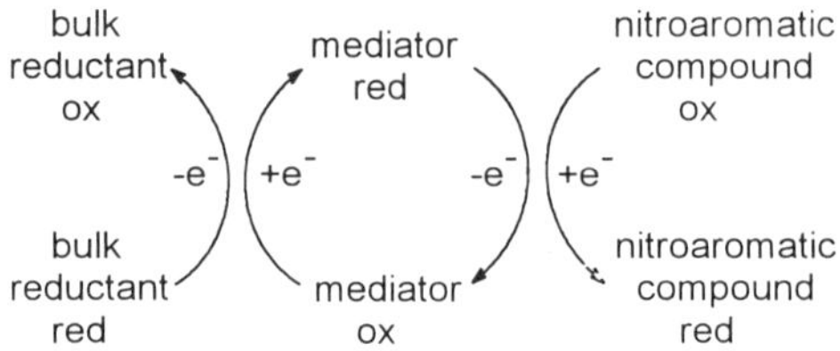

Figure 5. Abiotic reduction of aromatic nitro compounds via a mediator.

bacteria (40); and the combination of H_2-hydrogenase-ferredoxin, which has been described by several authors (3, 12, 35, 39).

O'Brien and Morris (35) observed that *Clostridium acetobutylicum* was able to detoxify chloramphenicol by reducing the nitro group via a ferredoxin-dependent reaction with pyruvate as electron donor. Chen and Blanchard (12) made use of the hydrogenase-mediated reduction of metronidazole to develop an assay for the quantification of ferredoxin. Angermaier and Simon (3) investigated the involvement of hydrogenase and/or ferredoxin in the reduction of 2-nitroethanol and 4-nitrobenzoate. Preparations of hydrogenase from *Clostridium kluyveri* in combination with purified ferredoxin from *Clostridium* sp. La1 or spinach ferredoxin catalyzed the uptake of molecular hydrogen with nitroethanol and 4-nitrobenzoate. Two moles of molecular hydrogen were taken up at a constant rate, whereas the supposed four-electron reduction intermediate, 4-hydroxylaminobenzoate, is a very poor substrate for the reoxidation of reduced ferredoxin (3). The involvement of hydrogenase in combination with ferredoxin in the reduction of TNT will be discussed in section 3.

Nitro Reduction by Reduced Electron Carriers

The observations mentioned above led to the assumption that nitro reduction could be catalyzed by reduced electron carriers in the absence of the enzyme. Indeed, nitro reduction can be spontaneously mediated by reduced flavins ($FADH_2$) after reduction by NADPH-cytochrome c-reductase (22). Similarly, it was to be expected that ferredoxin could act as a nonspecific reductant. Reports by Lindmark and Müller (29) describe the reduction of metronidazole by dithionite-reduced ferredoxin.

The role of ferredoxin as an electron carrier in the reduction of TNT will be discussed below.

3. ANAEROBIC REDUCTION OF 2,4,6-TRINITROTOLUENE

The investigation of nitro reduction with 2,4,6-trinitrotoluene is especially interesting in two respects: First, TNT is of environmental concern as one of the main contaminants at former military sites. Second, many questions about anaerobic nitro reduction and its implications can be elucidated with TNT as a model compound. Since the three nitro groups of the aromatic ring are sequentially reduced, the chemical properties of the molecule will change, thus causing different conditions for each subsequent reduction step.

Influence of the Reducing Potential of the Environment

The susceptibility of the three nitro groups of the TNT molecule to reduction is different with respect to the reducing potential in the environment. The reduction of one nitro group is catalyzed by a great variety of anaerobic as well as aerobic microorganisms as described, for example, by McCormick et al. (32), Parrish (37), Amerkhanova et al. (1), Kaplan and Kaplan (23), and Schackmann and Müller (42).

The subsequent reduction of the resulting aminodinitrotoluene with almost exclusive formation of 2,4-diamino-6-nitrotoluene is mediated by several mainly facultatively or strictly anaerobic bacteria, as described by Naumova et al. (34) and McCormick et al. (32). Moreover, Naumova et al. (34) observed that the further reduction of aminodinitrotoluene to diaminonitrotoluene was enhanced under conditions of nitrate respiration as compared to aerobic conditions.

The further conversion of 2,4-diamino-6-nitrotoluene to triaminotoluene seems to be confined to anaerobic microorganisms. Funk et al. (15) described the formation of triaminotoluene by a bacterial consortium in a reduced environment with a redox potential of -200 mV or less. The complete reduction of TNT to triaminotoluene in pure cultures has been observed so far with very few organisms (32, 39, Crawford, this volume). McCormick et al. (32) indirectly demonstrated the complete reduction by measuring the uptake of molecular hydrogen during the reduction with cell suspensions of *Veillonella alcalescens*. From the stoichiometry of hydrogen uptake, the authors concluded that TNT must have been reduced to triaminotoluene. Another strictly anaerobic microorganism, *Clostridium pasteurianum*, was also capable of catalyzing the complete reduction (32). In contrast to our own investigations (39), McCormick et al. (32) observed complete reduction in crude extracts or cell suspensions rather than in growing cultures. Boopathy and Kulpa (5) and Boopathy et al. (6) reported the biodegradation of TNT by a *Desulfovibrio* strain and described the complete reduction of TNT prior to elimination of the amino groups from the ring and formation of toluene. However, triaminotoluene was postulated rather than measured as the product of the last reduction step.

Anaerobic Transformation of TNT with a *Desulfovibrio* Strain

We studied the course of sequential reduction of the nitro groups of TNT with a sulfidogenic bacterium that was isolated with pyruvate and sulfate as carbon and energy sources and with TNT as the nitrogen source (39). The strategy of using nitroaromatic compounds as the nitrogen source has been applied successfully by Jensen and Lautrup-Larsen (21) and by Bruhn et al. (9). In the case of anaerobic metabolism, this strategy is especially advisable because of the low solubility of TNT. Anaerobic organisms have to catabolically degrade 80 to 90% of their growth substrate to obtain sufficient ATP for the synthesis of biomass. If TNT was supplied as the carbon and energy source, a concentration of at least 6 mM (i.e., 1.36 g/liter) would be required to gain measurable optical densities corresponding to about 100 mg/liter dry weight. With TNT as the sole nitrogen source, however, much lower concentrations will be sufficient, as it is only degraded to meet the anabolic nitrogen requirement.

According to its morphological and some physiological properties, the isolated strain was affiliated with the genus *Desulfovibrio*. Most representatives of the genus *Desulfovibrio* are capable of fixing molecular nitrogen. Therefore, a nitrogen-free atmosphere was maintained in the enrichment cultures.

The growth medium used for the enrichment of TNT-degrading organisms contained sodium sulfide as an oxygen scavenger at a concentration of 5 mM (39, according to 4). TNT was abiotically reduced to a considerable extent under these conditions. However, it is unclear whether TNT can be directly reduced by sulfide or whether a mediator, such as Fe(II), which was also present in the growth medium, is required for the reduction. Fe(II) in the absence of sulfide did not reduce TNT. The main product of this abiotic reduction was 2,4-diamino-6-nitrotoluene, but approximately 30-40% of the TNT initially added could not be recovered and might have been reduced to unknown products. No triaminotoluene was ever produced abiotically under the conditions described above.

The isolated *Desulfovibrio* strain is capable of catalyzing the complete reduction of TNT to triaminotoluene in cell suspensions in the absence of any reducing agent (39). It should be noted, however, that during centrifugation of a growing culture of *Desulfovibrio*, insoluble ferrous sulfide is sedimented together with the cells. Therefore, abiotic processes might have contributed to the reduction reactions observed, but no triaminotoluene was formed with pasteurized cell suspensions. Both aminodinitrotoluenes and 2,4-diamino-6-nitrotoluene could be identified as the intermediates during the reduction which yielded 2,4,6-triaminotoluene as the final product (Fig. 6). This complete reductive pathway has been described by some authors (15, 32, 39).

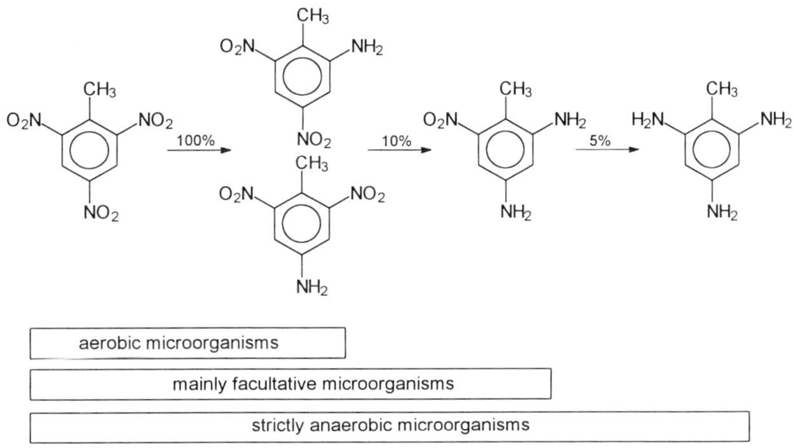

Figure 6. Pathway of complete reduction of trinitrotoluene to triaminotoluene and characteristics of the sequential reduction steps. The reduction rate is indicated in percent of the rate of TNT reduction to the aminodinitrotoluenes.

The reduction rates of the sequential steps catalyzed by cell suspensions of the *Desulfovibrio* strain differ markedly. Whereas the reduction of the first nitro group proceeds at high rates, yielding both aminodinitrotoluene isomers in almost equal amounts (39), the second reduction step is considerably slower. The reduction of 2,4-diamino-6-nitrotoluene is the rate-limiting step in the overall sequence (Table 2).

A plausible explanation for the sequentially decreasing reduction rates of the three nitro groups of the TNT molecule was outlined above: through reduction of a nitro group to an amino group, a substituent with a reduction rate-accelerating effect (-NO_2) is converted into one with the opposite effect (-NH_2) (32), thus "deactivating" the TNT molecule and the remaining nitro groups for further reduction.

Both aminodinitrotoluene isomers are products of the first reduction step in cell suspensions of *Desulfovibrio* (39). However, most publications on TNT reduction indicate that the nitro group *para* to the methyl group is reduced first, with 4-amino-2,6-dinitrotoluene being the main reduction product (23, 32, 37). Only Naumova et al. (33) observed the reduction of TNT mainly to 2-amino-4,6-dinitrotoluene by a *Pseudomonas* strain. Figure 6 illustrates the pathway of complete TNT reduction and summarizes the characteristics of the sequential reduction steps.

Microbial Enzymes Catalyzing TNT Reduction

Nitro reducing activity with TNT as the substrate has been observed in some aerobic microorganisms. A nitroaryl reductase able to reduce several nitrobenzenes, 3,5-dinitrobenzoic acid, and TNT has been purified from *Neurospora crassa* (52). Enzymes in crude

Table 2. Reduction of TNT and its reduced derivatives in cell suspensions of a *Desulfovibrio* strain (39)

Derivative	TNT	ADNT-isomers	DANT
Rate (μmol · l^{-1}.min^{-1})	23.2	2.3	1.2
%	100	10	5

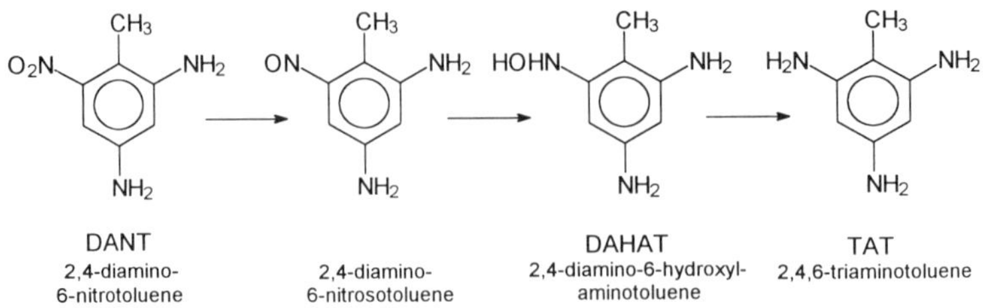

Figure 7. Tentative scheme for the reduction of 2,4-diamino-6-nitrotoluene, assuming that the reduction proceeds in two-electron reduction steps.

extracts of *Pseudomonas* CBS3 reduced *p*-nitrobenzoate and TNT with NADPH as a cofactor (42). Bryant and DeLuca (10) isolated a nitroreductase from *Enterobacter cloacae* that used NADH or NADPH as a cofactor and was also able to reduce TNT.

Little is known about enzymes that may be involved in the reduction of TNT by strictly anaerobic bacteria (32, 39). The reduction of TNT by enzymes from anaerobic organisms has been investigated by McCormick et al. (32). The authors postulated that *Veillonella alcalescens* nitroreductase consisted of hydrogenase and a ferredoxin-like material that could be separated by ion exclusion chromatography.

Since the reduction of 2,4-diamino-6-nitrotoluene to triaminotoluene is the rate-limiting step in reductive TNT transformation and can only be achieved under anaerobic conditions, this reaction is of special interest and will be described in more detail in the next section. Figure 7 shows a tentative scheme of the reduction of 2,4-diamino-6-nitrotoluene to triaminotoluene, assuming that the reduction proceeds in two-electron steps.

4. REDUCTION OF 2,4-DIAMINO-6-NITROTOLUENE (DANT)

The reduction of 2,4-diamino-6-nitrotoluene was investigated in cell suspensions of two clostridial strains, *Clostridium pasteurianum* and *Clostridium thermoaceticum* (39). The former organism contains high activities of hydrogenase; therefore, hydrogen is a feasible electron donor for the reduction. The key enzyme of *Clostridium thermoaceticum* is carbon monoxide dehydrogenase, and consequently, CO may serve as the electron donor. Both organisms reduce DANT to triaminotoluene via the four-electron reduction product 2,4-diamino-6-hydroxylaminotoluene (2,4-DAHAT) as the intermediate. The transient accumulation of 2,4-DAHAT during the course of the reduction was taken to indicate that the further reduction of 2,4-DAHAT is rate-limiting under these conditions (39).

The Involvement of Ferredoxin-Reducing Enzymes

The reduction of 2,4-diamino-6-nitrotoluene was therefore investigated with hydrogenase purified from *Clostridium pasteurianum* and with carbon monoxide dehydrogenase partially purified from *Clostridium thermoaceticum* (39). In contrast to methods described in earlier publications, a test system was applied by which the nitro-reducing activity could be directly measured: the consumption of DANT was assayed spectrophotometrically at 325 nm.

The reduction of DANT is catalyzed by purified hydrogenase from *Clostridium pasteurianum* and also by partially purified CO-dehydrogenase from *Clostridium thermoaceticum* in combination with methyl viologen and/or ferredoxin (39). During the whole purification procedure, nitro-reducing activity coeluted with both enzymes in all steps, thus presenting further evidence that ferredoxin-reducing enzymes are capable of catalyzing the reduction of DANT via methyl viologen and/or ferredoxin as the electron carrier. Obviously, the standard midpoint redox potential of ferredoxin (E_0 = - 420 mV) and of methyl viologen (E_0 = - 440 mV) is sufficiently low for the reduction of 2,4-diamino-6-nitrotoluene.

2,4-Diamino-6-nitrotoluene can also be reduced in the absence of extract or enzyme with a tenfold stoichiometric excess of methyl viologen or ferredoxin, stoichiometrically reduced by sodium dithionite prior to DANT addition (39). The rate of nitro reduction is, however, significantly higher in the presence of the ferredoxin-reducing enzymes. A possible explanation for this finding is the immediate regeneration of reduced ferredoxin, so that a constant negative redox potential is maintained.

The product of 2,4-diamino-6-nitrotoluene reduction by hydrogenase or CO-dehydrogenase in the presence of ferredoxin is 2,4-DAHAT rather than triaminotoluene. This is in accordance with the reports of Angermaier and Simon (3) on the reduction of 4-nitrobenzoate with hydrogenase and ferredoxin. In the presence of hydrogenase from *Clostridium kluyveri* together with *Clostridium pasteurianum* ferredoxin, two mol of molecular hydrogen per mol of *p*-nitrobenzoate were taken up at a nearly constant rate. The further reduction of the supposed four-electron reduction intermediate, *p*-hydroxylaminobenzoate, proceeded only slowly (3).

The elucidation of the pathway leading to 2,4-DAHAT requires further research. As described by Angermaier and Simon (3), the nitro radical anion was the first product of the reduction of *p*-nitrobenzoate. This nitro radical anion is subject to various dismutation and other chemical reactions (Fig. 4), until the final reduction product *p*-hydroxylaminobenzoate is formed. Since DANT is reduced by the ferredoxin/hydrogenase system (39), the possibility cannot be excluded that a radical anion is involved and that the formation of 2,4-DAHAT from this radical anion is due to dismutation and not to biologically mediated reactions.

In Figure 8 the clostridial pathway of the reduction of 2,4-diamino-6-nitrotoluene to triaminotoluene with the intermediates that are known so far is depicted. Since whole cells of *Clostridium pasteurianum* and *Clostridium thermoaceticum* performed the complete reduction to triaminotoluene (39), other enzyme systems that are not yet known must be involved in the reduction of 2,4-DAHAT.

DANT reduction was also investigated in cell suspensions of *Desulfovibrio* sp. The reduction rate was dependent on the addition of an external electron donor. Different substrates that supported growth of the *Desulfovibrio* strain were tested for their potential to serve as electron donors for DANT-reduction. Hydrogen, carbon monoxide, and pyruvate could serve as electron donors, whereas propionate, butyrate, and cysteine were not effective. Cells contained catabolic activities of hydrogenase and pyruvate-ferredoxin-oxidoreductase, as well as anabolic activities of carbon monoxide dehydrogenase. Hence it can be assumed that in sulfidogenic organisms as in the clostridia, DANT reduction is mediated by ferredoxin-reducing enzymes (39).

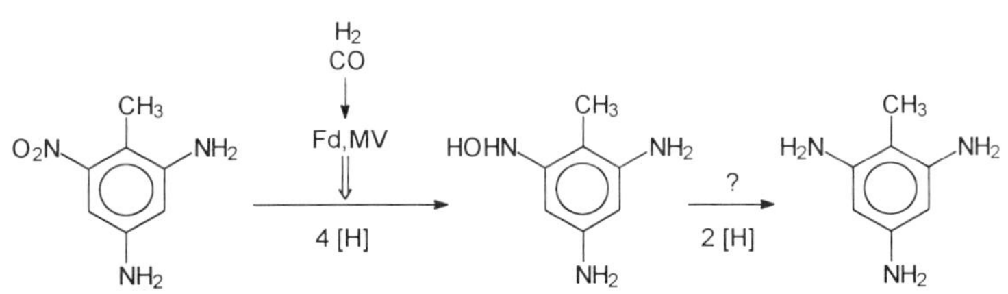

Figure 8. Pathway of the reduction of 2,4-diamino-6-nitrotoluene catalyzed by clostridia (Fd = ferredoxin, MV = methyl viologen).

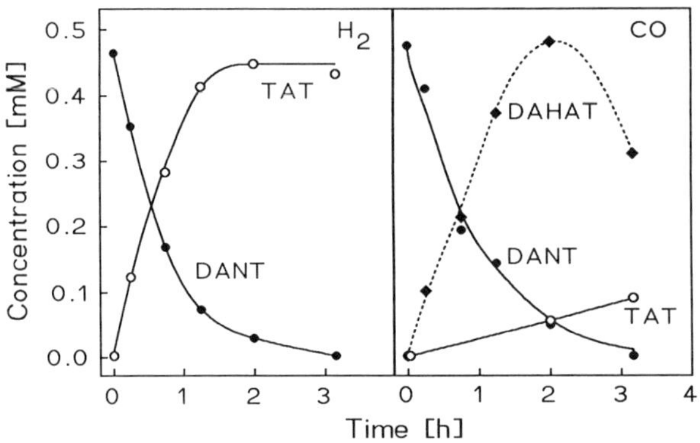

Figure 9. Progress of DANT-reduction in the presence of different electron donors (H_2, CO) in cell suspensions of *Desulfovibrio* sp. Adapted from Preuss et al. (39).

The kinetics of DANT reduction by *Desulfovibrio* sp. depend on the electron donor. With hydrogen or pyruvate, DANT is stoichiometrically reduced to triaminotoluene without transient accumulation of 2,4-DAHAT, as illustrated in Figure 9. With carbon monoxide as electron donor, however, DANT-reduction leads to the intermediary formation of 2,4-DAHAT. This is confirmed by comparison of the reduction kinetics when pyruvate is used as the electron donor in the absence and in the presence of carbon monoxide. In the presence of CO, 2,4-DAHAT transiently accumulates, which indicates that carbon monoxide inhibits the further reduction of the hydroxylaminotoluene (39).

Thus carbon monoxide seems to play two roles in the reduction sequence: on the one hand it serves as an electron donor for the reduction of 2,4-diamino-6-nitrotoluene, on the other hand it acts as an inhibitor of the last step, the reduction of the hydroxylamino-derivative. This suggested that the rate-limiting reduction of 2,4-DAHAT to triaminotoluene in *Desulfovibrio* is catalyzed by an enzyme system that is characteristic for sulfidogenic bacteria and is inhibited or inactivated by carbon monoxide (39). When 2,4-DAHAT instead of diaminonitrotoluene is directly applied as the substrate, its conversion to triaminotoluene by cell suspensions of *Desulfovibrio* is severely inhibited in the presence of carbon monoxide. Therefore, Preuss et al. (39) suggested that the enzyme system responsible for the effective reduction of 2,4-DAHAT is sulfite reductase.

Since there is no simple assay available for sulfite reductase activity, and the electron donor for sulfite reduction is not known (50), only indirect evidence exists for the involvement of sulfite reductase. Reduction of 2,4-diamino-6-hydroxylaminotoluene by cell suspensions of the isolate is inhibited by carbon monoxide and also by hydroxylamine, which is known to be a substrate for the sulfite reductase of *Desulfovibrio vulgaris* (26). Because not all sulfite reductases in sulfidogenic bacteria are CO-sensitive, the inhibition of sulfide formation from sulfite by carbon monoxide and hydroxylamine was tested. Indeed, carbon monoxide as well as hydroxylamine inhibited sulfide formation in cell suspensions of the isolate. Assuming that diaminohydroxylaminotoluene and/or DANT can be alternative, non-physiological substrates of sulfite reductase, an inhibition of sulfide formation by 2,4-DAHAT and/or DANT can be envisaged and was actually confirmed (39).

The reduction of DANT by NADH-dependent enzymes similar to the pyridine-nuleotide-dependent reduction of 4-nitrobenzoate described by Angermaier and Simon (2) could be excluded for the *Desulfovibrio* isolate (39). Figure 10 summarizes the complete reduction of trinitrotoluene in pure cultures of strictly anaerobic bacteria, including the main processes and enzymes involved.

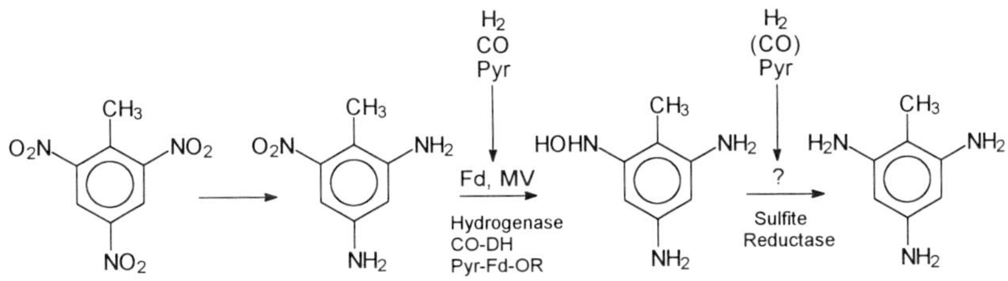

Figure 10. Summary of the most important metabolites during complete reduction of trinitrotoluene by strictly anaerobic bacteria (Fd = ferredoxin, MV = methyl viologen).

5. ANAEROBIC DEGRADATION OF TRIAMINOTOLUENE

The product of complete TNT reduction is triaminotoluene; therefore, the complete degradation of TNT under anaerobic conditions has to involve the elimination of the amino groups from triaminotoluene.

Chemical Instability of Triaminotoluene

The symmetrical substitution of the molecule with three amino groups greatly enhances the electron density of the aromatic ring. As a consequence, the molecule is unstable under aerobic conditions; triaminotoluene is easily susceptible to autoxidation and polymerization. This instability in the presence of oxygen is greatly increased by trace elements. Manganese (II) and nickel (II) in particular significantly accelerate the aerobic conversion of triaminotoluene (39).

For the complete degradation of nitrobenzene, a combined anaerobic-aerobic process has been proposed: first, nitrobenzene is reduced in an anaerobic stage to aniline, which is then completely degraded in the following aerobic reactor (14). Because of the instability of triaminotoluene, the combination of anaerobic reduction of the nitro groups and further aerobic degradation of the reduction product triaminotoluene is not feasible for complete biodegradation of TNT. Further biodegradation of triaminotoluene can therefore only be achieved under anaerobic conditions. This conclusion is confirmed by Funk et al. (15), who suggested a combined process for bioremediation of TNT contaminated soils, where the elimination of the amino groups occurs in the anaerobic stage.

But even under anaerobic conditions there are some limitations with respect to the chemical stability of triaminotoluene. Under acidic conditions, triaminotoluene is abiotically converted to products that are still unknown. The conversion occurs at pH values of 5 or below, with a pH-dependent rate. The conversion of triaminotoluene that can be observed in anaerobic cell suspensions of fermenting organisms like *Pantoea* sp., *E.coli*, or *Citrobacter amalonaticus* under acidogenic conditions is due to this pH sensitivity of triaminotoluene, as proposed by Preuss et al. (39). This can also be deduced from the fact that triaminotoluene conversion by these organisms is strictly dependent on the presence of glucose as a fermentable and acidogenic substrate. In cultures of *Citrobacter amalonaticus*, triaminotoluene conversion was studied during growth on glucose under acidogenic conditions which resulted in a pH shift from 7.6 to 5, and on formate plus fumarate at a constant pH of 7.6. Under acidogenic conditions, triaminotoluene was completely transformed, while no conversion could be observed under pH-neutral conditions. During incubation of triaminotoluene in buffer at pH 4, about one-third of the nitrogen is liberated as ammonia (39). Thus, triaminotoluene is probably subject to partial acidic hydrolysis. The complete hydrolysis would yield methylphloroglucinol (2,4,6-trihydroxytoluene) as the product; however this was never formed chemically under the conditions applied. This is contradictory to the suggestion of Funk et al. (15) who postulated that under physiological conditions triaminotoluene will be in equilibrium with methylphloroglucinol. Furthermore, Funk et al. (15) reported that other reduction products formed as intermediates in the course of TNT biodegradation polymerized at alkaline pH. Therefore, the productive degradation of TNT may occur just within a narrow pH range.

Elimination of Amino Groups from Aromatic Systems

Reductive elimination of ammonia from the aromatic ring by strictly anaerobic bacteria has been demonstrated with aniline and 4-aminobenzoate (44) and also with 2-aminobenzoate (30). To our knowledge, no publications exist so far on biodegradation of diaminobenzene or triaminobenzene.

Due to the high electron density of aminoaromatic compounds, these molecules have to be activated prior to reductive elimination of the amino groups. The mechanism for the elimination of the amino group from aniline is depicted in Figure 11. Aniline is carboxylated in *para*-position to the amino group, yielding 4-aminobenzoate, which is then activated through condensation with coenzyme A to 4-aminobenzoyl-CoA. The activated thioester is reductively deaminated to benzoyl-CoA. This mechanism is explained by the strong negative field and resonance effect of the CoA ester, which reduces the electron density at the 2- and 4-position of the aromatic ring, thus facilitating the reductive attack at these positions. This also explains the reductive elimination of ammonia from 2-aminobenzoate (30). The anaerobic degradation of 3-aminobenzoate as well includes an activation reaction yielding 3-aminobenzoyl-CoA; however, further reduction of this compound yielded not benzoyl-CoA but a different product, which was tentatively described as a reduced CoA ester (45).

The mechanism of reductive elimination of electron-donating substituents from a thioester-activated benzoic acid has also been described for the degradation of phenol (17), which, like aniline, is degraded by carboxylation and condensation with coenzyme A (48) prior to reductive elimination of the hydroxyl group. For polyphenolic compounds, the situation is different. The aromatic system of resorcinol and phloroglucinol is destabilized by keto-enol-tautomerism of the *meta*-hydroxyl groups. Reduction leads to the loss of aromaticity, followed by ring cleavage, and not to a reductive elimination of a hydroxyl group (27, 43).

Perspectives for Anaerobic Degradation of Triaminotoluene

There are very few reports on anaerobic degradation of triaminotoluene. Reductive elimination of the amino groups from triaminotoluene has been postulated by Boopathy and Kulpa (5), though the authors did not detect triaminotoluene as the reduction intermediate. After 45 days of incubation of a grown culture, high amounts of toluene were detected in

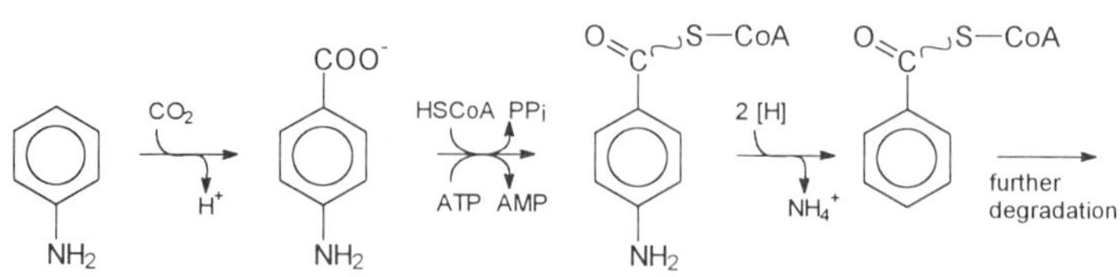

Figure 11. Mechanism for the anaerobic degradation of aniline by *Desulfobacterium anilini* according to Schnell and Schink (44).

the growth medium. Similarly, reductive elimination of the amino groups after complete reduction has been postulated for 2,4-dinitrotoluene, 2,6-dinitrotoluene, 2,4-dinitrophenol, and nitrobenzene, leading to the deaminated products toluene, phenol, and benzene (7). However, none of the mechanisms described above, which involve CoA ester activation, keto-enol-tautomerism or imin-enamin-tautomerism, respectively, applies to the detected metabolites.

In contrast to the observations of Boopathy and Kulpa (5), our *Desulfovibrio* strain is not capable of reductively deaminating triaminotoluene; at least no toluene was ever detected in cell suspensions or growing cultures of the isolate. However, the conversion of triaminotoluene to an unknown metabolite by a *Pseudomonas* strain under denitrifying conditions could be observed (39). C. Windfuhr, from our group, showed that this conversion is strictly dependent on the presence of nitrate and is stimulated by the addition of glucose. The metabolite is currently being characterized. It seems to be aromatic and stable in the presence of oxygen (Windfuhr, unpublished).

Another pathway of further anaerobic degradation of triaminotoluene, which involves methylphloroglucinol and *p*-cresol as the intermediates, has been postulated by Funk et al. (15). According to these authors, triaminotoluene under physiological conditions is in equilibrium with methylphloroglucinol, which is further converted to *p*-cresol. However, this equilibrium seems quite unlikely, as it would require an ammonolytic transformation of methylphloroglucinol, i.e. a nucleophilic substitution of all three hydroxyl groups by the released ammonium at pH 7. As the ammonium cation does not exhibit nucleophilic properties, this reaction does not seem feasible, from a chemical point of view. Therefore, the mechanism of the elimination of ammonia is not yet clear. Elucidation of the mechanism of formation of *p*-cresol from methylphloroglucinol also requires further research.

ACKNOWLEDGEMENTS

We thank Prof. Dr. Gabriele Diekert for helpful discussions and for critically reading this manuscript, and Claudia Windfuhr for providing us with unpublished results of her diploma thesis.

REFERENCES

1. Amerkhanova, N. N., and R. P. Naumova. 1978. 2,4,6-Trinitrotoluene as a source of nutrition for bacteria. Microbiologiya 47:393-395.
2. Angermaier, L., F. Hein, and H. Simon. 1981. Investigations on the reduction of aliphatic and aromatic nitro compounds by *Clostridium* species and enzyme systems, p. 266-275. *In* Bothe, H., and A. Trebst, (ed.) Biology of inorganic nitrogen and sulfur. Springer, Berlin.
3. Angermaier, L., and H. Simon. 1983. On the reduction of aliphatic and aromatic nitro compounds by *Clostridia*, the role of ferredoxin and its stabilization. Hoppe-Seyler's Z. Physiol. Chem. 364:961-975.
4. Balch, W. E., S. Schoberth, R. S. Tanner and R. S. Wolfe. 1977. *Acetobacterium*, a new genus of hydrogen-oxidizing, carbon-dioxide-reducing anaerobic bacteria. Int. J. Syst. Bacteriol. 27:355-361.
5. Boopathy, R., and C. F. Kulpa. 1992. Trinitrotoluene (TNT) as a sole nitrogen source for a sulfate-reducing bacterium *Desulfovibrio* sp. (B strain) isolated from an anaerobic digester. Curr. Microbiol. 25:235-241.
6. Boopathy R., C. F. Kulpa, and M. Wilson. 1993. Metabolism of 2,4,6-trinitrotoluene (TNT) by *Desulfovibrio* sp. (B strain) Appl. Microbiol. Biotechnol. 39:270-275.
7. Boopathy R., and C. F. Kulpa. 1993. Nitroaromatic compounds serve as nitrogen source for *Desulfovibrio* sp. (B strain) Can. J. Microbiol. 39:430-433
8. Brock, T. D., and M. T. Madigan. 1991. Biology of microorganisms. 6th ed. Prentice Hall, New Jersey.
9. Bruhn, C., H. Lenke, and H.-J. Knackmuss. 1987. Nitroaromatic compounds as nitrogen source for bacteria. Appl. Environ. Microbiol. 53:208-210.

10. Bryant, C., and M. DeLuca. 1991. Purification and characterization of an oxygen-insensitive NAD(P)H nitro reductase from *Enterobacter cloacae*. J. Biol. Chem. 266:4119-4125.
11. Bueding, E., and N. Jollife. 1946. Metabolism of trinitrotoluene (TNT) in vitro. J. Pharmacol. Exp. Ther. 88:300-312.
12. Chen, J.-S., and D. K. Blanchard. 1979. A simple hydrogenase-linked assay for ferredoxin and flavodoxin. Anal. Biochem. 93:216-222.
13. Clarke, E. D., K. H. Goulding, and P. Wardman. 1982. Nitroimidazoles as anaerobic electron acceptors for xanthine oxidase. Biochem. Pharmacol. 31:3237-3242
14. Dickel, O., W. Haug and H.-J. Knackmuss. 1993. Biodegradation of nitrobenzene by a sequential anaerobic-aerobic process. Biodegradation 4:187-194.
15. Funk, S. B., D. J. Roberts, D. L. Crawford, and R. L. Crawford. 1993. Initial-phase optimization for bioremediation of munition compound-contaminated soils. Appl. Environ. Microbiol. 59:2171-2177.
16. Glaus, M. A., C. G. Heijman, R. P. Schwarzenbach, and J. Zeyer. 1992. Reduction of nitroaromatic compounds mediated by *Streptomyces* sp. exudates. Appl. Environ. Microbiol. 58:1945-1951.
17. Glöckler, R., A. Tschech, and G. Fuchs. 1989. Reductive dehydroxylation of 4-hydroxybenzoyl-CoA to benzoyl-CoA in a denitrifying, phenol-degrading *Pseudomonas* species. FEBS Letters 251:237-240.
18 Gorontzy, T., J. Küver, and K.-H. Blotevogel. 1993. Microbial transformation of nitroaromatic compounds under anaerobic conditions. J. Gen. Microbiol. 139:1331-1336.
19. Harada, N., and T. Omura. 1980. Participation of cytochrome P-450 in the reduction of nitro compounds by rat liver microsomes. J. Biochem. 87:1539-1554.
20. Heijman, C. G., C. Holliger, M. A. Glaus, R. P. Schwarzenbach, and J. Zeyer. 1993. Abiotic reduction of 4-chloronitrobenzene to 4-chloroaniline in a dissimilatory iron-reducing enrichment culture. Appl. Environ. Microbiol. 59:4350-4353.
21. Jensen, H. L., and G. Lautrup-Larsen. 1967. Microorganisms that decompose nitroaromatic compounds, with special reference to dinitro-*ortho*-cresol. Acta Agric. Scand. 17:115-126.
22. Kamm, J. J., and J. R. Gillette. 1963. Mechanism of stimulation of mammalian nitroreductase by flavins. Life Sci. 4:254-260.
23. Kaplan, D. L., and A. M. Kaplan. 1982. Thermophilic biotransformations of 2,4,6-trinitrotoluene under simulated composting conditions. Appl. Environ. Microbiol. 44:757-760.
24. Kato, R., T. Oshima, and A. Takanaka. 1969. Studies on the mechanism of nitro reduction by rat liver. Mol. Pharmacol. 5:487-498.
25. Kinouchi, T., and Y. Ohnishi. 1983. Purification and characterization of 1-nitropyrene nitroreductases from *Bacteroides fragilis*. Appl. Environ. Microbiol. 46:596-604.
26. Kobayashi, K., Y. Seki, and M. Ishimoto. 1974. Biochemical studies on sulfate-reducing bacteria. J. Biochem. 75:519-529.
27. Krumholz, L. R., R. L. Crawford, M. E. Hemling, and M. P. Bryant. 1987. Metabolism of gallate and phloroglucinol in *Eubacterium oxidoreducens* via 3-hydroxy-5-oxohexanoate. J. Bacteriol. 169:1886-1890.
28. Lenke, H., and H.-J. Knackmuss. 1992. Initial hydrogenation during catabolism of picric acid by *Rhodococcus erythropolis* HL 24-2. Appl. Environ. Microbiol. 58:2933-2937.
29. Lindmark, D. G., and M. Müller. 1976. Antitrichomonad action, mutagenicity, and reduction of metronidazole and other nitroimidazoles. Antimicr. Agents Chemother. 10:476-482.
30. Lochmeyer, C., J. Koch, and G. Fuchs. 1992. Anaerobic degradation of 2-aminobenzoic acid (anthranilic acid) via benzoyl-coenzyme A (CoA) and cyclohex-1-enecarboxyl-CoA in a denitrifying bacterium. J. Bacteriol. 174:3621-3628.
31. Mason, R. P., and J. L. Holtzman. 1975. The role of catalytic superoxide formation in the O_2 inhibition of nitroreductase. Biochem. Biophys. Res. Comm. 67:1267-1274.
32. McCormick, N. G., F. E. Feeherry, and H. S. Levinson. 1976. Microbial transformation of 2,4,6-trinitrotoluene and other nitroaromatic compounds. Appl. Environ. Microbiol. 31:949-958.
33. Naumova, R. P., N. N. Amerkhanova, and V. A. Shaikhutdinov. 1979. Study of the first stage of the conversion of trinitrotoluene. Priklad. Biokhim. Mikrobiol. 15:45-50.
34. Naumova, R. P., S. Y. Selivanovskaya, and I. E. Cherepneva. 1988. Conversion of 2,4,6-trinitrotoluene under conditions of oxygen and nitrate respiration of *Pseudomonas fluorescens*. Appl. Biochem. Microbiol. 24:409-413.
35. O'Brien, R. W., and J. G. Morris. 1971. The ferredoxin-dependent reduction of chloramphenicol by *Clostridium acetobutylicum*. J. Gen. Microbiol. 67:265-271.
36. Oren, A., P. Gurevich, and Y. Henis. 1991. Reduction of nitrosubstituted aromatic compounds by the halophilic anaerobic eubacteria *Haloanaerobicum praevalens* and *Sporohalobacter marismortui*. Appl. Environ. Microbiol. 57:3367-3370

37. Parrish, F. W. 1977. Fungal transformation of 2,4-dinitrotoluene and 2,4,6-trinitrotoluene. Appl. Environ Microbiol. 34:232-233.
38. Peterson, F. J., R. P. Mason, J. Hovsepian, and J. L. Holtzman. 1979. Oxygen-sensitive and -insensitive nitroreduction by *Escherichia coli* and rat hepatic microsomes. J. Biol. Chem. 254:4009-4014.
39. Preuss, A., J. Fimpel, and G. Diekert. 1993. Anaerobic transformation of 2,4,6-trinitrotoluene (TNT). Arch. Microbiol. 159:345-353.
40. Rafii, F., W. Franklin, R. H. Heflich, and C. E. Cerniglia. 1991. Reduction of nitroaromatic compounds by anaerobic bacteria isolated from the human gastrointestinal tract. Appl. Environ. Microbiol. 57:962-968.
41. Rieger, P.-G., A. Preuss, V. Sinnwell, W. Francke, H. Lenke, and H.-J. Knackmuss. 1994. H^--additions as initial steps of aerobic bacterial degradation of 2,4,6-trinitrophenol (picric acid), abstr. Q-120, p. 409. Abstr. 94th Annu. Meet. Am. Soc. Microbiol. 1994.
42. Schackmann, A., and R. Mller. 1991. Reduction of nitroaromatic compounds by different *Pseudomonas* species under aerobic conditions. Appl. Microbiol. Biotechnol. 34:809-813.
43. Schink, B., and N. Pfennig. 1982. Fermentation of trihydroxybenzenes by *Pelobacter acidi-gallici* gen. nov. sp. nov., a strictly anaerobic, non-sporeforming bacterium. Arch. Microbiol. 133:195-201.
44. Schnell, S., and B. Schink. 1991. Anaerobic aniline degradation via reductive deamination of 4-amino-benzoyl-CoA in *Desulfobacterium anilini*. Arch. Microbiol. 155:183-190.
45. Schnell, S., and B. Schink. 1992. Anaerobic degradation of 3-aminobenzoate by a newly isolated sulfate reducer and a methanogenic enrichment culture. Arch. Microbiol. 158:328-334.
46. Schwarzenbach, R.P., R. Stierli, K. Lanz, and J. Zeyer. 1990. Quinone and iron porphyrin mediated reduction of nitroaromatic compounds in homogeneous aqueous solutions. Environ. Sci. Technol. 24:1566-1574.
47. Tatsumi, K., N. Koga, S. Kitamura, H. Yoshimura, P. Wardman, and Y. Kato. 1979. Enzymic *cis-trans* isomerization of nitrofuran derivatives - isomerizing activity of xanthine oxidase, lipoyl dehydrogenase, DT-diaphorase and liver microsomes. Biochim. Biophys. Acta 567:75-87.
48. Tschech, A., and G. Fuchs. 1989. Anaerobic degradation of phenol via carboxylation to 4-hydroxyben-zoate: in vitro study of isotope exchange between $^{14}CO_2$ and 4-hydroxybenzoate. Arch. Microbiol. 152:594-599.
49. Vorbeck, C., H. Lenke, P. Fischer, and H.-J. Knackmuss. 1994. Identification of a hydride-Meisenheimer complex as a metabolite of 2,4,6-trinitrotoluene by a *Mycobacterium* strain. J. Bacteriol. 176:932-934.
50. Widdel, F., and T. A. Hansen. 1992. The dissimilatory sulfate- and sulfur-reducing bacteria, p. 583-624. *In* Balows, A., H. G. Trüper, M. Dworkin, W. Harder, K.H. Schleifer (ed.), The Prokaryotes. Vol. I. Springer, Berlin.
51. Wolpert, M. K., J. R. Althaus, and D. G. Johns. 1973. Nitroreductase activity of mammalian liver aldehyde oxidase. J. Pharmacol. Exp. Ther. 185:202-213.
52. Zucker, M., and A. Nason. 1955. Nitroaryl reductase from *Neurospora crassa*. Methods Enzymol. 2:406-411.

6

BIODEGRADATION OF NITRATED MUNITION COMPOUNDS AND HERBICIDES BY OBLIGATELY ANAEROBIC BACTERIA

Ronald L. Crawford

Center for Hazardous Waste Remediation Research and
Dept. of Microbiology, Molecular Biology, and Biochemistry
University of Idaho
Moscow, Idaho 83844

INTRODUCTION

Research performed during the 1970s (15, 18) generally indicated that complete biomineralization of 2,4,6-trinitrotoluene (TNT) and similar highly nitrated compounds did not occur. Biological reductions ($R\text{-}NO_2 \rightarrow R\text{-}NO \rightarrow R\text{-}NHOH \rightarrow R\text{-}NH_2$) and polymerization reactions appeared to occur, but actual degradation of aromatic nuclei was not observed. However, this work involved studies of aerobic systems such as activated sludge and thermophilic composts, and pure cultures of aerobic fungi and bacteria such as pseudomonads. Pure cultures of some anaerobic bacteria such as *Veillonella alcalescens* (35) were examined, with similar results. Boopathy and Kulpa (2) recently isolated a *Desulfovibrio* that used TNT as a sole source of nitrogen, producing toluene as an end product. A *Pseudomonas* that produced dinitrotoluene, mononitrotoluene, and toluene from TNT, perhaps by hydride additions, was isolated by Duque et al. (10). These are still incomplete degradations of the parent molecule. Since the *Desulfovibrio* strain required obligately anaerobic conditions to produce toluene from TNT, and the *Pseudomonas* did so aerobically, the TNT transformation process may be mechanistically different in the two microorganisms. This should be a fruitful area for future research.

It has recently been found that obligately anaerobic microbial consortia can mineralize many recalcitrant chemicals (toluene, chloroform, benzene, chlorophenols, etc.) that had been considered essentially nonbiodegradable in the absence of oxygen (19, 20, 21). Extensive work at the University of Idaho with obligately anaerobic microbial consortia indicates that these systems are capable of complete biodegradation of nitroaromatic pollutants. Work with 2-*sec*-butyl-4,6-dinitrophenol (dinoseb) has shown complete fermentation of this nitroaromatic pollutant in soils by anaerobic consortia without buildup of aromatic biotransformation products (16, 17). Similar results were observed with a variety of nitrotoluenes and munitions residues, including TNT and trimethylenenitramine (RDX) under appropriately controlled conditions (12, 13). In the work with TNT and RDX, hydroxyaromatics were observed as metabolites of TNT. Their presence indicates a complex series of reductive and hydrolytic reactions that led to the conversion of TNT to products that can be fermented by members of the consortia. Additional work with radiolabeled TNT is required to clarify these pathways and provide overall mass balances for the various TNT biotransformation products

DEGRADATION OF NITROAROMATIC COMPOUNDS BY AN ANAEROBIC CONSORTIUM

Although several anaerobic microbial systems have been described for degrading aromatic chemicals, little practical information has been available on using such cultures to bioremediate nitroaromatic-contaminated soils. The method of Funk et al. (12, 13), developed, patented, and licensed at the University of Idaho, appears to be useful for biologically remediating nitroaromatic herbicide-contaminated soils and explosives-contaminated soils, such as those from agricultural chemical spill sites and military munitions depots (32, 33). For example, 12,000 mg/kg of TNT, 3,000 mg/kg of RDX, and 300 mg/kg of octahydro-1,3,5,7-tetranitro-1,3,5,7-tetrazocine (HMX) were found as contaminants in a soil used as a model system to develop the technology. A consortium of anaerobic bacteria transformed these contaminants to nontoxic, nonaromatic, and aerobically mineralizable end-products, including various organic acids. We have now begun the process of characterizing the microbiological components of this bioremediation system.

When a soil inoculum from this consortium was added to a buffer solution at near neutral pH containing a soil from a U.S. Army munitions site near Umatilla, Oregon, TNT was completely removed within 5 days, and RDX within 24 days (13). We tracked the fate of the TNT molecule to reduced aminonitro compounds, then to *para*-cresol and other yet to be identified intermediates. Nonaromatic volatile organic acids, such as acetate, were observed as ultimate products of anaerobic TNT degradation.

To investigate the pathway for the strictly anaerobic microbial degradation of TNT, an anaerobic bench-top reactor, initially established with a sewage sludge inoculum in 1989, was supplemented periodically with a mixture of TNT and other munitions compounds as a sole carbon source. The feed contained mostly TNT, but also significant amounts of RDX and HMX, and trace amounts of many other munitions-derived compounds. These compounds were supplied in a defined medium containing essential inorganic nutrients. As the consortium of anaerobes adapted over time to utilize the TNT, the reactor went through sequential stages of adaptation characterized by the accumulation of TNT metabolites which were identified by high perfomance liquid chromatography (HPLC) , gas chromatography-mass spectrometry (GC/MS), and ^{14}C-radiolabeled tracer studies (12).

The first metabolites to accumulate in culture supernatants were the reduced intermediates 4-amino-2,6-dinitrotoluene and 2,4-diamino-6-nitrotoluene. A second stage of adaptation led to the disappearance of these intermediates and the accumulation of 2,4,6-trihydroxytoluene. In a third stage of adaptation, 2,4,6-trihydroxytoluene gave way to *p*-cresol (4-methylphenol). After three years of selective pressure for growth on munitions compounds, the reactor accumulated acetate in the supernatant and the headspace as the first identifiable nonaromatic TNT degradation product. This adapted culture was used to establish optimal conditions to maximize TNT degradation rates. When fed the munitions mixture (final TNT concentration, 125 mg/liter), the reactor accumulated detectable amounts of *p*-cresol in 14 days. This bench-top reactor has proven useful in verifying the pathway of anaerobic TNT degradation to mineralizable products. The probable pathway is presented in Figure 1. Since this pathway is based only on results of characterizations of TNT

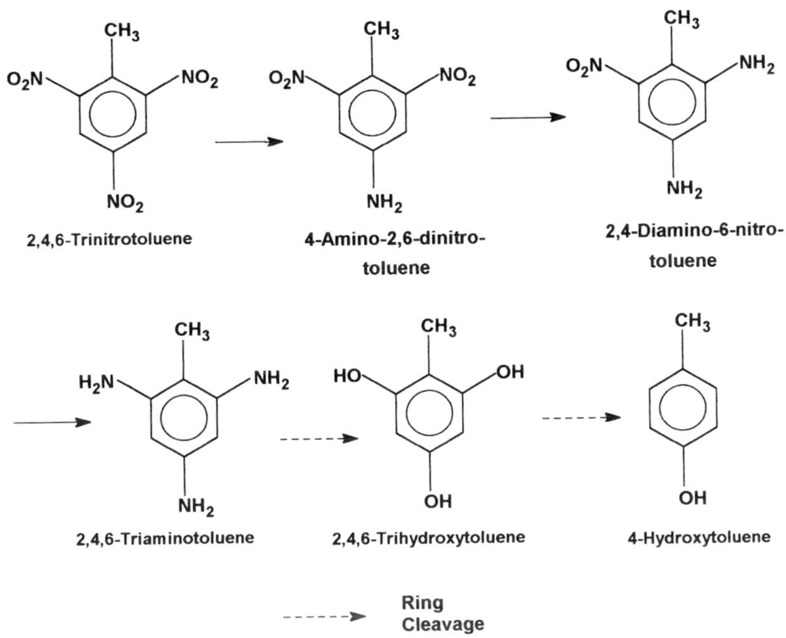

Figure 1. Proposed pathway for the anaerobic biodegradation of TNT. Solid lines, confirmed sequence of reactions; dashed lines, other, as-yet-unidentified intermediates (12).

metabolites, it must be considered as tentative. It remains to be determined which metabolites represent primary pathway intermediates, and whether multiple pathways are occurring simultaneously. Future studies with pure cultures isolated from the consortium will help answer these questions.

We also examined the culture parameters that affect the anaerobic bioremediation of soils contaminated with TNT, RDX, and HMX, in efforts to optimize the physical and chemical conditions promoting their degradation. We flooded the soils with 50 mM phosphate buffer, added starch as a supplemental carbon substrate, and incubated the soils under static conditions. Aerobic heterotrophs, naturally present in the soil or added as an inoculum, quickly removed the oxygen from the static cultures, creating anaerobic conditions. As observed previously, TNT was removed from the soil cultures by the strictly anaerobic microflora within 4 days. The reduced intermediates formed from TNT and RDX were removed from the cultures within 24 days. Optimal conditions were pH, 6.5 to 7.0; temperature, 25 to 35°C; nitrogen source, 25 mM; and redox potential, -100 to -300 mV. Degradation rates were improved when 25 mM ammonium chloride was added to cultures buffered with 50 mM potassium phosphate, but 50 mM ammonium phosphate buffer completely inhibited TNT reduction. When soils were incubated under aerobic conditions or under anaerobic conditions at alkaline pH, the TNT biodegradation intermediates polymerized. Polymerization was not observed at neutral to slightly acidic pH under anaerobic conditions. Polymerized TNT compounds are insoluble in the aqueous phase, so it is crucial to determine whether true degradation of TNT is taking place, or if only simple products of polymerization, which are still toxic or mutagenic, are being formed. The completion of the first stage of remediation of munitions-contaminated soils resulted in aqueous supernatants that contained no munitions residues or aminoaromatic compounds (13).

Our early work with dinoseb, a nitrophenolic herbicide commonly found as a soil contaminant, showed that under microaerophilic conditions, it is transformed to persistent multimeric forms that remain toxic, while under well-aerated conditions, no degradation occurs (29). However, in studies pre-dating our munitions work, we enriched an anaerobic consortium that fermented dinoseb and other nitroaromatic compounds under methanogenic conditions (16, 17). These initial observations ultimately led to our treatment of soils containing complex mixtures of TNT, dinitrotoluenes, mononitrotoluenes, nitrobenzoates, and related compounds (33), which showed that all contaminants could be removed to below detection limits of gas chromatography/mass-spectrometry. Even though biological treatment of several of these compounds in well-aerated cultures has been described, many of them are subject to polymerization reactions under microaerophilic conditions, which are almost certain to occur in soil treatment systems that are not maintained absolutely anoxic (18).

We have determined (17) that the optimal *in vitro* conditions for dinoseb biodegradation differ slightly from those for TNT: carbon source (carbohydrates, particularly starch), 5 g/liter; nitrogen source (ammonium chloride), 4 g/liter; temperature, 25°C (range 10-40°C); pH, 7.0-7.5; and redox potential, <-200 mV (range -200 to -500 mv). Data collected during a demonstration project in which 40 tons of dinoseb-contaminated soil underwent bioremediation indicated that the dinoseb level was reduced to below the analytical detection limit (0.5 mg/kg) in less than 25 days (32). Summary data were reported in a U.S. Environmental Protection Agency Superfund Innovative Technology Evaluation bulletin (32).

ISOLATION OF PURE CULTURES THAT DEGRADE TNT

Pure cultures of obligate anaerobes that degrade TNT have now been isolated by three researchers in our group. A gram-positive, rod-shaped bacterium capable of growth in nutrient media containing 30 ppm TNT and 50 ppm RDX (27) was shown to degrade TNT and RDX sequentially like our consortia. An API AN-IDENT culture identification test indicated *Clostridium bifermentans* as the closest match, an identification supported by urease and motility tests and an rRNA gene sequence. Other *Clostridium* spp. are known to reduce nitroaromatic compounds (25, 26), but do not catabolize TNT beyond simple reduction pathways. The TNT and RDX pathways employed by *C. bifermentans* have yet to be elucidated.

Biodegradation of TNT by a strictly anaerobic bacterial co-culture was originally observed in our lab in 1993. The co-culture was isolated from our three-year-old anaerobic bioreactor that had been fed only munitions compounds, by streaking samples from the bioreactor onto agar plates containing a mineral salts medium plus 100 ppm TNT. Preliminary evidence suggested that the co-culture degrades TNT fermentatively, producing volatile organic acids, and that it probably metabolized TNT efficiently only when a co-substrate (e.g., glucose or yeast extract) was provided (unpublished data). Apparently the agar in the original isolation medium provided contaminating nutrients sufficient for formation of colonies whose growth was primarily at the expense of TNT.

The co-culture consisted of two gram-positive, strictly anaerobic rods with heat-resistant endospores, apparently a *Clostridium* species and another similar coccoid bacterium. The culture grew on plates as a mixture of white and clear colonies, but both colony types contained co-cultures of the two cell types, as shown by scanning electron micrographs (SEM) of a white colony, and transmission electron micrographs (TEM) of a clear colony. As described below, co-cultures obtained from stocks derived from both the white and clear

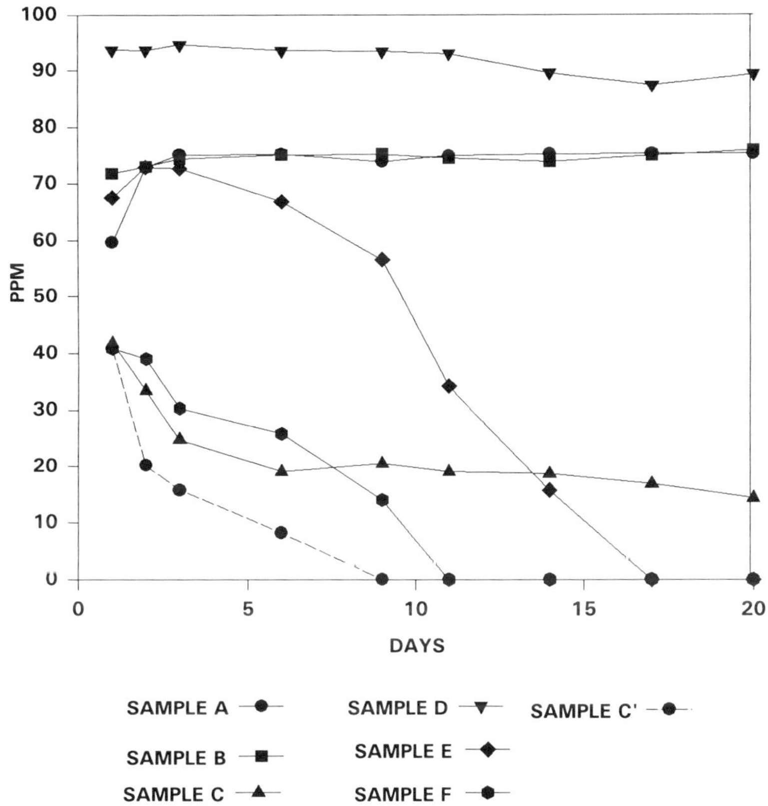

Figure 2. TNT degradation in soil by co-cultures. A (control), no inoculum, sterile (autoclaved) soil, 2% starch added; B (control), 10% inoculum of white colonies killed by 10% formaldehyde, sterile soil, 2% starch added; C, 10% inoculum of white colonies, sterile soil; D (control), no inoculum, non-sterile soil, 2% starch added; E, 10% inoculum of clear colonies, sterile soil; F, 10% inoculum of clear colonies, sterile soil, 2% starch added; C', (dashed line), 10% inoculum of white colonies, sterile soil, 2% starch added.

colony types behaved similarly in TNT degradation experiments (28). Very recently, working with the white co-culture, we have successfully separated the two cell types from one another.

To examine TNT degradation by the co-cultures, we used standard techniques employing anaerobic serum bottles (28). For the degradation experiments, a vitamin-supplemented minimal medium (11) was used, with or without additional starch (2% wt/vol). Slurried TNT-contaminated soil from the Umatilla, Oregon, site was used in place of pure TNT. Initial TNT concentrations were determined after dissolution of TNT from the soil matrix. Residual TNT and the appearance and disappearance of TNT degradation intermediates were monitored by HPLC. The co-cultures degraded TNT to nonaromatic products within 24 hours, in a process stimulated by starch. Co-cultures from both white and clear colonies were separately examined for TNT degradation in the presence and absence of starch. Use of killed controls allowed us to examine whether TNT biodegradation in the Umatilla soil could be mediated by enrichment of the natural microflora, or whether inoculation with the TNT-degrading clostridial culture was required for TNT degradation.

The results of this experiment are summarized in Figure 2. The autoclaved, uninoculated soil (sample A) and the formaldehyde-killed uninoculated soils (sample B) showed no TNT degradation, nor did the nonsterile Umatilla soil, even upon addition of starch as a

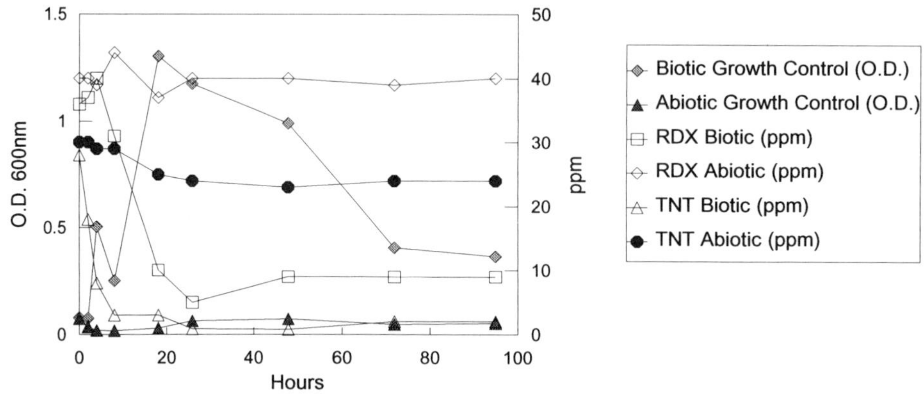

Figure 3. Growth curves of strain KMR-1 with concomitant biotransformation of the explosives in brain-heart infusion medium reduced with cysteine. E_h = -210 mv. Cysteine = 0.08% (wt/vol).

cosubstrate (sample D). Thus, this soil did not contain an anaerobic microflora that could be readily enriched to degrade the TNT, a conclusion we have reported in the past (13). Previous bioremediations of Umatilla soil have always involved inoculation of the soil with a competent microflora from another soil. When nonsterile soil was inoculated with either the white or clear clostridial co-cultures, we observed TNT degradation, which was generally more rapid in the presence of added starch. In recent experiments we have found that a 10% clostridial inoculum with a 3% (w/v) addition of starch resulted in complete removal of 100 ppm TNT from the soil in 24 hours at room temperature. Transient intermediates detected as a result of TNT degradation include 4-amino-2,6-dinitrotoluene (4A-2,6-DNT) and 2,4-diamino-6-nitrotoluene (2,4-DA-6-NT). These intermediates accumulated during the early phases of TNT degradation, but were then removed. Thus, both of the clostridial co-cultures were capable of bioremediating Umatilla soil containing 100 ppm of TNT. Higher concentrations have not yet been examined. However, soils from Weldon Spring, Missouri, containing >2000 ppm TNT were treated successfully in field trials employing anaerobic consortia (33). Since the solubility of TNT in aqueous solution is about 100 mg/liter, it must be solubilized as it is degraded. Thus, very high concentrations of TNT (above solubility) apparently were not toxic to the clostridia in our system.

We have also examined the bioconversion of HMX and RDX, which are aliphatic compounds common in explosives. After demonstrating a sequential bioconversion of TNT, and then RDX by an anaerobic consortium (13), we investigated this sequential process further (27), beginning with the isolation from our mixed munitions-degrading consortium (12, 13) of an obligately anaerobic bacterial isolate capable of removing RDX from its environment. Of several dissimilar isolates, the one chosen for further study was KMR-1, a strain of *C. bifermentans* that degraded the most RDX in 72 hours. Electron micrographs showed it to be a pure culture. The organism, a motile, urease-negative, gram-positive, anaerobic bacillus, is similar to strains previously isolated by our group. The growth of strain KMR-1 was monitored by A_{600} in the presence of explosives under a variety of reducing conditions. In nonreduced media containing no explosives, strain KMR-1 grew rapidly and maintained a very high A_{600}, whereas on media containing explosives it grew much more slowly, regardless of the reductant. In all cases there was a significant lag time before growth of strain KMR-1 in media containing explosives. Data on the biotransformation of explosives by strain KMR-1 during growth in explosive-containing media under different reducing conditions are shown in Figures 3-5.

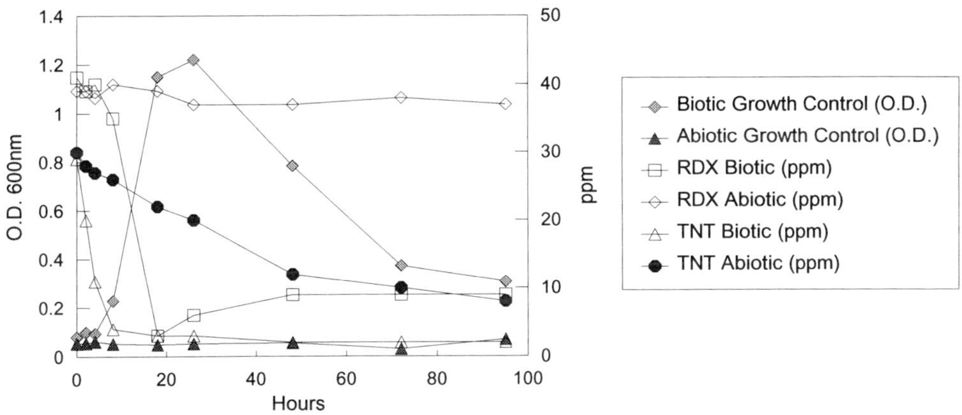

Figure 4. Growth curves of strain KMR-1 with concomitant biotransformation of the explosives in brain-heart infusion medium reduced with dithionite. $E_h < -600$ mv. Dithionite = 0.001% (wt/vol).

In all three media, the transformation of TNT occurred prior to the transformation of RDX. Cell numbers did not increase until TNT concentrations approached their minimum. Once the TNT concentrations dropped, an increase in cell density occurred, with concomitant co-metabolism of RDX. As the concentration of RDX reached its minimum, the cell density in the culture reached a maximum. In abiotic controls, concentrations of RDX did not decrease in any of the reducing treatments, but concentrations of TNT decreased in all treatments. A trend was seen in the amount of abiotically transformed TNT and the theoretical E_h of the media. The more reduced the medium, the greater the amount of abiotic reduction of TNT. In all treatments the amount of biotically transformed explosive was greater than the amount of abiotically transformed explosive.

Funk et al. (13) reported that the biotransformation of TNT occurred over an experimental time frame of 4 days, and RDX over 24 days. This long period required for the observed biotic transformation may be due to the limited amount of growth substrate supplied to the consortium in these experiments. The comparatively short time required for

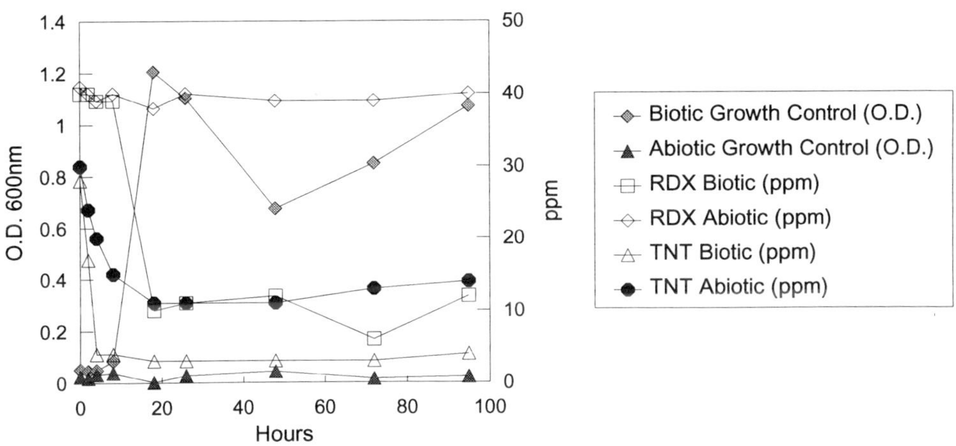

Figure 5. Growth curves of strain KMR-1 with concomitant biotransformation of the explosives in non-reduced brain-heart infusion medium. E_h was not determined.

biotic transformation of TNT and RDX in the pure culture, 4 hours and 23 hours respectively, may be due to the rich substrate supplied to strain KMR-1. In both these groups of experiments (13, 27), the biotransformation of TNT took place before that of RDX, separated by a short lag time of approximately 2 hours. This time lag may represent the time required for the bacteria to develop the cellular machinery to biotransform the RDX. Thus, our accumulated evidence indicates that the organisms responsible for the biotransformation of both RDX and TNT in our enrichments probably are *C. bifermentans* and closely related strains. The bioconversion of RDX and TNT occurs under anaerobic conditions both in the consortium and in pure culture without the need of an added reductant, and TNT is degraded before RDX. The presence of a readily metabolizable carbon source accelerated both biotransformations. *C. bifermentans* was readily and reproducibly isolated from all our enrichments, which contain very high numbers of clostridial endospores (9, 27).

PHYSIOLOGY OF ANAEROBIC TNT BIODEGRADATION

As discussed above, an anaerobic TNT-degrading mixed culture was established by combining 1 liter of an anaerobic, dinoseb-degrading microbial consortium (29) with fresh anaerobic sewage sludge. Neither the sludge nor the enrichment could transform TNT beyond reducing the 4- and 2-nitro groups to amino groups. The consortium was maintained by periodic feeding with a munitions feed solution that contained TNT, RDX, and HMX as the major contaminants (13), and allowed to acclimate to the munitions compounds for 13 months. Intermediates other than the amino-substituted toluenes accumulated sequentially, e.g., methyl phloroglucinol and *p*-cresol (12). Fresh anaerobic sludge was added to the culture twice during the first year of acclimation to restore a lost methanogenic population. The consortium appeared to stabilize after the first year. Removal of the amino-substituted intermediates to below detectable limits usually occurred within 18 to 28 days of feeding. After two years, the consortium was modified by adding a portion to fresh sewage sludge and fresh medium (11). At the same time, fresh medium was added to the original consortium. Both consortia continued to produce methane from the munitions feed. The subenrichment that had received sewage sludge was used for the isolation of culture LJP-1.

Through electron microscope examinations, LJP-1 originally appeared to be a co-culture of two heat-resistant bacteria, a rod and a coccus. It was purified to a single strain, the rod-shaped bacterium, by heat treatment at 82°C for 20 min. The bacterium was a gram-positive (negative in older cultures), non-motile, strictly anaerobic rod with heat-stable spores, identified as a *Clostridium* strain (8).

Experiments were performed to determine whether LJP-1 could grow in a defined medium containing TNT as a sole carbon and nitrogen source. A control without TNT showed no increase in cell numbers after 24 h, whereas the culture with 50 ppm TNT increased from 10^3 to 10^5 cells/ml 24 h after inoculation, when the cells appeared largely as spores. The aqueous phases of the cultures were assayed by HPLC (20). LJP-1 produced and then removed several major TNT degradation intermediates, including 4-amino-2,6-dinitrotoluene, 2,4-diamino-6-nitrotoluene (DANT), and 2,4,6-trihydroxytoluene, within 24 h after inoculation. No nitroaromatic intermediates were seen after 24 h. The predominant organic acids present after 24 h were isovaleric, heptanoic, isocaproic, valeric, butyric, isobutyric, and formic acids. ^{14}C from ^{14}C-TNT was found in these fatty acids (unpublished data).

This was our first observation of a defined, obligately anaerobic culture that could degrade TNT provided as a source of carbon and nitrogen. The culture converted TNT to a mixture of volatile organic acids and a small amount of carbon dioxide via a fermentative pathway apparently differing from that of most *Clostridium* species by producing at least five different organic acids from the seven-carbon TNT molecule.

TRANSFORMATION AND DEGRADATION OF MUNITIONS COMPOUNDS BY STRICT ANAEROBES: OVERALL PERSPECTIVE

If we examine all of the available information concerning transformation and degradation of nitroaromatic compounds by strict anaerobes, some emerging patterns can be seen. Work by Boopathy and Kulpa (2), Boopathy et al. (4-7), Preuss and Rieger (this volume), and Preuss et al. (25) shows that sulfate-reducing bacteria such as *Desulfovibrio* strains can use 2,4,6-trinitrotoluene as an electron acceptor and/or nitrogen source while using compounds like lactate or pyruvate as carbon sources. TNT is reduced by these strains sequentially through monoamino- and diaminonitrotoluenes to triaminotoluene (TAT), which may accumulate or be reductively deaminated to toluene. Methanogenic bacteria also reduce nitrophenols (14) and TNT (4). Our work indicates that an alternative route for TNT degradation involves the ubiquitous reductions to monoamino-, diamino-, and probably triaminotoluene, but subsequent reactions involve hydrolytic displacements of the amino groups to form polyphenols. Polyphenols are then reductively dehydroxylated to yield simpler phenols like *p*-cresol. These alternative routes appear prominent among *Clostridium* strains isolated by enrichments on munitions compounds. These patterns are summarized in Figure 6.

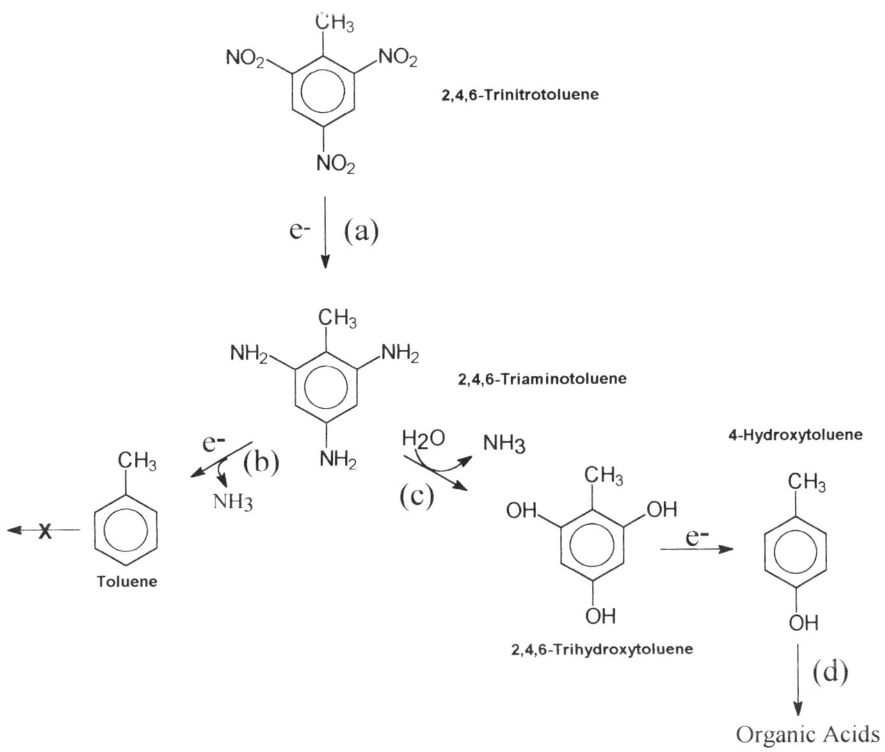

Figure 6. Degradation of 2,4,6-trinitrotoluene by strictly anaerobic bacteria: (a) Many steps, probably involving abiotic and biotic reactions; intermediates include nitroso- and hydroxylamine- derivatives; catalyzed by a variety of reductive enzymes (Preuss and Rieger, this volume); common in many anaerobes. (b) Observed in sulfate reducers. (c) Observed in *Clostridium* strains. (d) Probably employs traditional fermentative/reductive pathways.

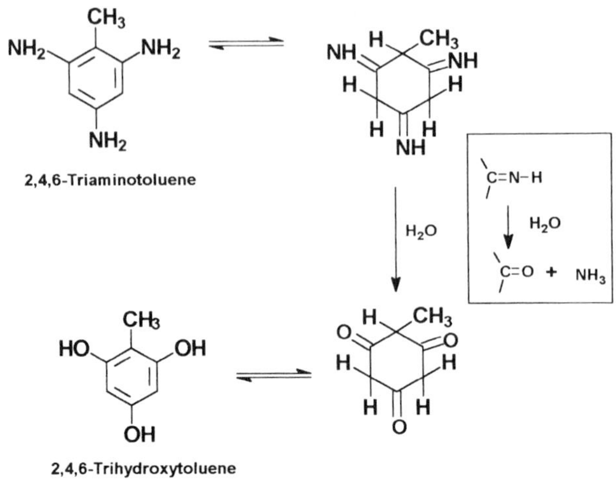

Figure 7. Proposed mechanism for non-enzymatic conversion of 2,4,6-triaminotoluene to 2,4,6-trihydroxytoluene.

There is a great deal of work still required to firmly establish the centrality among anaerobes of the pathways in Figure 6. Very few enzymes have been examined, though there have been some exciting initial studies of TNT and DANT reductases in some systems (Preuss and Rieger, this volume). The complexities of electron transport to TNT and various other intermediary substrates need attention. Many intermediates still must be identified; for example, those between triaminotoluene and trihydroxytoluene in the clostridial systems, and between triaminotoluene and toluene in the *Desulfovibrio* strains. A possible mechanism for the non-enzymatic conversion of TAT to trihydroxytoluene is shown in Figure 7. We have mentioned the likelihod of this conversion in a previous publication (13). In theory, all these reactions would be reversible, but the equilibirium would be expected to favor almost complete conversion to the deaminated products. Thus, such a process may explain how TAT is metabolized further under anaerobic conditions through the involvement of some non-enzymatic, but physiologically productive, reactions.

Mass balances of conversion of TNT to final products are still largely missing from the literature. How our *Clostridium* strains ferment compounds like *p*-cresol to mixtures of volatile organic acids is not known. Fortunately, with the variety of pure cultures now available, progress on these questions should be rapid. Techniques of modern enzymology (e.g., capillary electrophoresis to identify TNT-induced proteins) and molecular biology (e.g., transposon mutagenesis to produce blocked pathway mutants) should be particularly productive. Finally, we should not be surprised to encounter additional variations on TNT catabolism not suggested by work to date.

Other important munitions contaminants receiving study are RDX and HMX. Strict anaerobes appear to degrade these compounds efficiently (18, 22, 34). RDX appears to be reduced sequentially to its nitroso derivatives, and formaldehyde and methanol seem to be biodegradation products (22). The basis for the diauxic process by which both anaerobic consortia (13) and pure *Clostridium* strains (27) degrade TNT from munitions mixtures prior to RDX needs investigation. Are different reductive enzyme systems used for TNT and RDX, and are they induced sequentially? Do similar enzymes reduce both, but preferentially reduce the substrates and their partial reduction products according to their redox potentials? The reviews of Kaplan (18) and Walker and Kaplan (34) provide a summary of RDX and HMX degradation pathways. As with TNT, pure cultures of strict anaerobes that degrade RDX are available, so progress on understanding the details of these processes should also be rapid.

ACKNOWLEDGMENTS

Work reported here was supported by the Idaho State Board of Higher Education; the J. R. Simplot Company; the U. S. Environmental Protection Agency, Emerging Technologies and SITE Programs (award CR816818010); the U. S. Army Corps of Engineers (contract DACA 88-93-M-0810); a U.S. EPA Cooperative Agreement (award CR8200804); and the U.S. Air Force Office of Scientific Research (awards AFOSR-91-0315 and F49620-93-1-0464).

We thank Ms. Connie Bollinger for assistance in review and editing of our manuscript, Dr. Wendy Davis-Hoover and Dr. Eric Weber of the U.S. EPA for technical advice, and Dr. Gerald Mead and Mr. Dane Higdem of the J. R. Simplot Company of managing our SITE demonstrations.

REFERENCES

1. Boopathy, R. Transformation of nitroaromatic compounds by a methanogenic bacterium *Methanococcus* sp. (strain B). Arch Microbiol. 162:167-172.
2. Boopathy, R., and C. F. Kulpa. 1992. Trinitrotoluene (TNT) as a sole nitrogen source for a sulfate reducing bacterium *Desulfovibrio* sp. (B strain) isolated from an anaerobic digester. Curr. Microbiol. 25:235-241.
3. Boopathy, R., and C. F. Kulpa. 1993. Nitroaromatic compounds serve as nitrogen source for *Desulfovibrio* sp. (B strain). Can. J. Microbiol. 39:430-433.
4. Boopathy, R., and C. F. Kulpa. 1994. Biotransformation of 2,4,6-trinitrotoluene (TNT) by a *Methanococcus* sp. (strain B) isolated from a lake sediment. Can. J. Microbiol. 40:273-278.
5. Boopathy, R. C. F. Kulpa, and M. Wilson. 1993. Metabolism of 2,4,6-trinitrotoluene (TNT) by *Desulfovibrio* sp. (B strain). Appl. Microbiol. Biotechnol. 39:270-275.
6. Boopathy, R., J. Manning, C. Montemagno, and C. F. Kulpa. 1994. Metabolism of 2,4,6-trinitrotoluene (TNT) by soil bacteria isolated from TNT-contaminated soil. Biores. Technol. 47:19-24.
7. Boopathy, R., M. Wilson, and C. F. Kulpa. 1993. Anaerobic removal of 2,4,6-trinitrotoluene under different electron-accepting conditions: laboratory study. Wat. Environ. Res. 65:271-275.
8. Cato, P. E., W. L. George, and S. M. Finegold. 1986. The clostridium, p. 1141-1200. *In* P. H. A. Sneath (ed.), Bergey's manual of systematic bacteriology, vol. 2. Williams & Wilkins, Baltimore.
9. Crossley, K. M., D. J. Roberts, R. H. Kaake, R. L. Crawford and D. L. Crawford. Effects of bioremediation on the microbial population of a soil contaminated with dinoseb. Submitted for publication.
10. Duque, E., A. Haïdour, F. Godoy, and J. L. Ramos. 1993. Construction of a *Pseudomonas* hybrid that mineralizes 2,4,6-trinitrotoluene. J. Bact. 175:2278-2283.
11. Fedorak, P. M., J. M. Foght, and W. S. Westlake. 1982. A method for monitoring mineralization of ^{14}C-labeled compounds in aqueous samples. Water Res. 16:1285-1290.
12. Funk, S. B., D. J. Roberts, D. L. Crawford, and R. L. Crawford. 1993. Degradation of trinitrotoluene (TNT) and sequential accumulation of metabolic intermediates by an anaerobic bioreactor during its adaptation to a TNT feed, abstr. Q 410, p. 421. Abstr. Annu. Meet. Am. Soc. Microbiol. 1993.
13. Funk, S. B., D. J. Roberts, D. L. Crawford, and R. L. Crawford. 1993. Initial phase optimization for bioremediation of munition-compound-contaminated soils. Appl. Environ. Microbiol. 59:2171-2177.
14. Gorontzy, T., J. Kuver, and K. H. Blotevogel. 1993. Microbial transformation of nitroaromatic compounds under anaerobic conditions. J. Gen. Microbiol. 139:1331-1336.
15. Higson, F. K. 1992. Microbial degradation of nitroaromatic compounds. *In* S. L.Neidleman and A. I. Laksin (ed.), Advances in applied microbiology, vol 37. Academic Press, New York.
16. Kaake, R. H., D. J. Roberts, T. O. Stevens, R. L. Crawford, and D. L. Crawford. 1992. Bioremediation of soils contaminated with 2-*sec*-butyl-4,6-dinitrophenol (dinoseb). Appl. Environ. Microbiol. 58:1683-1689.
17. Kaake, R. H., D. L. Crawford, and R. L. Crawford. 1994. Optimization of an anaerobic bioremediation process for soil contaminated with the nitroaromatic herbicide dinoseb (2-*sec*-butyl-4,6-dinitrophenol),

p. 337-341. *In* R. E. Hinchee, D. B. Anderson, F. B. Metting Jr., and G. D. Sayles (ed.), Applied biotechnology for site remediation. Lewis Publishers, Boca Raton.

18. Kaplan, D. L. 1992. Biological degradation of explosives and chemical agents. Curr. Opin. Biotechnol. 3:253-260.
19. Kuhn, K., T. Townsend, and J. Suflita. 1990. Effect of sulfate and organic supplements on reductive dehalogenation of chloroanilines in anaerobic aquifer slurries. Appl. Environ. Microbiol. 56:2630-2637.
20. Lovley, D. and D. Lonergan. 1990. Anaerobic oxidation of toluene, phenol, and *p*-cresol by the dissimilatory iron-reducing microorganism GS-15. Appl. Environ. Microbiol. 56:1858-1864.
21. Lovley, D., M. Baedecker, D. Lonergan, I. Cozzarelli, E. Phillips, and D. Siegel. 1989. Oxidation of aromatic contaminants coupled to microbial iron reduction. Nature 339:297-299.
22. McCormick, N. G., J. H. Cornell, and A. M. Kaplan. 1981. Biodegradation of hexahydro-1,3,5-trinitro-1,3,5-triazine. Appl. Environ. Microbiol. 42:817-823.
23. McCormick, N., F. E. Feeherry, and H. S. Levinson. 1976. Microbial transformation of 2,4,6-trinitrotoluene and other nitroaromatic compounds. Appl. Environ. Microbiol. 31:949-958.
24. Naumova, P. R., S. Y. Selivanovskay, and F. A. Mingatina. 1986. Possibility of deep bacterial destruction of 2,4,6-trinitrotoluene. Mikrobiologiya 57:218-222.
25. Preuss, A., J. Fimpel, and G. Diekert. 1993. Anaerobic transformation of 2,4,6-trinitrotoluene (TNT). Arch. Microbiol. 159:345-353.
26. Rafii, F., W. Franklin, R. H. Heflich, and C. E. Cerniglia. 1991. Reduction of nitroaromatic compounds by anaerobic bacteria isolated from the human gastrointestinal tract. Appl. Environ. Microbiol. 57:962-968.
27. Regan, K. M., and R. L. Crawford. Characterization of *Clostridium bifermentans* and its biotransformation of 2,4,6-trinitrotoluene (TNT) and 1,3,5-triaza-1,3,5-trinitrocyclohexane (RDX). Biotechnol. Lett. In press.
28. Shin, K.-R., S. B. Funk, and D. L. Crawford. Treatment of TNT-contaminated soil by inoculation with a *Clostridium* sp. Submitted for publication.
29. Stevens, T. O., R. L. Crawford, and D. L. Crawford. 1990. Biodegradation of dinoseb (2-*sec*-butyl-4,6-dinitrophenol) in several Idaho soils with various dinoseb exposure histories. Appl. Environ. Microbiol. 56:133-139.
30. Stevens, T. O., R. L. Crawford, and D. L. Crawford. 1991. Selection and isolation of bacteria capable of degrading dinoseb (2-*sec*-butyl-4,6-dinitrophenol). Biodegradation 2:1-13.
31. Tschech, A., and B. Schink. 1988. Methanogenic degradation of anthranilate (2-aminobenzoate). Syst. Appl. Microbiol. 11:9-12.
32. U.S. Environmental Protection Agency. 1994. Demonstration of the J. R. Simplot ex-situ bioremediation technology for treatment of nitroaromatic contaminants at the Weldon Spring Ordnance works site in Weldon Spring, Missouri: TNT. March 1994
33. U.S. Environmental Protection Agency. 1994. Ex-situ anaerobic bioremediation system: dinoseb. EPA Superfund Innovative Technology Evaluation Demonstration Bulletin. EPA/540/MR-94/508. April 1994.
34. Walker, J. E., and D. L. Kaplan. 1992. Biological degradation of explosives and chemical agents. Biodegradation 3:369-385.
35. Woolfolk, C. A. 1963. Ph.D. dissertation. University of Washington, Seattle.

7

REDUCTIVE METABOLISM OF NITROAROMATIC AND NITROPOLYCYCLIC AROMATIC HYDROCARBONS

Carl E. Cerniglia* and Charles C. Somerville†

*National Center for Toxicological Research
U.S. Food and Drug Administration
Jefferson, Arkansas 72079
†Armstrong Laboratory
U.S. Air Force
Tyndall AFB, Florida 32403

INTRODUCTION

Nitro-PAHs are polycyclic aromatic hydrocarbon derivatives that contain one or more nitro groups covalently bound at chemically reactive positions on the aromatic ring. Mixtures of nitrated PAHs are generated either by reactions of PAHs with nitrogen oxides or as byproducts of the incomplete combustion of fossil fuels (65). A wide variety of nitro-PAHs have been isolated from environmental sources, such as coal fly ash, diesel emission particulates, cigarette smoke and carbon black photocopier toners (29, 52, 63, 74, 75, 86, 87, 88). Structures of representative nitro-PAHs isolated from the environment are shown in Figure 1.

More than twenty years ago, research in Bruce Ames's laboratory (51) showed that nitro-PAHs are potent mutagens in the *Salmonella typhimurium* reversion assay, and interestingly, they do not require metabolic activation to be genotoxic (3, 92). This direct mutagenic activity contrasts with the response observed with PAHs, which normally require mammalian post-mitochondrial supernatant (S9) to induce mutations (28, 39, 55). Fu and co-workers have postulated that nitro-PAHs can be divided into two groups, based on whether the nitro substituent is co-planar (for example, 1-nitropyrene) or perpendicular (for example, 6-nitrobenzo[*a*]pyrene) to the aromatic ring (26, 27, 28, 41). A nitro-PAH with the nitro substituent in a perpendicular orientation and relatively high first-wave reduction potential usually exhibits weak or no direct-acting mutagenicity in *Salmonella typhimurium*.

Many nitro-PAHs are mutagenic to bacteria and mammalian cells and are a potential carcinogenic risk to humans (40). The genotoxic activity of many nitro-PAHs is dependent upon the reduction of the nitro group to *N*-arylhydroxylamines by either mammalian or bacterial nitroreductases (3, 54, 76). Nitro reduction can occur as a series of two-electron transfers which yield the nitroso, hydroxylamino, and amino derivatives of the parent

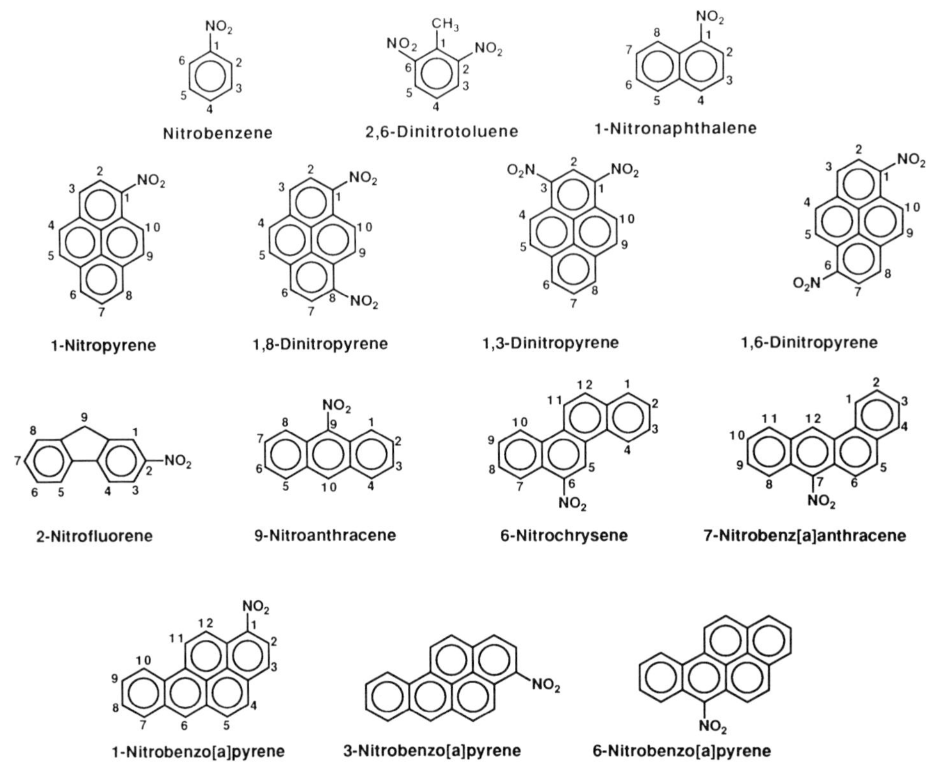

Figure 1. Chemical structures of nitropolycyclic aromatic hydrocarbons detected in environmental samples.

compound, or through one-electron transfers resulting in the nitroanion radical and nitroso compounds (Fig. 2). These enzymatic mechanisms are discussed in detail by Corbett and Corbett (this volume). After esterification the *N*-hydroxylarylamines can react with the guanine moiety of DNA to form a DNA adduct (Fig. 2). It appears that nitro-PAHs are activated to DNA damaging species *in vivo* by nitroreduction and subsequent *O*-esterification, by a combination of intestinal microflora and mammalian enzymes (32, 33, 53). The structures of the DNA adducts formed from incubation of nitro-PAHs with *Salmonella* support this conclusion (31, 38). When *Salmonella typhimurium* tester strains deficient in nitroreductase were exposed to nitro-PAHs, no genotoxic intermediates were formed and mutagenicity was not induced. Nitroreduction by intestinal microflora has a major influence on the disposition and metabolic activation of nitro-PAHs in experimental animals. In addition, oxidative metabolism *via* mammalian cytochrome P-450 mediated reactions has been shown to be important in the metabolic activation of nitro-PAHs in experimental animals (3, 28) (Fig. 2).

METABOLISM OF NITRO-PAHS BY GASTROINTESTINAL MICROFLORA

Intestinal microflora are an essential component of human physiology because they act as a barrier against colonization of the gastrointestinal tract by pathogenic bacteria. They also play an important role in the digestion of food and in the metabolism of xenobiotics,

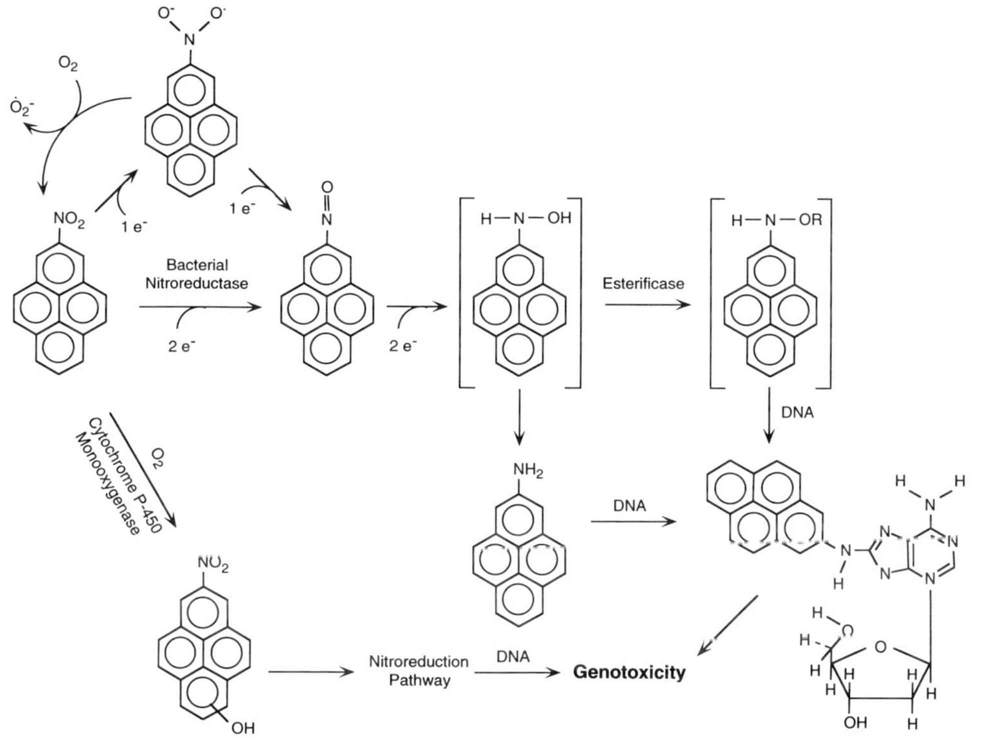

Figure 2. Metabolic activation pathways for nitropolycyclic aromatic hydrocarbons.

drugs and nutrients (19). Humans are continuously exposed to nitroaromatic compounds in consumer products, because they are important intermediates in the manufacture of drugs, plastics, agricultural chemicals and dyes (40). The human gastrointestinal tract is an extremely complex community, which contains more than 400 bacterial species (23). The number of microorganisms per gram of feces can reach a concentration of 10^{11} organisms, with the predominant genera being *Bacteroides, Bifidobacterium, Clostridium, Fusobacterium, Peptococcus* and *Peptostreptococcus*. Genotoxic compounds like nitro-PAHs can gain entrance to the intestinal tract either directly *via* ingestion of food, drugs and water or indirectly *via* the enterohepatic circulation or inhalation of airborne particulates. Therefore, there are plenty of opportunities for nitro-PAHs to enter the gastrointestinal tract and be metabolized by intestinal microflora.

Bacteria within the gastrointestinal tract perform a wide variety of biochemical reactions (77). Since the lower part of the gastrointestinal tract is anaerobic, the biochemical reactions performed by intestinal microflora are usually reductive and hydrolytic in nature. Nitro-PAH metabolism by intestinal microflora has been demonstrated both *in vitro* and *in vivo* (Table 1, 2). It is well established that the nitroreductases of intestinal microflora are responsible for the metabolism and bioactivation of nitro-PAHs. However, it should be noted that the intestinal microflora are not separate from the host metabolic processes and that synergistic metabolic interactions between enzymes of the gut mucosa, hepatic tissue and microflora are important in the metabolic activation of nitro-PAHs. El-Bayoumy et al. (25) and Kinouchi et al. (45) demonstrated that the predominant metabolites formed from 1-nitropyrene-treated germ-free animals (intestinal microflora were absent) were primary ring-hydroxylated derivatives of

Table 1. Anaerobic microbial metabolism of nitropolycyclic aromatic hydrocarbons

Compound	Organisms	Metabolites	Reference
1-Nitropyrene	*Bacteroides* sp., *Bacteroides thetaiotaomicron, Bacteroides fragilis, Bifidobacterium infantis, Clostridium* sp., *Clostridium perfringens, Clostridium clostridiiforme, Clostridium leptum, Citrobacter* sp., *Peptococcus anaerobium, Peptostreptococcus productus, Salmonella typhimurium, Escherichia coli, Lactobacillus* sp., *Lactobacillus acidophilus,* Rat and Human intestinal microflora.	1-Aminopyrene, 1-N-Acetyl-1-aminopyrene, 1-N-Formyl-1-aminopyrene.	53, 36, 37, 13, 12, 45, 41, 44, 43, 64, 65, 10.
1,8-Dinitropyrene	*Salmonella typhimurium*, Human, Rhesus Monkey and Rat intestinal microflora.	1-Amino-8-nitropyrene, 1,8-Diaminopyrene, 1-N-Acetyl-8-nitropyrene, N-N-Diacetyl-1,8-diaminopyrene.	9, 31, 32, 33, 15.
2-Nitrofluorene	Rat intestinal microflora	2-Aminofluorene.	56.
7-Nitrobenz[*a*]anthracene	*Bacteroides* sp., *Butyrivibrio* sp., *Clostridium clostridiiforme, Clostridium nexile, Clostridium paraputrificum, Clostridium perfringens, Eubacterium* sp., *Eubacterium hadrum*, Human, Mouse, Pig and Rat intestinal microflora.	7-Aminobenz[*a*]anthracene, Benz[*a*]anthracene7,12-dione.	57.
6-Nitrobenzo[*a*]pyrene	*Bacteroides thetaiotaomicron*, *Bifidobacterium infantis, Clostridium* sp., *Citrobacter* sp., *Clostridium perfringens, Peptostreptococcus productus, Escherichia coli*, Rat and Human intestinal microflora.	6-Aminobenzo[*a*]pyrene, 6-Nitrosobenzo[*a*]pyrene.	13, 12, 68, 27.
3-Nitrobenzo[*a*]pyrene	Rat intestinal microflora	3-Aminobenzo[*a*]pyrene.	68, 81.
1-Nitrobenzo[*a*]pyrene	Rat intestinal microflora	1-Aminobenzo[*a*]pyrene, 1-Nitrosobenzo[*a*]pyrene.	68, 81.
6-Nitrochrysene	*Bifidobacterium infantis, Peptococcus magnus, Eubacterium lentum, Clostridium perfringens, Bacteroides fragilis, Peptostreptococcus productus, Bacteroides thetaiotaomicron, Lactobacillus acidophilus, Citrobacter* sp., *Salmonella typhimurium*, Human, Mouse and Rat intestinal microflora.	6-Aminochrysene, N-Formyl-6-aminochrysene, 6-Nitrosochrysene, N-Acetyl-6-aminochrysene.	46, 47, 20.

Table 2. Examples of the importance of intestinal microflora nitroreductases in the toxicity of nitroaromatic compounds

Compound	Experimental Approach	Toxicological Effects	Reference
Nitrobenzene	Conventional but not germ-free or antibiotic treated animals showed methemoglobin levels of 30-40% higher after exposure to nitrobenzene.	Methemoglobinemia	67
Dinitrotoluene	Nitroreduction and covalent binding to DNA was decreased in germ-free versus conventional animals. No induction of unscheduled DNA synthesis in germ-free animals.	Hepatic genotoxicity	55, 22, 69
1-Nitropyrene	Conventional but not germ-free or antibiotic treated rats metabolized 1-nitropyrene to 1-aminopyrene. Urine from 1-nitropyrene-treated conventional rats was more mutagenic than that from germ-free animals.	Carcinogenic in experimental animals	43
2-Nitrofluorene	Intestinal microflora reduced the excretion of mutagenic metabolites in conventional versus germ-free animals.	Carcinogenic in experimental animals	56
6-Nitrochrysene	Intestinal microflora can reduce the nitro group to form 6-aminochrysene. A comparison of levels of DNA adducts formed from germ free and conventional mice receiving an i.p. dose of 6-nitrochrysene indicates that mammalian reductases are primarily responsible for metabolic activation.	Carcinogenic in experimental animals.	20

1-nitropyrene, whereas the conventional animal used both oxidative and reductive pathways to form 1-aminopyrene, acetylated and ring-hydroxylated metabolites. In addition, the excreta collected from germ-free animals were less mutagenic than those from conventional animals, indicating that intestinal microflora play an important role in the bioactivation of 1-nitropyrene. These results were in agreement with studies with pure cultures of intestinal microflora representative of the human gastrointestinal tract and from mixed microflora, from rat and human intestinal samples, which reduced 1-nitropyrene to 1-aminopyrene (Table 1).

Nitroreductases of intestinal microflora are important in the metabolic activation of nitro-PAHs, such as nitrobenzene, dinitrotoluenes, 2-nitrofluorene, and 6-nitrochrysene to genotoxic intermediates (Table 2). In these studies, as previously described for 1-nitropyrene, the major metabolites from germ-free animals are primarily products of oxidative hepatic enzymes, whereas conventional animals utilize a reductive pathway mediated by bacterial nitroreductases. The extent of nitroreduction of nitro-PAHs to the corresponding arylamines by cultures of intestinal microflora depends upon the geometric structure of the nitro-PAH, the position of the nitro substituent, the source of intestinal microflora and diet (60). Generally, the nitroreductive capacities of mixed populations of intestinal microflora are greater than those observed in pure culture, demonstrating that microbial interactions may play an important role in governing the potential hazards of nitro-PAHs in the gastrointestinal tract.

ENVIRONMENTAL DEGRADATION OF NITRO-PAHS

Nitro-PAHs are released into the environment from anthropogenic sources and *via* chemical nitration of PAHs formed from incomplete combustion of organic material and

Table 3. Aerobic microbial metabolism of nitropolycyclic aromatic hydrocarbons

Compound	Organisms	Metabolites	Ref.
1-Nitropyrene	*Cunninghamella elegans*	1-Nitropyrene-6-a-D-glucopyranoside	14
	Mycobacterium sp. PYR-1	1-Nitropyrene-8-a-D-glucopyranoside	35
		1-Nitropyrene *cis*-4,5-dihydrodiol	
		1-Nitropyrene *cis*-9,10-dihydrodiol	
		1-Aminopyrene	
3-Nitrofluoranthene	*Cunninghamella elegans*	3-Nitrofluoranthene-8-sulfate	66
		3-Nitrofluoranthene-9-sulfate	
6-Nitrobenzo[*a*]pyrene	*Cunninghamella elegans*	6-Nitrobenzo[*a*]pyrene-1-α -D-glucopyranoside	63
		6-Nitrobenzo[*a*]pyrene-3-α -D-glucopyranoside	54

photochemical reactions of PAHs with NO_x. In contrast to the number of studies on the metabolism of nitro-PAHs by intestinal microflora (Table 1), less is known about the metabolism of these contaminants by aquatic and terrestrial microorganisms (Table 3). Although a wide variety of bacteria, fungi and algae have been shown to degrade PAHs containing two to five aromatic rings (16, 17), nitro-substituted PAHs are slowly degraded by indigenous microorganisms and may persist in soils and sediments (36). Interestingly, in soils with high pyrene biodegradative ability, 1-nitropyrene was degraded slowly under aerobic conditions, so that less than 1% of the added 1-nitropyrene was mineralized to $^{14}CO_2$ after 8 weeks of incubation. Under anaerobic conditions, the 1-nitropyrene was reduced to 1-aminopyrene. The recalcitrance of high molecular weight nitro-PAHs is due in part to the strong adsorption to soil organic matter, low solubility, large molecular size and hydrophilic character of the nitro group. In addition, the nitro substituent is an electron-withdrawing group which makes the aromatic moiety of 1-nitropyrene less accessible for oxidation and therefore a relatively poor substrate compared to the parent compound, pyrene. Heitkamp and Cerniglia (35) isolated *Mycobacterium* sp. PYR-1 from oil-contaminated sediments and showed that it was capable of metabolizing pyrene, phenanthrene, and fluoranthene completely to CO_2. The metabolism of 1-nitropyrene in microcosms containing pyrene-induced *Mycobacterium* sp. PYR-1 cultures was limited to approximately 11% of the total 1-nitropyrene added. The major ethyl acetate extractable metabolites were identified as 1-nitropyrene *cis*-9,10- and 4,5-dihydrodiols (Table 3, Fig. 3). The low mineralization of 1-nitropyrene, as compared to the parent compound pyrene, could be due to the nitro substituent in the C-1 position, which could sterically block enzymatic oxidation. Recently, Rafii et al. (70) showed that *Mycobacterium* sp. PYR-1 cells grown without pyrene had nitroreductase activity and were capable of reducing 1-nitropyrene to the corresponding arylamine, 1-aminopyrene (Table 3). The nitroreductase activity, similar to those of *Enterobacter cloacae* and *Escherichia coli* (5, 84), was oxygen-insensitive and intracellular.

Most non-ligninolytic fungi, like the zygomycete *Cunninghamella elegans*, initially oxidize PAHs *via* a cytochrome-P-450 monooxygenase to arene oxides, which isomerize to form phenols or are enzymatically hydrated to form *trans*-dihydrodiols. The dihydrodiols are optically active and are predominantly in the *S,S* configuration. The phenols can be subsequently conjugated with sulfate, glucose, xylose or glucuronic acid to form detoxified products (Fig. 3). These reactions are similar to those reported for PAH oxidation by mammalian systems (16, 17, 18). *Cunninghamella elegans*, which metabolizes two- to five-ring PAHs to dihydrodiols, phenols, quinones, dihydrodiol-epoxides and conjugates of the hydroxylated derivatives, has also been shown to metabolize the nitro-PAHs 1-nitropyrene, 3-nitrofluoranthene, and 6-nitrobenzo[*a*]pyrene (Table 3). *C. elegans* metabolized approximately 55% of the added nitropyrene to glucoside conjugates of 6- and 8-hydroxy-1-nitropy-

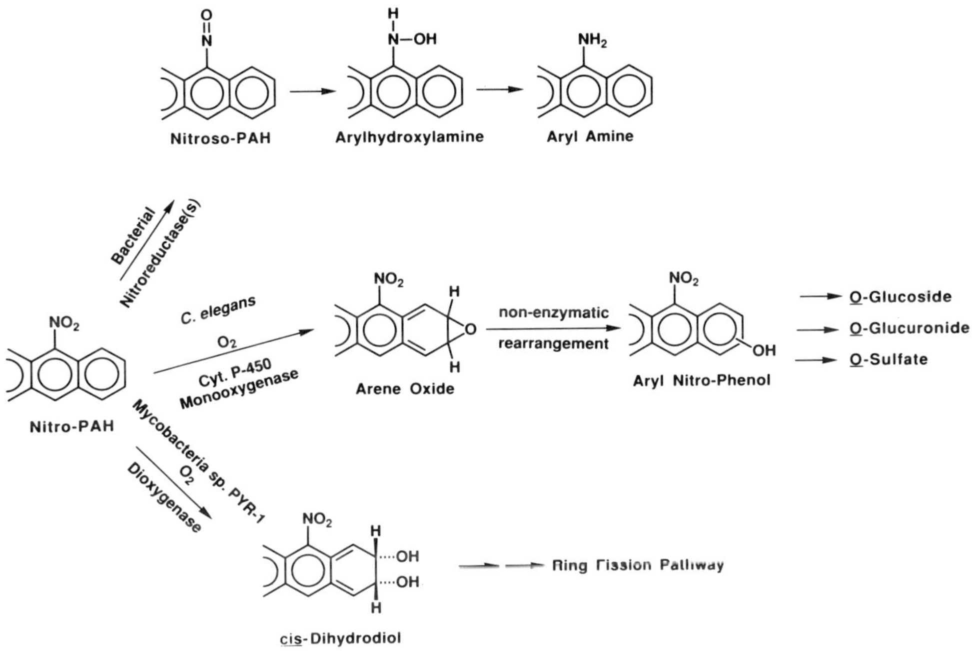

Figure 3. Diverse metabolic pathways for the microbial metabolism of nitropolycyclic aromatic hydrocarbons.

rene (14). No intermediates resulting from reduction of the nitro group were detected. Although free nitropyrenols are mutagenic in the *Salmonella* reversion assay, the two glucoside conjugates formed by *C. elegans* were not mutagenic, which further substantiates the importance of fungi in the detoxification of nitro-PAHs in the environment. Similar results were also found when *C. elegans* was incubated with 6-nitrobenzo[*a*]pyrene and 3-nitrofluoranthene (57, 66). *C. elegans* metabolizes 6-nitrobenzo[*a*]pyrene to glucoside conjugates of 1- and 3-hydroxy-6-nitrobenzo[*a*]pyrene and a sulfate conjugate of 1-hydroxy-6-nitrobenzo[*a*]pyrene (Table 3). The nitro group at the C-6 position blocks benzo[*a*]pyrene metabolism at the C-7 and C-8 positions, which are *peri* to the nitro-substituent, and shifts the metabolism to the C-1 and C-3 positions (57). A similar nitro substituent effect was also found in the metabolism of 3-nitrofluoranthene by *C. elegans* (66).

OXYGEN-INSENSITIVE BACTERIAL NITROREDUCTASES

A wide variety of enzymes in both prokaryotic and eukaryotic cells can catalyze the reduction of the nitro group of nitroaromatic compounds. Although these enzymes have few common characteristics, they can be placed in one of two functional groups based on their ability to catalyze nitroreduction in the presence of oxygen. Type I (oxygen-insensitive) enzymes reduce the nitro group in two-electron increments *via* nitroso- and hydroxylamino-intermediates. The nitroso group is sufficiently stable to allow reduction to proceed in the presence of oxygen. Type II (oxygen-sensitive) enzymes catalyze a one-electron reduction of the nitro group yielding a nitro anion radical. The nitro anion radical reacts with oxygen to form superoxide radical and regenerate the parent nitro compound (Fig. 2). This expenditure of reducing power without achieving net reduction of the nitro group is the basis for oxygen sensitivity (64).

Eukaryotic enzymes with nitroreductase activity include aldehyde oxidase, cytochrome *c* reductase, cytochrome P-450 reductase, glutathione reductase, hepatic cytochrome P-450s, hepatic NAD(P)H:quinone reductase, hydrogenosomal pyruvate:ferredoxin oxidoreductase, succinic dehydrogenase, xanthine dehydrogenase and xanthine oxidase (6). The reduction of aromatic nitro groups is probably not the physiological role of these enzymes, but the characteristics of the nitro group make it a facile electron acceptor. Of these enzymes, only hepatic NAD(P)H:quinone reductase and xanthine dehydrogenase have oxygen-insensitive nitroreductase activity (6). Bacteria also have both Type I and Type II nitroreductase activities. Oxygen-sensitive enzymes with nitroreductase activity have been described (67, 69), but most of the enzymes that have been thoroughly characterized are oxygen-insensitive. Little is known about the physiological role of these enzymes in the bacterial host, but a review of the research on bacterial Type I enzymes does provide some clues to their functions and origins.

Early observations of bacterial nitroreductase activity were made with cells capable of reducing chloramphenicol and/or *p*-nitrobenzoic acid to the corresponding aryl amines (78, 78, 80, 81, 82). From this body of work there emerged an early image of a bacterial nitroreductase as a soluble metalloflavoprotein that uses reduced pyridine nucleotide as an electron donor and catalyzes at least one two-electron reduction (Fig. 4).

Yamashina and co-workers (94, 95) showed that azoxy-, azo- and hydrazo-compounds were not reduced by bacterial nitroreductase. Their results supported the observations of Saz and coworkers (78, 78, 80, 81) on the occurrence of individual nitro and nitroso reductase activities, and inhibitor studies indicated the involvement of metals in enzymatic activity. Furthermore, these studies contradicted the earlier work of Egami et al. (24) which indicated that nitroreductase activity could be attributed to nitrite reductase.

Cartwright and Cain (11) studied reduction of the nitro group of nitrobenzoic acids by enzymes in cell-free extracts of *Nocardia erythropolis*. Detection of the nitroso and hydroxylamino derivatives of nitrobenzoate indicated that they were intermediates in reduction to aminobenzoate. This study attributed nitroreductase activity to a single soluble metalloflavoprotein that catalyzes the entire reduction from the nitro to the amino group. Glucose grown cells reduced nitrobenzoate to aminobenzoate, which suggested that the reaction was not catalyzed by a specific inducible nitroreductase.

The first purified nitroreductase was isolated from *Nocardia V* (89). The enzyme involved in the reduction of *p*-dinitrobenzene was purified to homogeneity, and shown to catalyze the reduction of *p*-dinitrobenzene to *p*-nitroaniline. The enzyme preparation contained a single component with a molecular weight of 24 kDa as measured by ultracentrifugal analysis. Unfortunately, the enzyme activity was unstable and too little material was recovered for detailed characterization. Still, three key features of the reductase were identified; the molecular weight of the enzyme, the ability to catalyze the entire series of reductions from a nitro to an amino group, and the use of NADH as an electron donor (Table 4).

Several studies have shown that bacteria possess multiple nitroreductase activities (2, 8, 21, 43, 61, 84). Although results vary from strain to strain, the "typical" Type I enzyme was found to be an FMN-containing flavoprotein that uses NAD(P)H as an electron donor and is markedly inhibited by *p*-chloromercuribenzoate, *o*-iodosobenzoic acid, and dicumarol. Dicumarol is a specific inhibitor of menadione reductase and DT diaphorase, suggesting that the Type I enzymes serve as quinone reductases *in vivo* (1, 21, 84). These studies gave no evidence for the involvement of a metal cofactor in the reduction process.

The late 1980's and early 1990's were an active time in the study of bacterial nitroreductase enzymes, and particularly in the study of the "classical" nitroreductase of

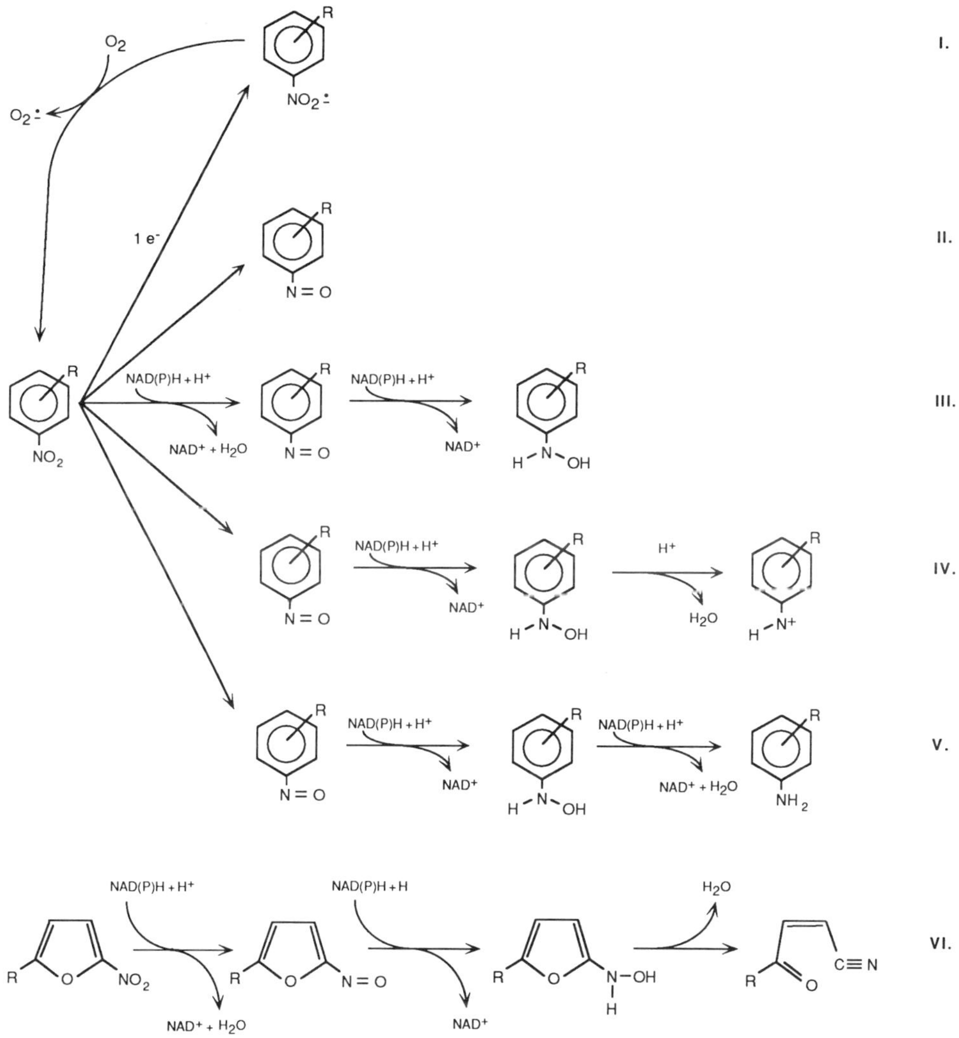

Figure 4. The major enzymatic reduction reactions of aromatic nitro groups. I. Type II (oxygen-sensitive) nitroreductases catalyze a one-electron reduction of the nitro group and produce the nitro anion radical. If oxygen is present, the anion radical reacts to produce superoxide and the parent nitro compound (64). In the absence of oxygen the radical anions disproportionate nonenzymatically to yield the nitroso and nitro compounds (not shown) (49). II. A single two-electron reduction of the nitro group yields a nitroso compound. Some bacterial nitroreductases catalyze only this reaction (78, 79, 80, 81, 95). III. Nitroreductase enzymes from *Comamonas acidovorans* (30), *E. coli* (1) and *Pseudomonas pseudoalcaligenes* (83) reduce nitroaromatic compounds to the corresponding hydroxylamino derivatives but not to the amino compounds. IV. Reduction of a nitroaromatic compound to the hydroxylamino derivative may be followed by an acid-catalyzed decomposition to a nitrenium ion (34). A similar reaction can occur following *O*-acetylation of the hydroxylamino group. In this case the decomposition reaction yields acetate and the nitrenium ion (6). V. Several bacterial nitroreductases catalyze the reduction of aromatic nitro groups to the corresponding amino compounds (e.g. 4, 5, 43, 89, 90). VI. The reduction of oxygen-containing heterocycles can lead to the formation of a nitrile and concurrent ring cleavage (5, 50).

Table 4. Characteristics of purified bacterial Type I nitroreductase enzymes.

Organism	MW (kDa)	Quaternary Structure	Flavin co-factor (molar ratio)	Electron Donor	-NO2 group reduced to:	Refs.
Bacteroides fragilis GAI0624, NRase I	52	monomer	FMN (N.R.)	NADH	$-NH_2$	42
Enterobacter cloacae	27	monomer	FMN (1:1)	NAD(P)H	-NHOH or-NH_2	5
Escherichia coli B	24	monomer	(FMN (N.R.)	NAD(P)H	-NHOH	1
Nocardia sp. V.	24	monomer	N.R.	NADPH	$-NH_2$	85
Pseudomonas pseudoalcaligenes JS45	29-33	monomer	FMN (2:1)	NADPH	-NHOH	79
Rhodobacter capsulatus E1F1	54	homodimer	FMN (N.R.)	NAD(P)H	$-NH_2$	4
Salmonella typhimurium	24-28	N.R.	N.R.	N.R.	$-NH_2$	86
Vibrio fischeri	25	monomer	FMN (N.R.)	NAD(P)H	-N.R.	92

N.R. = Not Reported

enteric bacteria. During this time enzymes from the genera *Salmonella*, *Enterobacter* and *Escherichia* were characterized (1, 5, 7, 90, 91). Furthermore, molecular sequence data became available for the first time, allowing assessment of the relationships among enzymes with nitroreductase activity and between the nitroreductases and other proteins in the data bases.

Watanabe et al. (90) studied nitroreductase activity in *Salmonella typhimurium* by constructing a gene library and assaying for transformants with increased sensitivity to 2-nitrofluorene. They isolated a hypersensitive transformant carrying a recombinant plasmid, and demonstrated that sensitivity was conferred on insensitive cells after transformation with the plasmid. They found increased nitroreductase activity in cell extracts of the sensitive transformant, and theorized that the strain carried a construct encoding the classical nitroreductase. DNA sequence analysis (91) identified an open reading frame of 651 nucleotides corresponding to a putative translation product of 24.0 kDa. Maxicell analysis revealed a protein of approximately 28 kDa (91), indicating that the reductase was active as a monomer.

Bryant and DeLuca (5) purified and characterized a Type I nitroreductase from *Enterobacter cloacae*. The protein is a monomer of approximately 27 kDa, it has a loosely bound FMN cofactor and uses NAD(P)H as an electron donor. The substrate range of the enzyme includes nitrofurans, nitrobenzenes, nitrotoluenes and quinones. The enzyme appears to produce the hydroxylamino derivative of nitrofurazone under aerobic conditions, but the product was not thoroughly characterized. Under anaerobic conditions the product is the amine. The authors did not test the effect of metals on enzyme activity, and no inhibition studies were reported. Therefore, it is not known if this enzyme is a metalloprotein. The gene encoding the *Enterobacter* reductase was cloned and sequenced (7). The authors found a 651 base pair open reading frame corresponding to a protein of 23.9 kDa. Comparison of the *Enterobacter* and *Salmonella* amino acid sequences revealed 88% sequence identity between the two proteins (7).

Anlezark et al. (1) followed shortly thereafter with a report on the purification of a classical nitroreductase from *Escherichia coli* B. The enzyme is an FMN-containing flavoprotein of approximately 24 kDa that uses NAD(P)H as the electron donor. The enzyme specifically reduces the anti-tumor agent CB1954 (5-[aziridin-1-yl]-2,4-dinitrobenzamide) to the 4-hydroxylamino derivative (Fig. 4). The isoelectric point of the protein is 5.1 which is similar to the values of 5.3 and 5.6 calculated for the *Salmonella* and *Enterobacter* enzymes, respectively. The sequence of 31 amino acids at the N-terminus of the *E. coli* enzyme was determined, and a further 41 residues were obtained from a peptide generated by cyanogen bromide digestion (1). The high degree of similarity among the *E. coli*, *Salmonella* and *Enterobacter* amino acid sequences (Table 3 in 1) indicates that they share a recent common ancestor. Comparison of these sequences to protein sequence data bases found no other matches with significant homology (1, 7).

It is clear that the "classical" nitroreductase enzymes are highly conserved among the enteric bacteria, but their physiological role is still unknown. The *E. coli* enzyme and DT diaphorase share the ability to reduce both menadione and CB1954 and both are inhibited by dicumarol. However, there is no sequence similarity between the two and the reductase does not cross react with polyclonal antibodies raised against DT diaphorase (1). The fact that menadione is a good substrate for the *E. coli* enzyme suggests that they may serve as quinone reductases *in vivo*.

A strain of *Comamonas acidovorans* able to grow on 4-nitrobenzoate was recently isolated by Groenewegen et al. (30). It had been generally accepted that the biodegradation of nitroaromatic compounds could proceed *via* one of two routes: oxidation of the ring with the release of nitrite, or reduction of the nitro group to yield the corresponding amine. In *Comamonas,* however, hydroxylaminobenzoate is on the degradation pathway, but aminobenzoate is not (Fig. 4). The purified reductase catalyzes the reduction of nitrobenzoate to hydroxylaminobenzoate under both aerobic and anaerobic conditions. Nitrosobenzoate was postulated as an intermediate in the reduction, but was not detected in reaction mixtures. The enzyme is induced by 4-nitrobenzoate and uses NAD(P)H as the electron donor. The authors found no enhancement of activity with the addition of FAD, FMN or metal ions, although this does not rule out the participation of tightly bound cofactors. The key points of this study are the inducible nature of the reductase and its relatively narrow substrate range, which indicate that the reduction of 4-nitrobenzoate is the physiological role of this enzyme.

Sequence data have recently been made available for an enzyme with nitroreductase activity from *Vibrio fischeri* (96). Translation of the DNA sequence predicts a protein of 218 amino acids with a calculated molecular weight of 24.6 kDa (96), and an isoelectric point of 5.7. All of these values are in good agreement with characteristics of the classical nitroreductase enzymes from enteric bacteria. An alignment of the *Vibrio* and *Enterobacter* nitroreductase amino acid sequences yields only a 30% overall identity, but regions of high amino acid sequence conservation suggest that the proteins share a common ancestor (data not shown). The degrees of similarity within the enteric group and between the enteric and *Vibrio* enzymes, correlates with the phylogeny of these bacteria based on 16S rRNA sequences (93), and support the hypothesis that "classical" nitroreductase is conserved among these bacteria.

Recently Blasco and Castillo (4) reported the characterization of a nitrophenol nitroreductase purified from *Rhodobacter capsulatus*. It is a soluble, FMN-containing flavoprotein that uses NAD(P)H as electron donor, and reduces 2,4-dinitrophenol to 2-amino-4-nitrophenol which is not further metabolized (Fig. 4). The native enzyme is a homodimer of 54 kDa. The inhibition of enzymatic activity by metal chelators, salicylhydroxamate, *p*-hydroxymercuribenzoate and Cu^{2+} indicates the importance of sulfhydryl groups in the stability and/or activity of the protein, and the possible involvement of non-heme iron in reductase activity. Nitrophenol reductase activity was induced by a variety of nitrophenols, but not by other nitroaromatic compounds. However, its induction was suppressed by the inclusion of ammonium in the medium, and its role in the cell is not clear.

Recent research with the nitroreductase of *Pseudomonas pseudoalcaligenes* grew out of the observation that this strain degrades nitrobenzene *via* hydroxylaminobenzene, but aniline is not produced in whole cells or crude extracts (62). A nitrobenzene-inducible, oxygen-insensitive nitroreductase was purified to homogeneity from crude extracts of *P. pseudoalcaligenes* (83). Nitrosobenzene appears to be produced as an intermediate in the reduction reaction but is not released into the medium. The purified enzyme reduces nitrobenzene to hydroxylaminobenzene and no further. Chelating agents had little effect on enzyme activity, yet incubation with 1 mM salicylhydroxamate, a non-heme iron reagent, caused nearly complete inhibition of enzyme activity. Rigorous analysis by plasma emmission spectroscopy indicated that purified nitrobenzene nitroreductase has no significant metal content (83). The structure of salicylhydroxyamate (*N*, 2-dihdroxybenzamide) suggests that it may inhibit the reductase by competition for the active site, but this hypothesis has not been tested. The induction of enzyme expression during growth on nitrobenzene clearly shows that the reduction of nitrobenzene is its physiological role.

Table 4 summarizes the physical characteristics of purified Type I bacterial nitroreductases. The enzyme described by Groenewegen et al. (30) has not been included here because of a lack of detailed physical characterization. All of the enzymes in Table 4 have FMN prosthetic groups, and most are active as monomers from 24 to 30 kDa. However, the apparent physical similarities may give a false sense of relatedness among these enzymes. Preliminary analyses with partial amino acid sequence data from the *Pseudomonas* reductase (83) reveals no obvious relationship to the enteric enzymes (data not shown), and no sequence data are available for the *Bacteroides*, *Comamonas*, *Nocardia* or *Rhodobacter* nitroreductases. A comprehensive determination of the relationships among these enzymes must await more sequence data. The *Comamonas*, *E. coli* and *Pseudomonas* enzymes share the characteristic of reducing nitro groups only to the level of the corresponding hydroxylamino compound. It is not known what aspect of the protein structure determines whether the final product of the reaction will be a nitroso, hydroxylamino, or amino compound. Additional analysis of the enzyme structures and mechanisms will be necessary to answer this question.

The nitroreductases purified from *Comamonas* (30) and *Pseudomonas* (83) function in biodegradative pathways that are induced by their nitroaromatic substrates. Therefore, these enzymes can be considered "specific" nitroreductases because the reduction of nitro compounds is their physiological role. The "specific" and "classical" (enteric) nitroreductases share several characteristics: they are FMN-containing flavoproteins, they use reduced pyridine nucleotides as electron donors, they catalyze two-electron reductions of the nitro group yielding hydroxylamino- or amino-derivatives of the parent compound, they are active as monomers, and they have molecular weights between 24 and 30 kDa. Although the *Rhodobacter* enzyme is active as a homodimer, the monomer weight of 27 kDa is in good agreement with the other enzymes in the group. The role of metal ions in nitro group reduction is still unclear. Stimulation of nitroreductase activity in cell extracts by divalent cations has been reported (11, 79) and metal requirements have been inferred from inhibitor studies (4, 95). However, the majority of the studies to date give no specific information regarding metal cofactor requirements. Resolution of this issue awaits detailed mechanistic studies on individual nitroreductases.

CONCLUSIONS

This chapter has illustrated that higher molecular weight nitro-PAHs are a widely distributed class of environmental pollutants, some of which are potent bacterial mutagens and animal carcinogens. Additionally, nitro-PAHs have been isolated in air and soil samples and may be considered a human health risk, because these compounds are genotoxic, recalcitrant to microbial degradation (i.e., mineralization) and persist in the environment.

Nitroreduction is the critical reaction in the bioactivation of nitro-PAHs. Germ-free and conventional animal studies indicate that intestinal microflora nitroreductases are primarily involved in enzymatic reduction of the nitro-PAHs to form arylhydroxylamines, which bind to DNA and induce mutagenesis. Little is known about how diet can modulate nitroreductase activity or the biological effects of nitro-PAHs in humans.

A significant body of literature now exists on oxygen-insensitive bacterial nitroreductases. The "classical" enteric nitroreductases are oxygen-insensitive, FMN-containing flavoproteins that catalyze two-electron reductions of the nitro group to produce the hydroxylamino or amino derivatives of the parent nitro compound. These enzymes appear to be similar to the "specific" degradative nitroreductases isolated from bacteria that can grow at the expense of nitroaromatic compounds, but insufficient sequence data are available to determine whether or not they are homologous.

Future directions for research should include:

1. the effect of diet on inhibition of intestinal microflora nitroreductases that convert nitro-PAHs to genotoxic intermediates,
2. ways to inhibit growth in the gastrointestinal tract of bacteria that have high nitroreductase activity,
3. elucidation of the mechanisms of bacterial nitroreductases,
4. determination of the phylogeny of genes that encode nitroreductases, the mechanisms by which they become part of degradative operons, and their genetic regulation,
5. development of microorganisms with improved nitro-PAH degradative capabilities.

ACKNOWLEDGEMENTS

The authors thank John Sutherland, Jim Spain and Shirley Nishino for critically reading the manuscript and Pat Fleischer, Sandra Malone and Shirley Nishino for assistance in preparing the manuscript.

REFERENCES

1. Anlezark, G. M., R. G. Melton, R. F. Sherwood, B. Coles, F. Friedlos, and R. Knox. 1992. The bioactivation of 5-(aziridin-1-yl)-2,4-dinitrobenzamide (CB1954)-1. Purification and properties of a nitroreductase enzyme from *Escherichia coli* - a potential enzyme for antibody-directed enzyme prodrug therapy (ADEPT). Biochem. Pharmacol. 44:2289-2295.
2. Asnis, R. E. 1957. The reduction of furacin by cell-free extracts of furacin-resistant and parent-susceptible strains of *Escherichia coli*. Arch. Biochem. Biophys. 66:208-216.
3. Beland, F. A., R. H. Heflich, P. C. Howard, and P. P. Fu. 1985. The *in vitro* metabolic activation of nitropolycyclic aromatic hydrocarbons, p. 371-396. *In* R. G. Harvey (ed.), Polycyclic hydrocarbons and carcinogenesis. ACS Symposium Series 283. American Chemical Society, Washington, D.C.
4. Blasco, R., and F. Castillo. 1993. Characterization of a nitrophenol reductase from the phototrophic bacterium *Rhodobacter capsulatus* EIFI. Appl. Environ. Microbiol. 59:1774-1778.
5. Bryant, C., and M. DeLuca. 1991. Purification and characterization of an oxygen-insensitive NAD(P)H nitroreductase from *Enterobacter cloacae*. J. Biol. Chem. 266:4119-4125.
6. Bryant, C., and W. D. McElroy. 1991. Nitroreductases. *In* F. Muller (ed.), Chemistry and biochemistry of flavoenzymes - Vol. II. CRC Press, Boca Raton.
7. Bryant, C., L. Hubbard, and W. D. McElroy. 1991. Cloning, nucleotide sequence and expression of the nitroreductase gene from *Enterobacter cloacae*. J. Biol. Chem. 266:4126-4130.
8. Bryant, D. W., D. R. McCalla, M. Leeksma, and P. Laneuville. 1981. Type I nitroreductases of *Escherichia coli*. Can. J. Microbiol. 27:81-86.
9. Bryant, D. W., D. R. McCalla, P. Lultschik, M. A. Quilliam, and B. E. McCarry. 1984. Metabolism of 1,8-dinitropyrene by *Salmonella typhimurium*. Chem.-Biol. Interact. 49:351-368.

10. Campbell, W. L., W. Franklin, and C. E. Cerniglia. 1992. Validation studies on an *in vitro* semicontinuous culture system designed to simulate a bacterial ecosystem of the human intestine. J. Microbiol. Methods 16:239-252.
11. Cartwright, N. J., and R. B. Cain. 1959. Bacterial degradation of nitrobenzoic acids, 2. Reduction of the nitro group. Biochem. J. 73:305-314.
12. Cerniglia, C. E. 1985. Metabolism of 1-nitropyrene and 6-nitrobenzo[*a*]pyrene by intestinal microflora, p.133-137. *In* B. Wostmann, J. R. Pleasants, M. Pollard, B. A. Teah and M. Wagner (ed.), Progress in clinical and biological research, Vol. 181. Germfree research; microflora control and its application to the biomedical sciences. Alan R. Liss, Inc., New York.
13. Cerniglia, C. E., P. C. Howard, P. P. Fu, and W. Franklin. 1984. Metabolism of nitropolycyclic aromatic hydrocarbons by human intestinal microflora. Biochem. Biophys. Res. Commun. 123:262-269.
14. Cerniglia, C. E., J. P. Freeman, G. L. White, R. H. Heflich, and D. W. Miller. 1985. Fungal metabolism and detoxification of the nitropolycyclic aromatic hydrocarbon, 1-nitropyrene. Appl. Environ. Microbiol. 50:649-655.
15. Cerniglia, C. E., K. J. Lambert, G. L. White, R. H. Heflich, W. Franklin, E. K. Fifer, and F. A. Beland. 1987. Metabolism of 1,8-dinitropyrene by human, rhesus monkey and rat intestinal microflora. Toxic. Assess. Int. Q. 3:147-159.
16. Cerniglia, C. E. 1992. Biodegradation of polycyclic aromatic hydrocarbons. Biodegradation 3:351-368.
17. Cerniglia, C. E. 1993. Biodegradation of polycyclic aromatic hydrocarbons. Curr. Opin. Biotechnol. 4:331-338.
18. Cerniglia, C. E., J. B. Sutherland, and S. A. Crow. 1992. Fungal metabolism of aromatic hydrocarbons, p. 193-217. *In* G. Winkelmann (ed.) Microbial degradation of natural products. VCH Verlagsgesellschaft, Weinheim.
19. Chadwick, R. W., S. E. George, and L. R. Claxton. 1992. Role of gastrointestinal mucosa and microflora in the bioactivation of dietary and environmental mutagens or carcinogens. Drug Metab. Rev. 24:425-492.
20. Delclos, K. B., C. E. Cerniglia, K. L. Dooley, W. L. Campbell, W. Franklin, and R. P. Walker. 1989. The role of intestinal microflora in the metabolic activation of 6-nitrochrysene to DNA-binding derivatives in mice. Toxicology 60:137-150.
21. Doi, T., H. Yoshimura, and K. Tatsumi. 1983. Properties of nitrofuran reductases from *Escherichia coli B/r*. Chem. Pharm. Bull. 31:1105-1107.
22. Doolittle, D. J., J. M. Sherrill, and B. E. Butterworth. 1983. Influence of intestinal bacteria, sex of the animal, and position of the nitro group on the hepatic genotoxicity of nitrotoluene isomers *in vivo*. Cancer Res. 43:2836-2842.
23. Drasar, B. S., and B. I. Duerden. 1991. Anaerobes in the normal flora of man, p. 162-179. *In* B. I. Duerden and B. S. Drasar (ed.), Anaerobes in human disease. Wiley-Liss, New York.
24. Egami, F., H. Ebata, and R. Sato. 1951. Reduction of chloromycetin by a cell-free bacterial extract and its relation to nitrite reduction. Nature 167:118-119.
25. El-Bayoumy, K., C. Sharma, Y. M. Louis, B. Reddy, and S. S. Hecht. 1983. The role of intestinal microflora in the metabolic reduction of 1-nitropyrene to 1-aminopyrene in conventional and germ free rats and in humans. Cancer Lett. 19:311-316.
26. Fu, P. P., M. W. Chou, D. W. Miller, G. L. White, R. H. Heflich, and F. A. Beland. 1985. The orientation of the nitro substituent predicts the direct-acting bacterial mutagenicity of nitrated polycyclic aromatic hydrocarbons. Mutat. Res. 143:173-181.
27. Fu, P. P., C. E. Cerniglia, K. E. Richardson, and R. H. Heflich. 1988. Nitroreduction of 6-nitrobenzo[*a*]pyrene in humans. Mutat. Res. 209:123-129.
28. Fu, P. P. 1990. Metabolic activation of nitro-polycyclic aromatic hydrocarbons. Drug Metab. Rev. 22:209-268.
29. Gibson, T. L. 1983. Sources of direct-acting nitroarene mutagens in airborne particulate matter. Mutat. Res. 122:115-121.
30. Groenewegen, P. E. J., P. Breeuwer, J. M. L. M. van Helvoort, A. A. M. Langenhoff, F. P. de Vries, and J. A. M. de Bont. 1992. Novel degradative pathway of 4-nitrobenzoate in *Comamonas acidovorans* NBA-10. J. Gen. Microbiol. 138:1599-1605.
31. Heflich, R. H., E. K. Fifer, Z. Djuri , and F. A. Beland. 1985. DNA adduct formation and mutation induction by nitropyrenes in *Salmonella* and Chinese hamster ovary cells: relationships with nitroreduction and acetylation. Environ. Health Perspect. 62:135-143.
32. Heflich, R. H., Z. Djuri , E. K. Fifer, C. E. Cerniglia, and F. A. Beland. 1986a. Metabolism of dinitropyrenes to DNA-binding derivatives *in vitro* and *in vivo*, p. 185-197. *In* N. Ishinishi, A. Koizumi, R. O. McClellan, and W. Stîber (ed.), Carcinogenic and mutagenic effects of diesel engine exhaust. Elsevier, Amsterdam.

33. Heflich, R.H., E.K. Fifer, Z. Djuri , and F.A. Beland. 1986b. Mutation induction and DNA adduct formation by 1,8-dinitropyrene in Chinese hamster ovary cells, p. 265-273. *In* C. Ramel, R. Lambert, and J. Monson (ed.), Genetic toxicology of environmental chemicals, Part A: Basic principles and mechanisms of action. Alan R. Liss, New York.
34. Heflich, R. H., P. C. Howard, and F. A. Beland. 1985. 1-Nitrosopyrene: an intermediate in the metabolic activation of 1-nitropyrene to a mutagen in *Salmonella typhimurium* TA1538. Mutat. Res. 149:25-32.
35. Heitkamp, M. A., and C. E. Cerniglia. 1988. Mineralization of polycyclic aromatic hydrocarbons by a bacterium isolated from sediments below an oil field. Appl. Environ. Microbiol. 54:1612-1614.
36. Heitkamp, M. A., J. P. Freeman, D. W. Miller, and C. E. Cerniglia. 1991. Biodegradation of 1-nitropyrene. Arch. Microbiol. 156:223-230.
37. Howard, P. C., F. A. Beland, and C. E. Cerniglia. 1983. Reduction of 1-nitropyrene to 1-aminopyrene by rat intestinal bacteria. Carcinogenesis 4:985-990.
38. Howard, P. C., R. H. Heflich, F. E. Evans, and F. A. Beland. 1983. Formation of DNA adducts *in vitro* and in *Salmonella typhimurium* upon metabolic reduction of the environmental mutagen 1-nitropyrene. Cancer Res. 43:2052-2058.
39. IARC. 1984. IARC Monographs on the Evaluation of the Carcinogenic Risk of Chemicals to Humans, p. 171-178. Vol. 33. Polynuclear aromatic compounds, Part 2. Carbon blacks, mineral oils and some nitroarenes. IARC, Lyon.
40. IARC. 1989. IARC Monographs on the Evaluation of the Carcinogenic Risks to Humans, p. 31-458. Vol. 46. Diesel and gasoline engine exhausts and some nitroarenes. IARC, Lyon.
41. Jung, H., A. U. Shaikh, R. H. Heflich, and P. P. Fu. 1991. Nitro group orientation, reduction potential, and direct-acting mutagenicity of nitro-polycyclic aromatic hydrocarbons. Environ. Mol. Mutagen. 17:169-180.
42. Kinouchi, T., Y. Manabe, K. Wakisaki, and Y. Ohnishi. 1982. Biotransformation of 1-nitropyrene in intestinal anaerobic bacteria. Microbiol. Immunol. 26:992-1005.
43. Kinouchi, T., and Y. Ohnishi. 1983. Purification and characterization of 1-nitropyrene reductases from *Bacteroides fragilis*. Appl. Environ. Microbiol. 46:596-604.
44. Kinouchi, T., and Y. Ohnishi. 1986. Metabolic activation of 1-nitropyrene and 1,6-dinitropyrene by nitroreductases from *Bacteroides fragilis* and distribution of nitroreductase activity in rats. Microbiol. Immunol. 30:979-992.
45. Kinouchi, T., K. Nishifuji, and Y. Ohnishi. 1987. *In vitro* intestinal microflora-mediated metabolism of biliary metabolites from 1-nitropyrene-treated rats. Microbiol. Immunol. 31:1145-1159.
46. Manning, B. W., C. E. Cerniglia, and T. W. Federle. 1986. Biotransformation of 1-nitropyrene to 1-aminopyrene and *N*-formyl-1-aminopyrene by the human intestinal microbiota. Toxicol. Environ. Health 18:339-346.
47. Manning, B. W., W. C. Campbell, W. Franklin, K. B. Delclos, and C. E. Cerniglia. 1988. Metabolism of 6-nitrochrysene by intestinal microflora. Appl. Environ. Microbiol. 34:197-203.
48. Manning, B. W., T. W. Federle, and C. E. Cerniglia. 1987. Use of a semi-continuous culture system as a model for determining the role of human intestinal microflora in the metabolism of xenobiotics. J. Microbiol. Methods 6:81-94.
49. Mason, R. P., and J. L. Holtzman. 1975. The role of catalytic superoxide formation in the O_2 inhibition of nitroreductase. Biochem. Biophys. Res. Commun. 67:1267-1274.
50. Mattammal, M. B., T. V. Zenser, M. O. Palmier, and B. B. Davis. 1985. Renal reduced nicotinamide adenine dinucleotide phosphate:cytochrome *c* reductase-mediated metabolism of the carcinogen *N*-[4-(5-nitro-2-furyl)-2-thiazolyl]acetamide. Cancer Res. 45:149-156.
51. McCann, J., E. Choi, E. Yamasaki, and B. N. Ames. 1975. Detection of carcinogens as mutagens in the *Salmonella* microsome test: assay of 300 chemicals. Proc. Natl. Acad. Sci. USA 72:5135-5139.
52. McCartney, M. A., B. F. Chatterjee, E. C. McCoy, E. A. Mortimer, Jr., and H. S. Rosenkranz. 1986. Airplane emissions: a source of mutagenic nitrated polycyclic aromatic hydrocarbons. Mutat. Res. 171:99-104.
53. McCoy, E. C., H. S. Rosenkranz, and P. C. Howard. 1990. *Salmonella typhimurium* TA100Tn5-1012, a strain deficient in arylhydroxylamine *O*-esterificase, exhibits a reduced nitroreductase activity. Mutat. Res. 243:141-144.
54. McCoy, E. C., H. S. Rosenkranz, and R. Mermelstein. 1981. Evidence for the existence of a family of bacterial nitroreductases capable of activating nitrated polycyclics to mutagens. Environ. Mutagen. 3:421-437.
55. Mermelstein, R., D. K. Kiriazides, M. Butler, E. C. McCoy, and H. S. Rosenkranz. 1981. The extraordinary mutagenicity of nitropyrenes in bacteria. Mutat. Res. 89:187-196.

56. Messier, F., C. Lu, P. Andrews, B. E. McCarry, M. A. Quilliam, and D. R. McCalla. 1981. Metabolism of 1-nitropyrene and formation of DNA adducts in *Salmonella typhimurium*. Carcinogenesis 2:1007-1011.
57. Millner, G. C., and C. E. Cerniglia. 1986. Microbial transformation and detoxification of 6-nitrobenzo[*a*]pyrene. Toxicol. Environ. Health 19:519-530.
58. Mirsalis, J. C., T. E. Hamm, J. M. Sherrill, and B. E. Butterworth. 1982. Role of gut flora in genotoxicity of dinitrotoluene. Nature 295:322-323.
59. Mîller, L., M. Corrie, T. Midtvedt, J. Rafter, and J.-A. Gustafsson. 1988. The role of the intestinal microflora in the formation of mutagenic metabolites from the carcinogenic air pollutant 2-nitrofluorene. Carcinogenesis 9:823-830.
60. Morehead, M. C., W. Franklin, P. P. Fu, F. E. Evans, T. M. Heinze, and C. E. Cerniglia. 1994. Metabolism of 7-nitrobenz[*a*]anthracene by intestinal microflora. J. Toxicol. Environ. Health (In press).
61. Narai, N., S. Kitamura, and K. Tatsumi. 1984. A comparative study on 1-nitropyrene and nitrofurazone reductases in *Escherichia coli*. J. Pharm. Dyn. 7:407-413.
62. Nishino, S. F., and J. C. Spain. 1993. Degradation of nitrobenzene by a *Pseudomonas pseudoalcaligenes*. Appl. Environ. Microbiol. 59:2520-2525.
63. Ohnishi, Y., T. Kinouchi, Y. Manabe, H. Tsutsui, H. Otsuka, H. Tokiwa, and T. Otofuji. 1985. Nitro compounds in environmental mixtures and foods, p. 195-204. *In* M. D. Waters, S. S. Sandhu, J. Lewtas, L. Claxton, G. Strauss, and S. Nesnow (ed.), Short-term bioassays in the analysis of complex environmental mixtures. Vol. IV. Plenum Publishing Corporation, New York.
64. Peterson, F. J., R. P. Mason, J. Hovespian, and J. L. Holtzman. 1979. Oxygen-sensitive and -insensitive nitroreduction by *Escherichia coli* and rat hepatic microsomes. J. Biol. Chem. 254:4009-4014.
65. Pitts, J. N., Jr., K. A. Van Cauwenberghe, D. Grosjean, J. P. Schmid, D. R. Fitz, W. L. Belser, Jr., G. B. Knudson, and P. M. Hynds. 1978. Atmospheric reactions of polycyclic aromatic hydrocarbons: facile formation of mutagenic nitro derivatives. Science 202:515-518.
66. Pothuluri, J. V., F. E. Evans, T. M. Heinze, and C. E. Cerniglia. 1994. Metabolism of 3-nitrofluoranthene, a nitropolycyclic aromatic hydrocarbon, by the fungus *Cunninghamella elegans*. J. Toxicol. Environ. Health 42:209-218.
67. Preuss, A., J. Fimpel, and G. Diekert. 1993. Anaerobic transformation of 2,4,6-trinitrotoluene (TNT). Arch. Microbiol. 159:345-353.
68. Rafii, F., W. Franklin, R. H. Heflich, and C. E. Cerniglia. 1991. Reduction of nitroaromatic compounds by anaerobic bacteria isolated from the human gastrointestinal tract. Appl. Environ. Microbiol. 57:962-968.
69. Rafii, F., and C. E. Cerniglia. 1993. Comparison of the azoreductase and nitroreductase from *Clostridium perfringens*. Appl. Environ. Microbiol. 59:1731-1734.
70. Rafii, F., A. L. Selby, R. K. Newton, and C. E. Cerniglia. Reduction and mutagenic activation of nitroaromatic compounds by a *Mycobacterium* sp. Appl. Environ. Microbiol. 60:4263-4267.
71. Reddy, B. G., L. R. Pohl, and G. Krishna. 1976. The requirement of the gut flora in nitrobenzene-induced methemoglobinemia in rats. Biochem. Pharmacol. 25:1119.
72. Richardson, K. E., P. P. Fu, and C. E. Cerniglia. 1988. Metabolism of 1-, 3-, and 6-nitrobenzo[*a*]pyrene by intestinal microflora. J. Toxicol. Environ. Health 23:527-537.
73. Rickert, D. E., Butterworth, B. E., and J. A. Popp. 1984. Dinitrotoluene: Acute toxicity, oncogenicity, genotoxicity, and metabolism. CRC Crit. Rev. Toxicol. 13:217-234.
74. Rosenkranz, H. S., and D. R. Sanders. 1980. Nitropyrenes: isolation, identification, and reduction of mutagenic impurities in carbon black toners. Science 209:1039-1043.
75. Rosenkranz, H. S. 1982. Direct-acting mutagens in diesel exhausts: magnitude of the problem. Mutat. Res. 101:1-10.
76. Rosenkranz, H. S., and R. Mermelstein. 1983. Mutagenicity and genotoxicity of nitroarenes. All nitro-containing chemicals were not created equal. Mutat. Res. 114:217-267.
77. Rowland, I. R. 1988. Factors affecting metabolic activity of the intestinal flora. Drug Metab. Rev. 19:243-262.
78. Saz, A. K., and J. Marmur. 1953. The inhibition of organic nitro-reductase by aureomycin in cell-free extracts. Proc. Soc. Exp. Biol. Med. 82:783-784.
79. Saz, A. K., and M. L. Martinez. 1956. Enzymatic basis of resistance to aureomycin I. Differences between flavoprotein reductases of sensitive and resistant *Escherichia coli*. J. Biol. Chem. 223:285-292.
80. Saz, A. K. and R. B. Slie. 1954. Reversal of aureomycin inhibition of bacterial cell-free nitro reductase by manganese. J. Biol. Chem. 210:407-412.
81. Saz, A. K., and R. B. Slie. 1954. The inhibition of organic nitro reductase by aureomycin in cell-free extracts II. Cofactor requirements for the nitro reductase enzyme complex. Arch. Biochem. Biophys. 51:5-16.

82. Smith, G. N., and C. S. Worrel. 1949. Enzymatic reduction of chloramphenicol (chloromycetin). Arch. Biochem. 24:216-223.
83. Somerville, C. C., S. F. Nishino, and J. C. Spain. Isolation and characterization of nitrobenzene nitroreductase from *Pseudomonas pseudoalcaligenes* JS45. J. Bacteriol. (In press).
84. Tatsumi, K., T. Doi, H. Yoshimura, H. Koga, and T. Horiuchi. 1982. Oxygen-insensitive nitrofuran reductases in *Salmonella typhimurium* TA100. J. Pharm. Dyn. 5:423-429.
85. Thornton-Manning, J. R., W. L. Campbell, B. S. Hass, J. J. Chen, P. P. Fu, C. E. Cerniglia, and R. H. Heflich. 1989. The role of nitro-reduction in the synergistic mutational response induced by mixtures of 1- and 3- nitrobenzo[*a*]pyrene in *Salmonella typhimurium*. Environ. Mol. Mutagen. 13:203-210.
86. Tokiwa, H., and Y. Ohnishi. 1986. Mutagenicity and carcinogenicity of nitroarenes and their sources in the environment. CRC Crit. Rev. Toxicol. 17:23-60.
87. Tokiwa, H., T. Otofugi, R. Nakagawa, K. Horikawa, T. Maeda, N. Sano, K. Izumi, and H. Otsuka. 1986. Dinitro derivatives of pyrene and fluoranthene in diesel emission particulates and their tumorigenicity in mice and rats. Dev. Toxicol. Environ. Sci. 13:253-270.
88. Tokiwa, H., R. Nakagawa, K. Horikawa, and A. Ohkubo. 1987. The nature of the mutagenicity and carcinogenicity of nitrated aromatic compounds in the environment. Environ. Health Perspect. 73:191-199.
89. Villanueva, J. R. 1964. The purification of a nitro-reductase from *Nocardia V.* J. Biol. Chem. 239:773-776.
90. Watanabe, M., M. Ishidate, Jr., and T. Nohmi. 1989. A sensitive method for the detection of mutagenic nitroarenes: construction of nitroreductase-overproducing derivatives of *Salmonella typhimurium* strains TA98 and TA100. Mutat. Res. 216:211-220.
91. Watanabe, M., M. Ishidate, Jr., and T. Nohmi. 1990. Nucleotide sequence of *Salmonella typhimurium* nitroreductase gene. Nucleic Acids Res. 18:1059.
92. Watanabe, M., T. Nohmi, and M. Ishidate, Jr. 1987. New tester strains of *Salmonella typhimurium* highly sensitive to mutagenic nitroarenes. Biochem. Biophys. Res. Commun. 147:974-979.
93. Woese, C. R. 1987. Bacterial evolution. Microbiol. Rev. 51:221-271.
94. Yamashina, I. 1954. Enzymatic reduction of aromatic nitro compounds. Bull. Chem. Soc. Japan 27:85-89.
95. Yamashina, I., S. Shikata, and F. Egami. 1954. Enzymatic reduction of aromatic nitro, nitroso and hydroxyl amino compounds. Bull. Chem. Soc. Japan 27:42-45.
96. Zenno, S., S. Inouye, and H. Kanoh. 1993. Gene encoding enzyme having flavin reducing activity and nitroreductase activity. European Patent Application no. 0 547 876 A1.

8

BIODEGRADATION OF 2,4,6-TRINITROTOLUENE BY THE WHITE ROT FUNGUS *PHANEROCHAETE CHRYSOSPORIUM*

James D. Stahl and Steven D. Aust

Biotechnology Center
Utah State University
Logan, UT 84322-4705

INTRODUCTION

Trinitrotoluene (TNT) has become the predominant explosive worldwide. TNT wastes generated during TNT production and munitions manufacturing, as well as improper storage, use, and disposal have led to contamination of the environment (67, 94). Due to the solubility of TNT in water (~ 125 ppm at 20°C) large volumes of contaminated waste water can be generated during TNT production and munitions loading operations. As much as 500,000 gallons of TNT-contaminated water have been generated per day by a single munitions plant (91). In the past, the contaminated water was typically collected in large lagoons. Repetitive use of the lagoons has left high residual concentrations of TNT in the sediment (42, 67). Other contamination sites include missile production facilities, mining sites, military firing ranges and sites where outdated explosives were burned. Continual leaching of TNT from these soils over the years has resulted in the contamination of groundwater (42, 67). Typical contaminated sites may contain average concentrations of 10,000 mg/kg TNT in soil and 100 ppm TNT in water (30).

It has been recommended that a maximum allowable TNT concentration in drinking water be set at 140 ppb and the U.S. Environmental Protection Agency has calculated a lifetime health advisory of 2 ppb TNT in drinking water (70, 71). The cleanup criterion recommended by the U.S. Army is 5 ppm for TNT-contaminated soils at inactive army ammunition plants (69). Thus, there is a clear need to develop TNT remediation procedures for water and soil.

A number of feasibility studies for the bioremediation of TNT-contaminated water or soil have been performed. The earliest work was performed in the 1940's in municipal water treatment facilities. Schott et al. (75) and Ruchoft et al. (71) concluded that TNT (< 2 ppm in liquid waste) had an adverse effect on BOD removal by activated sludge systems. Carpenter et al. (18) showed that the TNT metabolites were mainly associated with the lipid and protein components of microflora in an activated sludge system. The majority of the metabolites were found as precipitates which had solubility properties and infrared spectra consistent with amide polymers. In a follow up study, Bell and Burrows (7) used a semicontinuous activated sludge system to completely remove TNT from the effluent. No TNT or TNT metabolites were detected in the effluent, sludge or supernatant from simulated landfill sludge. Hoffsommer et al. (34) used a 3,000-gallon reactor to treat aqueous solutions of TNT. They obtained the best results with activated sludge and supplemental nutrients. Under these conditions, TNT concentrations were reduced by 97 ± 2% on a continual basis with a combined yield of 12 ± 2% amino and diamino metabolites. Supporting studies with ^{14}C-TNT showed that the radioactivity was about equally distributed between the solids, bacterial floc, and aqueous phase of the reactor. However, TNT was not mineralized in any of these studies.

Composting has also been proposed for the bioremediation of munition contaminated soils. Demonstrations (36, 41, 58, 62, 95, 98) revealed that composting shows promise but requires long incubation times and the final products are unknown. More research is necessary to determine the toxicity of the final products. Also, little is known of mineralization of TNT in composting processes.

Use of white rot fungi for bioremediation has generated considerable interest in recent years. The interest arises from the ability of the fungi to degrade a diverse group of environmentally persistent and toxic chemicals. *Phanerochaete chrysosporium* is known to mineralize a variety of organopollutants, including TNT, under nutrient limiting conditions (14, 15, 17, 24,28-30, 88, 89). Under these conditions (carbon, nitrogen or sulfur limitation), the fungus secretes a number of enzymes and factors that together constitute the lignin degrading system. The system contains the critical enzymes responsible for the conversion of xenobiotic compounds to carbon dioxide. Components of this system include lignin peroxidases (LiP), manganese dependent peroxidases (MnP), oxidases, reductases, hydrogen peroxide, veratryl alcohol, oxalate, and quinones (3, 33, 44-47, 86, 93).

The use of white rot fungi for soil bioremediation has several potential advantages. First, the fungi use the same system to degrade organic pollutants as they use to degrade lignin which is a complex three dimensional, heterogeneous biopolymer (23). It contains no regular repeating subunits that can be easily cleaved. Therefore, the enzymes necessary to degrade lignin must be very non-specific. Because of the non-specific nature of the lignin degrading system, white rot fungi may be excellent candidates for the bioremediation of sites contaminated with multiple pollutants. Second, the lignin degrading system is induced in response to nutrient limitation, not by the presence of the pollutants. This may allow the fungi to degrade the pollutants to essentially nondetectable levels because the enzymes will still be produced when the concentration of chemicals is low. Third, because the system is extracellular, the fungi can degrade many chemicals that would be toxic to other organisms. Finally, the system should be applicable to insoluble chemicals because lignin is insoluble.

White rot fungi will grow on any number of ligninocellulose nutrient sources such as sawdust, corn cobs, etc. Therefore, any readily available ligninocellulose waste may be used for the fungal nutrient at the remediation site. Thus, the system could be quite economical. Also, methods have been developed to provide large quantities of the fungi.

The biodegradation of environmental pollutants by fungi has some disadvantages. The growth of fungi and their rate of biodegradation are rather slow. Also, a nutrient supply must be added to the contaminated soil. This may increase the total volume. But it should be noted that the added nutrients are transformed by the fungi so they may actually enrich the soil. The degradation of these materials results in the production of humus. Finally, it must be noted that some of the fungi which are good biodegraders are not indigenous to soil.

However, with all these considerations, the use of white rot fungi shows promise as a cost effective method for the bioremediation of TNT-contaminated water and soil. However, before this technology can be implemented, TNT toxicity and metabolism must be understood. Such understanding has led to our ability to remediate soil that otherwise is toxic to the fungi.

TOXICITY

The toxic and mutagenic effects of TNT are well documented (26, 37, 38, 61, 68, 96). TNT is toxic to fathead minnows (*Pimephales promelas*) and bluegills (*Leopomis macrochirum*) at concentrations of 2-4 ppm (35, 66, 78, 96). It is also toxic to certain green algae (*Selenastrum capricornutum, Microcystis aeruginosa, Chlamydomonas reinhardtii*), tidepool copepods (*Tigriopus californicus*), and oyster larvae (*Crassostrea gigas*) at concentrations less than 10 ppm TNT (35, 78, 96). Several studies on the actual toxic component(s) of TNT metabolism have been performed (2, 66, 72, 84, 96, 97). In one study, TNT was found to be highly toxic but reduced metabolites were not toxic to green algae, tidepool copepods, and oyster larvae (96). Amino derivatives of TNT have been found to be less toxic to some organisms and more toxic to others (2). Several studies have suggested that mixtures of TNT complexes (surfactant or metal chelated) may be more toxic and mutagenic than pure TNT or amino metabolites (42, 84). This may represent a synergistic effect between the chemicals or an effect of the metals typically contained in the mixtures.

In any case, it is generally agreed that the toxic effects of nitroaromatic compounds require enzymatic reduction to yield reactive intermediates (60). It has been suggested that the toxicity of nitroaromatic compounds may be due to a nitro anion free radical ($R\text{-}NO_2^{\cdot -}$) formed during sequential one electron reductions (54). The free radicals may react with oxygen and lead to the formation of potentially toxic reduced oxygen species ($O_2^{\cdot -}$ and $OH^{\cdot}$). Also highly reactive nitroso intermediates (R-NO) formed during the reduction are known to form complexes with iron (54, 76).

Toxicity of TNT to *P. chrysosporium* has also been demonstrated. Spiker et al. (79) observed little fungal growth from a spore inoculum in the presence of 20 ppm TNT. They also observed that when TNT was added to pregrown mycelia (7 days after inoculation) in fresh media, up to 100 ppm TNT could be degraded but further fungal growth was inhibited by 65 ppm TNT. In these studies inhibition of mycelial growth was used as the measure of toxicity.

We have reported similar results suggesting that TNT is toxic to fungal spores and mycelia (80). In these studies toxicity to the fungus was estimated by quantitating ^{14}C-glucose dependent respiration. This method measures the metabolic activity of the fungus in the presence of TNT. In these studies 22 μM (5 ppm) TNT had no effect on fungal growth from spores but respiration was inhibited about 50% by 44 μM TNT and 96% by 220 μM TNT (Fig. 1). When TNT was added to cultures grown for two days, respiration was inhibited 0%, 50%, 80% at 440, 660 and 880 μM TNT, respectively. As the amount of mycelium increased, i.e., cultures grown for six days, the amount of TNT necessary to inhibit respiration 50% increased to approximately 1000 μM TNT. This inverse relationship between mycelial mass and toxicity was also shown by incubating TNT with washed mycelia. In this case respiration was only reduced 6.6% when 4.4 mM TNT was incubated with 36 mg (dry weight)/ml mycelium compared with 62.6% inhibition with 0.55 mg/ml mycelium (80). Thus it may be possible to circumvent the toxic effects by increasing the amount of mycelium exposed at any given time.

The lignin degrading system did not appear to play a role in TNT toxicity. The toxicity of TNT was identical in both ligninolytic and nonligninolytic cultures (80). It would therefore appear that the toxic substance was not the substrate for the lignin degrading system. Also, the lignin degrading system did not produce a metabolite that contributed to toxicity. Because

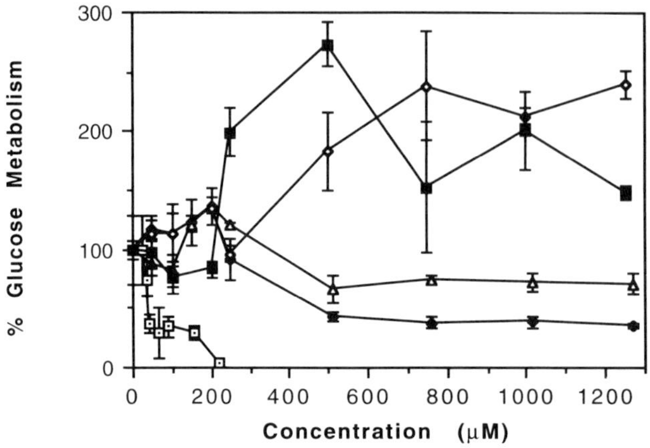

Figure 1. Toxicity of TNT, AmDNT and DAT to *P. chrysosporium*. Indicated amounts of TNT (open squares), 2-AmDNT (closed diamond), 4-AmDNT (open triangle), 2,4-DAT (closed square) or 2,6-DAT (open diamond) and [^{14}C]-glucose (0.05 μCi) were added to nutrient nitrogen-deficient media and fungal spores.

TNT was metabolized to amino and diamino metabolites under nonligninolytic conditions, the toxicities of these reduced metabolites were evaluated. The aminodinitrotoluenes (AmDNT) or diaminonitrotoluenes (DAT) exhibited much less toxicity to fungal growth from spores (Fig. 1). No inhibition of respiration was seen below 200 μM. At higher concentrations, the AmDNT were more toxic than the DAT. Respiration was inhibited 50% and 65% by the two isomers of AmDNT at 1250 μM, whereas respiration was actually stimulated 2 to 3-fold by the two DAT isomers. The differences in toxicity between the two AmDNT isomers may represent a difference in the rate of reduction of the *ortho* vs *para* nitro-group.

METABOLISM

The initial step in the metabolism of TNT by nearly every organism seems to be reduction (8-13, 18, 19, 27, 32, 51, 52, 59, 63, 64, 74, 87, 96, 97). Most often nitrotoluenes are biotransformed to aromatic amines (31, 40) through a series of reduction steps as described by Yamashima et al. (99). The *para* nitro group of TNT is usually reduced preferentially to the *ortho* group. A second series of reductions predominantly leads to the formation of 2,4-diaminonitrotoluene (2,4-DAT). Reduction of the third nitro group usually only occurs under anaerobic conditions (55). Under aerobic conditions, hydroxylamino metabolites are sufficiently long lived to lead to the formation of azoxy compounds or polymers which are products of condensation of hydroxylamino and nitroso dinitrotoluenes (56).

Extensive biodegradation of TNT and red water (a spent-sellite waste solution produced during the TNT purification process) by *P. chrysosporium* has been observed (29, 81). A large portion of the TNT was mineralized in these studies. The TNT mineralization rates were found to be proportional to the TNT concentration which may be related to the pseudo-first order kinetics observed for free radical mechanisms used by the fungus. This effect has also been observed for AmDNT (56) which suggests a free radical mechanism for the degradation of the AmDNT. When the concentration of TNT in cultures (both liquid and soil) approached that typically found in the environment, i.e., 10,000 mg/kg in soil and 100 mg/liter in water, approximately 20% of the TNT was converted to CO_2 and 15% was AmDNT at 90 days (30). In both these cases, the TNT completely disappeared.

Fungal spores first reduced TNT to the 4-AmDNT and 2-AmDNT metabolites (Fig. 2) (80). This reduction has been shown to produce a transient concentration of

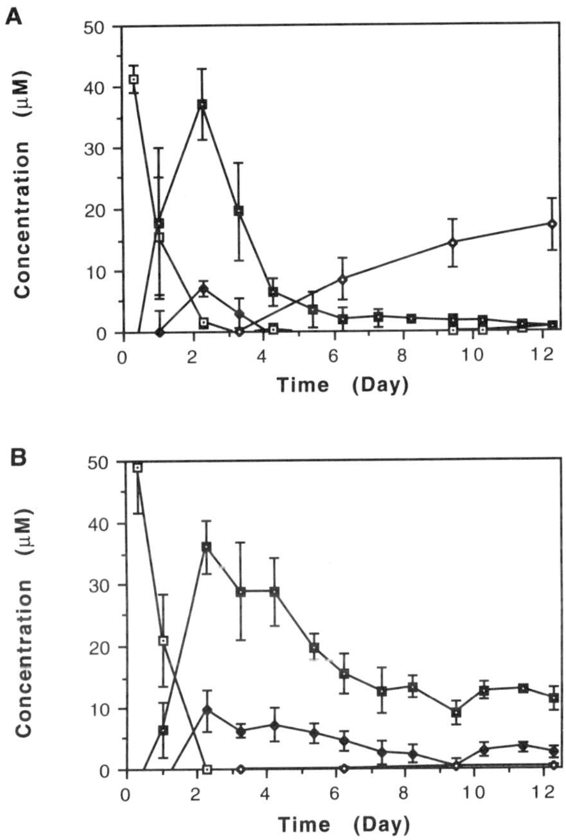

Figure 2. TNT metabolism in nutrient nitrogen-deficient (A) and sufficient (B) cultures of *P. chrysosporium*. TNT (44 μM, 0.05 μCi) was added to quadruplicate 10 ml, 2.4 mM and 24.0 mM nutrient nitrogen-containing media and fungal spores. TNT disappearance (open square), 4-AmDNT appearance (closed diamond), 2-AmDNT (closed square) and TNT mineralization (open diamond) are shown (80).

hydroxylamino metabolites (56). In most experiments, the 4-AmDNT was preferentially formed but occasionally the 2-AmDNT was the dominant amine. Valli et al. (88) reported that 2-aminonitrotoluene was the predominant metabolite formed during 2,4-dinitrotoluene (2,4-DNT) degradation.

Under nonligninolytic conditions, the AmDNT metabolites were slowly reduced to DAT metabolites. Insubstantial amounts of azoxy dimers were also formed under both these conditions. Mass balances showed that all the TNT metabolites could be accounted for in the extracellular fluid until the cultures became ligninolytic (Figs. 3, 4). The AmDNT metabolites were quickly degraded at the onset of manganese dependent peroxidase activity (Figs. 2, 3) (81). Mineralization occurred when the lignin peroxidases were produced by the fungus.

Even when the fungus is grown in media with no manganese, TNT is mineralized at 50% of the rate observed with manganese containing cultures. Under these conditions there can be no manganese dependent peroxidase activity. The mechanism for mineralization under these conditions is unknown.

REDUCTION

As previously discussed, TNT is initially reduced to AmDNT metabolites by most organisms. Channon et al. (19) showed that tissue homogenates of rabbit livers, kidneys, or

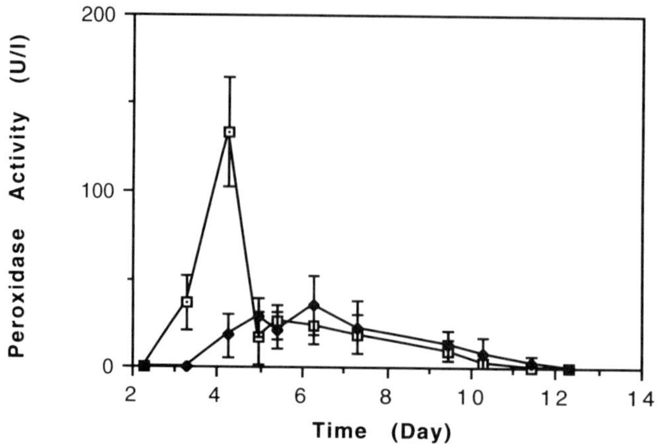

Figure 3. Peroxidase activity in nutrient nitrogen deficient cultures of *P. chrysosporium*. Manganese dependent (open square) and lignin (closed diamond) peroxidase activities were determined spectrophotometrically by following guaiacol and veratryl alcohol oxidation, respectively.

hearts all were capable of reducing TNT. Parrish et al. (64) reported that of 190 fungi representing 98 genera, the mycelia of 183 species were able to reduce TNT.

Valli et al. (88) suggested that intracellular nitroreductases may be involved in 2,4-DNT reduction by *P. chrysosporium*. Even though intracellular nitroreductases of *P. chrysosporium* probably exist, there is no current evidence that supports their involvement in TNT reduction. Conversely, there are a number of experiments which suggest that TNT is reduced by a plasma membrane redox system in *P. chrysosporium* (81). TNT reduction requires live, intact mycelia. Any condition that disrupts the integrity of the plasma membranes (freeze-thawing or grinding) destroys the reduction activity. Also, no significant reduction is observed with either supplemented (NADPH, NADH, or ATP) or unsupplemented extracellular or intracellular fractions of the culture, under aerobic or anaerobic conditions (81).

Classical membrane impermeable chemicals such as ferricyanide or tetrazolium dyes are typically used to study these membrane redox systems (1, 22, 53, 65). Some of these compounds inhibit TNT reduction (81). Reduction was inhibited by potassium ferricyanide and the easily reducible tetrazolium dyes (3-(4,5-dimethylthiazol-2-yl)-2,5-diphenyl tetrazolium and tetranitroblue tetrazolium) but not the more difficult to reduce dyes (nitroblue tetrazolium, 2,3,5-triphenyl tetrazolium and blue tetrazolium). In these experiments, either the TNT reduction or the dye reduction is inhibited depending on the relative ease of

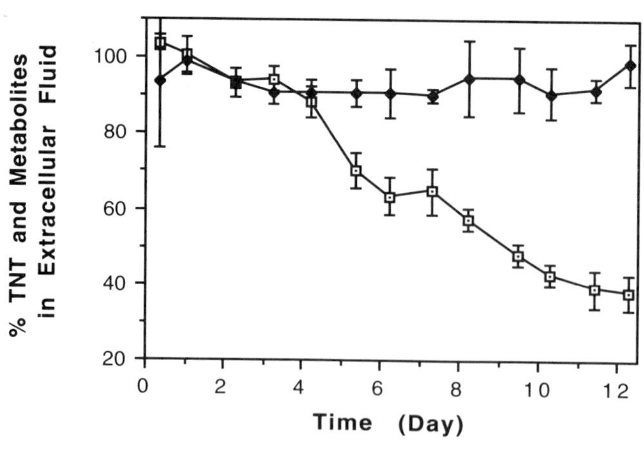

Figure 4. Percent of TNT and TNT metabolites in extracellular fluid of *P. chrysosporium* cultures. TNT (44 μM, 0.05 μCi) was added to 10 ml of 2.4 mM (open square) or 24 mM (closed square) nutrient nitrogen-containing media and fungal spores.

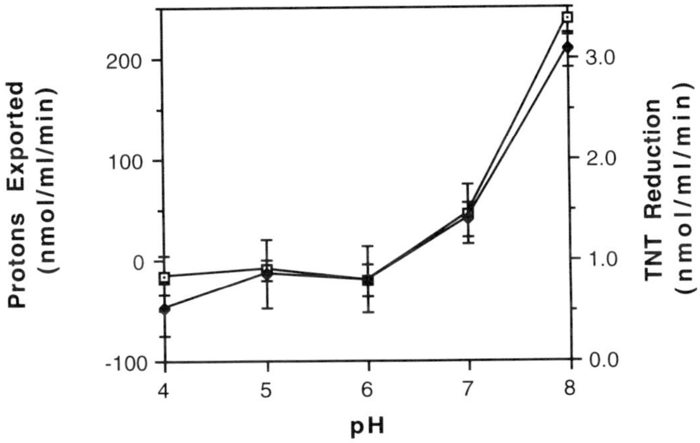

Figure 5. Effect of pH on proton excretion and TNT reduction rates. Washed mycelia (5 g) in potassium phosphate buffer (10 ml, 1.0 M, pH 4, 5, 6, 7, 8) were incubated with TNT (130 μM). Initial rates of protons generated (open square) and TNT reduction (closed diamond) were calculated from concentration changes during the first 30 minutes (81).

reduction which would be a reflection of the redox potentials. Thus, it appears that TNT, dyes and ferricyanide are reduced by the same mechanism.

Protonophores such as carbonyl cyanide trifluoromethoxyphenyl hydrazone (FCCP) and 2,4-dinitrophenol (2,4-DNP) and metabolic inhibitors such as sodium azide are also known to inhibit these membrane redox systems (39, 57). Protonophores act by dissipating the proton motive force at the membrane which at low concentrations actually stimulates electron flow across membranes (4, 22). A slight stimulation in TNT reduction was observed with FCCP, between 0 and 50 μM, but was not statistically significant. In general, TNT reduction was inhibited by these compounds at concentrations known to inhibit membrane redox systems.

The rates of proton export and TNT reduction could be correlated (Fig. 5). Incubating fungal mycelia in buffers of pH 6.0 and higher stimulated both proton flow across the membrane and increased TNT reduction. *P. chrysosporium* maintains an extracellular physiological pH of approximately 4.5 (50). We believe the plasma membrane redox system helps to maintain this pH. When mycelia were placed in buffers with pH significantly above this physiological pH, protons appear to have been exported by the mycelia (81). At pH 8.0, over 200 nmol of protons/ml/minute were exported by 0.5 g (wet weight) of mycelia. Even though the rates of reduction and proton export correlate, they appear to be uncoupled above pH 6.0. This apparent uncoupling is due in part to phosphate uptake by the mycelia. Replacment of the phosphate buffer in the assay with Tris or HEPES buffer (200 mM) decreased the proton export rate ~ 50-60%. A basal rate of protons were exported by the mycelia in the absence of any electron acceptor, however, after correction for the basal rate, stoichiometric rates of proton and electron generation were observed using Tris or HEPES buffer. Thus the protons exported may help to maintain the fungal extracellular pH and the electrons may be used to reduce chemicals.

PEROXIDASE REACTIONS

Lignin (LiP) and manganese dependent (MnP) peroxidases of *P. chrysosporium* are secreted in conjunction with a number of other enzymes and chemicals under nutrient deficient conditions (3, 33, 44-47, 86, 92, 93). The peroxidases are known to catalyze a number of free radical reactions. Research on these enzymes has been extensively reviewed recently (5, 49, 85). The "resting" enzyme (ferric-heme protein) is initially oxidized by two electrons in a reaction with hydrogen peroxide (Fig. 6). The resulting intermediate, com-

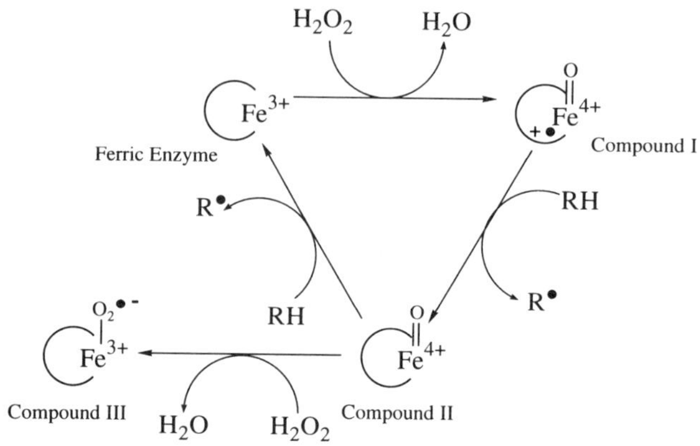

Figure 6. The peroxidase catalytic cycle (85).

pound I, is an oxoferryl (FeIV) heme protein with a cation radical stabilized in the heme porphyrin ring. A good electron donor, such as veratryl alcohol, can reduce this intermediate in two sequential one-electron reductions to return the enzyme to the ferric state. In the absence of a good electron donor, the enzyme will still abstract one electron from other reductants (perhaps from H_2O_2) to form compound II (ferryl enzyme). Compound II can be reduced again to regenerate ferric enzyme or react with hydrogen peroxide to form a catalytically inactive species, compound III (ferric-superoxide protein). Alternatively, a compound that is oxidized directly by the enzyme can oxidize other chemicals. Both direct and indirect oxidations of chemicals by the enzymes are known (4).

In addition to catalyzing the oxidation of many compounds, LiP is also able to catalyze reductive reactions in the presence of electron donors such as EDTA or oxalate (Fig. 7) (6, 77). Veratryl alcohol is a free radical mediator in these reactions. The electron donors appear to be oxidized by a LiP generated veratryl alcohol cation radical. The resulting anion radical can catalyze the reduction of good electron acceptors such as cytochrome *c*, nitroblue tetrazolium, and oxygen. Evolution of CO_2 from EDTA or oxalate effectively drives the reductive reactions. Similar reactions have also been observed with the manganese dependent peroxidases in the presence of quinones (20). Early work performed in our laboratory showed that these reductive mechanisms are not involved in TNT reduction. However, these reactions may be involved in other steps in TNT metabolism.

Bumpus and Tatarko (16) suggested that AmDNT were not substrates or inhibitors of the LiP oxidative reactions. On the other hand, Michels and Gottschalk (56) found that hydroxylaminodinitrotoluenes (NHOH-DNT) are good substrates for LiP (56). They demonstrated that a transient decrease in LiP activity in cultures of the fungus correlated with NHOH-DNT concentrations. Bumpus and Tatarko found that culture fluid (after 24 hours incubation of the fungus with TNT) inhibited veratryl alcohol oxidase activity. They concluded that the inhibition was due to NHOH-DNT. Both groups found that inhibition of veratryl alcohol oxidation by LiP isozyme H8 was characterized by a concentration-dependent lag period, after which veratryl alcohol oxidase activity was completely restored. These results are consistent with the hypothesis that oxidation of alternate LiP substrates inhibit veratryl aldehyde formation. Michels and Gottschalk (56) presented evidence that the NHOH-DNT could protect the LiP from inactivation by H_2O_2. If NHOH-DNT was not a good substrate, compound III would be formed quickly and the enzyme inactivated. They suggested that once NHOH-DNT reaches an equilibrium between oxidized and reduced forms, veratryl aldehyde formation begins. Since the

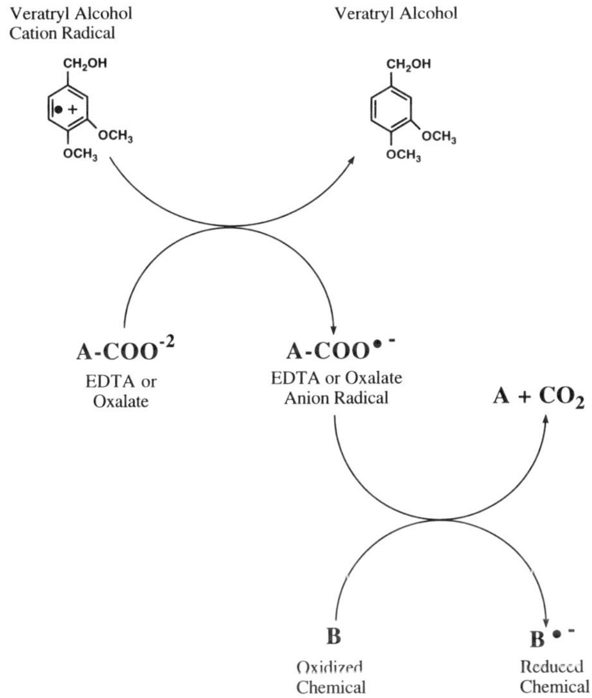

Figure 7. Mechanism of lignin peroxidase catalyzed reduction (6, 77).

enzyme turnover is not inhibited, near complete veratryl alcohol activity is restored. Michels and Gottschalk (56) found that at low NHOH-DNT concentrations, increasing veratryl alcohol concentration could shorten the lag period. This effect was saturable above 2.8 mM veratryl alcohol. These results are consistent with the mediation effect of veratryl alcohol seen in other indirect oxidation reactions.

Although the degradation of TNT metabolites by the peroxidases has not been studied in detail, potentially useful information can be inferred from the reactions of 2,4-dinitrotoluene metabolites with the peroxidases. Valli et al. (88) reported that the initial metabolites in 2,4-DNT degradation by the fungus were aminonitrotoluenes (AmNT). These compounds appear to be substrates for the peroxidases. The 2-AmNT was oxidized by MnP to yield 4-nitro-1,2-benzoquinone and methanol (Fig. 8). The quinone was reduced and methylated in cultures of the fungus. The nitro group was then released as nitrite by either MnP or LiP. MnP oxidizes the 4-nitro-1,2-hydroquinone to 2-hydroxy-1,4-benzoquinone which then can be reduced by cultures. LiP oxidized the methylated compound (1,2-dimethoxy-4-nitrobenzene) to yield 2-methoxy-1,4-benzoquinone. The methoxy group can then be removed by either LiP or MnP. The 1,2,4-trihydroxybenzene is a known substrate for intracellular 1,2-dioxygenases (25). In cell-free extracts, this chemical was shown to undergo ring cleavage to produce, after reduction, ß-ketoadipic acid.

With this pathway in mind, it is interesting to note that no research has been published to date on the potential reactions of MnP with TNT metabolites. The Valli et al. (88) work represents the first report on release of nitrite in MnP reactions. Release of methanol from methoxy benzene compounds has been shown by a number of researchers since it was first published by Kersten et al. (48).

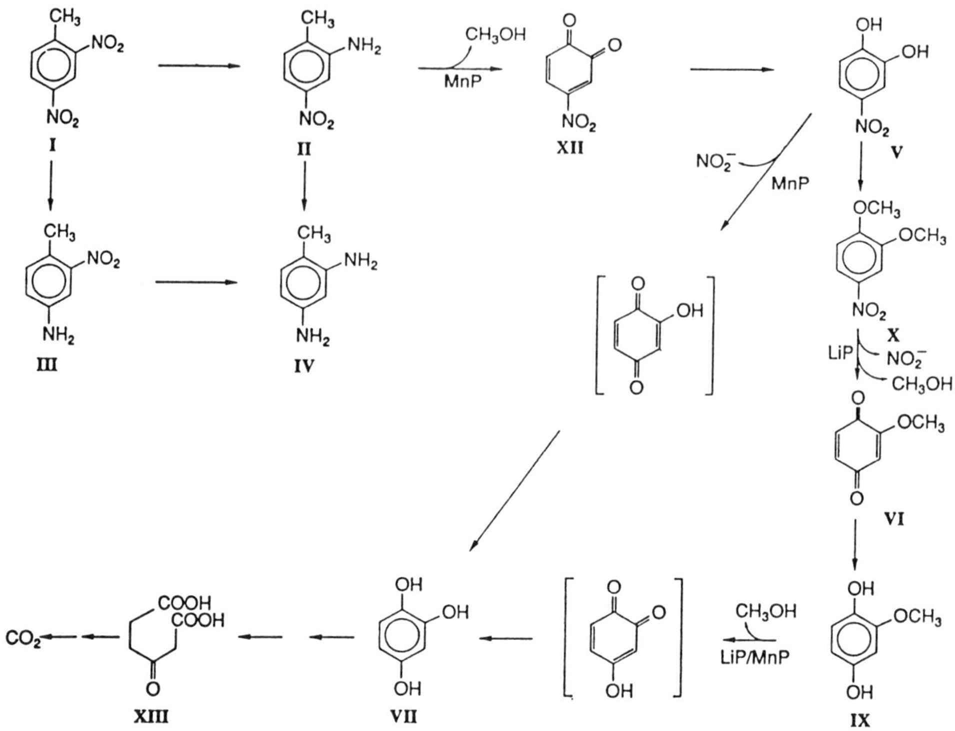

Figure 8. Proposed pathway for the degradation of 2,4-dinitrotoluene by *P. chrysosporium* (88).

REMEDIATION STUDIES

Various types of reactors have been developed for the remediation of TNT-contaminated water. Sublette et al. (83) used a rotating biological contactor (RBC) to develop a cost effective treatment for pink water. Pink water is an aqueous effluent generated during TNT manufacture from plant cleanup and scrubbing processes. The major components of this waste water are TNT and cyclotrimethylene trinitramine (RDX) at concentrations of 120-175 ppm and 25 ppm, respectively. Greater than 90% removal of TNT from a simulated pink water (120-175 ppm TNT) was observed in a continuous RBC with a residence time of about 24 hours. The RBC disk area required was about 10,000 ft^2/GPM feed. Sublette et al. (83) reported that at high TNT concentration (and/or short residence times), an apparent intermediate(s) of TNT degradation accumulated in the reactor. These intermediates were probably a mixture of AmDNT metabolites which have subsequently been shown to be mineralized by *P. chrysosporium* (16, 80).

Bumpus and Tatarko (16) showed that a 2-liter fixed-film silicone membrane bioreactor first developed by Venkatadri and Irvine (90) was able to degrade TNT and AmDNT metabolites under ligninolytic conditions. The reactor was able to degrade 30 ppm TNT to less than 60 ppb within 96 hours under batch conditions. Both 4-AmDNT and 2-AmDNT degraded to unidentified yellow metabolites that were water soluble and could not be extracted into methylene chloride. These could be hydroxylaminodinitrotoluene conjugated gluconides but there are no data to suggest that the fungus can catalyze glucuronidation. It would have been interesting if the Webster (68) test had been performed on the metabolites.

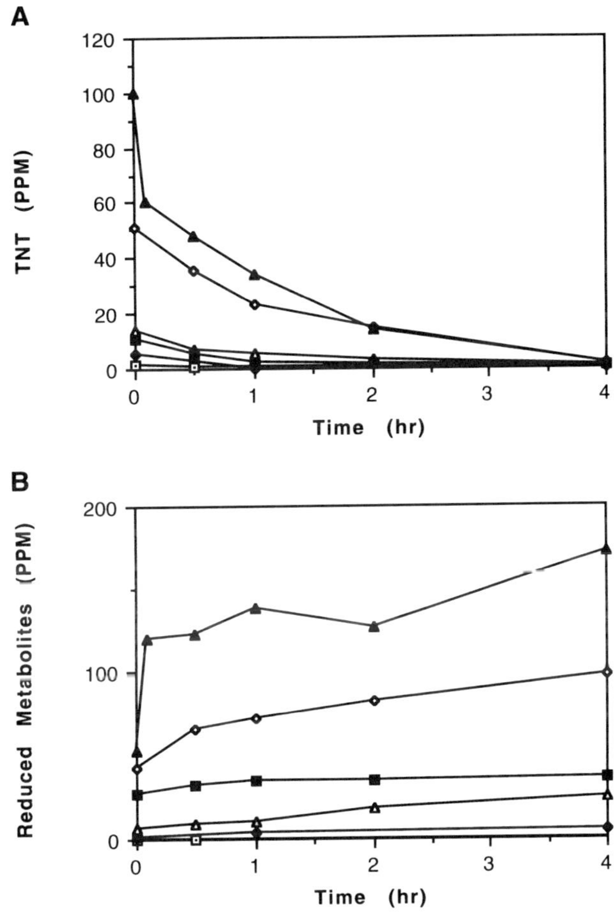

Figure 9. TNT metabolism by *P. chrysosporium* in liquid 2-liter reactor. Mycelia (100 g wet weight) were added to 2.0 liters of nitrogen-deficient media. Indicated concentrations of TNT were added to the reactor at 10 day intervals. (1st (open square), 2nd (closed diamond), 3rd (open triangle), 4th (closed square), 5th (open diamond), and 6th additions (closed triangle)). TNT concentration (A) and cumulative concentrations of AmDNT and DAT (B) were determined by gas chromatography.

We have been studying the potential use of continuously stirred reactors and air-lift reactors for the remediation of TNT-contaminated water (82). We propose that the toxicity of TNT can be circumvented by increasing the mycelial mass in the reactor. This has the additional benefit of increasing the initial rate of TNT reduction. We found that up to 100 ppm TNT could be transformed to less than 140 ppb within 4 hours in a 2-liter stirred reactor with approximately 50 g/liter (wet weight) of mycelia under batch conditions (Fig. 9). Under these conditions, mechanical agitation was necessary to maintain aerobic conditions. Due to the mechanical agitation, the peroxidases were not active and AmDNT and DAT metabolites were not degraded. Ligninolytic conditions could be maintained in an air lift reactor containing approximately 8 g/liter (wet weight) of mycelia. Under these conditions, 100 ppm TNT was degraded within 6 days, lignin peroxidase activity was maintained, and the reduced metabolites were completely degraded within 10 days.

This has led us to propose a two stage reactor system. The first stage can be a nonligninolytic stirred reactor containing large amounts of fungus which would rapidly reduce TNT to the amino metabolites, and thus decrease toxicity. The reduced metabolites would then be completely degraded in a second stage ligninolytic reactor (fixed-film or air-lift designs). The second stage reactor would obviously need to be much larger than the first in order to optimize degradation rates.

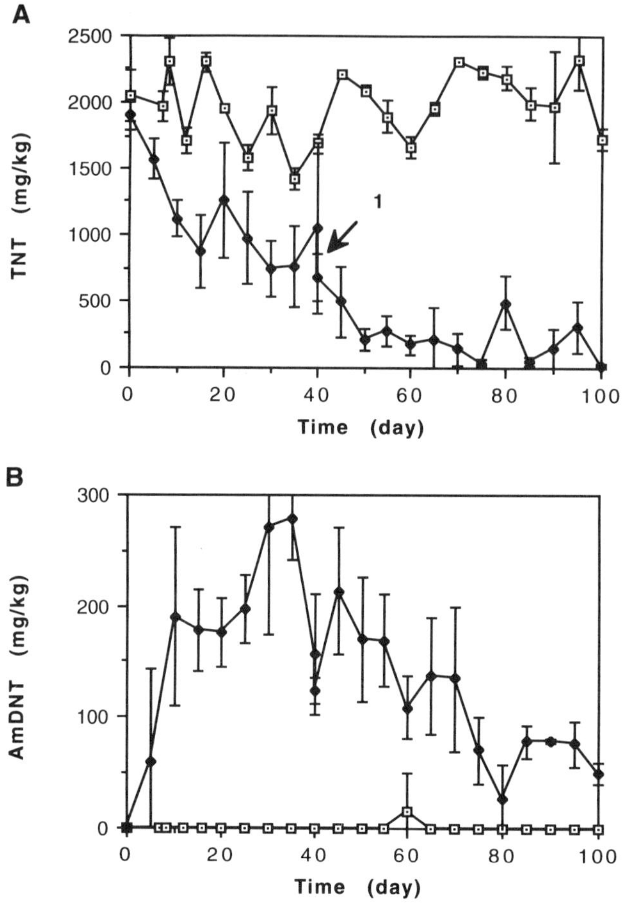

Figure 10. TNT metabolism in soil (2000 mg/kg TNT) by *P. chrysosporium*. TNT-contaminated soil (2.0 kg) was incubated (at 39°C) with sawdust (20%) preinoculated with *P. chrysosporium* (closed diamond). Contaminated soil (0.5 kg) incubated (at 39°C) without sawdust was used as controls (open square). Inoculated sawdust (10%) was added and the microcosms mixed at 40 days (arrow 1). Concentrations of TNT (A) and AmDNT (B) were determined by HPLC.

Remediation of TNT-contaminated soil by *P. chrysosporium* has also been observed (30, 82). In a bench scale (1.0-2.0 kg soil) feasibility study (82), TNT was degraded from 200 and 2,000 mg/kg (Fig. 10) to nondetectable levels within 14 and 100 days, respectively. In another microcosm, the TNT concentration was reduced from 10,000 to 3,500 mg/kg in 100 days. The principal metabolites were the isomeric aminodinitrotoluenes, which reached maximum concentrations of 35, 280 and 2,600 mg/kg on days 14, 35 and 55 for the 200, 2,000 and 10,000 mg/kg microcosms. These metabolites decreased to less than 10 mg/kg by day 28 in the 200 mg/kg microcosms and to 80 and 1,000 mg/kg by day 100 in the 2,000 and 10,000 mg/kg microcosms, respectively. The ultimate fate of the TNT in the microcosms was not explored. Because the microcosms were open systems, no radioactive labeling was possible. However, Fernando et al. (30) observed that in soil contaminated with 10,000 mg/kg TNT approximately 20% of the TNT was converted to CO_2 within 90 days. They reported that 11.5% of the radioactivity was bound to a soil-corn cob-fungal matrix.

The major problem encountered in the bench scale remediation study mentioned above was a decrease in the rate of TNT degradation after 20 days (82). The effectiveness of mixing the microcosm, addition of fungal inoculum, or addition of inoculated substrate to the microcosms containing 2,000 and 10,000 mg/kg TNT was evaluated when TNT metabolism slowed. Mixing the microcosms did not result in the resumption of biodegradation. Addition of inoculum or inoculated substrate was effective in the microcosms containing 2,000 mg/kg TNT (i.e., degradation resumed at near its original rate) but only the addition of inoculated substrate proved effective in the microcosms containing 10,000 mg/kg TNT.

Based on the experiments described above, we suggest that bioremediation of TNT-contaminated soil by *P. chrysosporium* is feasible. Toxicity can be circumvented by growing the fungus first on an appropriate nutrient substrate then incubating the substrate with the contaminated soil after growth of the fungus. This would reduce or eliminate the toxicity of TNT to the fungus. A second addition of inoculated substrate might be necessary to complete TNT metabolism in highly contaminated soils. It should be noted, however, that research into maintaining degradation rates beyond 20 days needs to be conducted.

CONCLUDING REMARKS

Due to the persistence, toxicity, and mutagenicity of TNT, there is considerable interest in remediation of TNT-contaminated water and soil. Metabolism of TNT by many organisms has been extensively studied. In most cases reduction represents the whole of TNT metabolism with AmDNT and DAT being the principal end products. A few organisms use TNT as sole carbon and/or nitrogen sources (8, 27, 87). However, in all cases, only small amounts of TNT were mineralized by these organisms. Several of these organisms and various consortia have been used to develop remediation procedures. However, potential problems in degrading TNT to nondetectable levels will have to be overcome. Also, the potential toxicity of intermediates and/or products must be evaluated.

White rot fungi show promise for the remediation of TNT because the fungi can significantly mineralize TNT. Multiple systems for the remediation of TNT-contaminated water and soil by *P. chrysosporium* are being developed. Currently, research is being conducted to optimize these systems.

ACKNOWLEDGEMENTS

This publication was made possible by grant number ES04922 from the National Institutes of Environmental Health Sciences, National Institutes of Health, Global Environmental Solutions, Inc., Magna, Utah, and Intech One-Eighty Corporation, Logan, Utah. We thank Terri Maughan for preparing this manuscript and for technical assistance.

REFERENCES

1. Altman, F. P. 1969. The use of eight different tetrazolium salts for a quantitative study of pentose shunt dehydrogenation. Histochemie 19:363-374.
2. Amerkhanova, N. N. and R. P. Naumova. 1979. Comparative study of the acute toxicity of alpha-trinitrotoluol and products of its biodegradation for hydrobiots. Biol. Navkl. 2:26-28.
3. Ander, P., C. Mishra, R. L. Farrell, and K.-E. Eriksson. 1990. Redox reactions in lignin degradation: interactions between laccase, different peroxidases and cellobiose:quinone oxidoreductase. J. Biotechnol. 13:189-198.
4. Bakker, E. P., J. C. Arents, J. P. M. Hoebe, and H. Terada. 1975. Surface potential and the interaction of weak acidic uncouplers of oxidative phosphorylation with liposomes and mitochondria. Biochim. Biophys. Acta. 387:491-506.
5. Barr, D. P. and S. D. Aust. 1994. Mechanisms white rot fungi use to degrade pollutants. Environ. Sci. Technol. 28:78A-87A.
6. Barr, D. P., M. M. Shah, T. A. Grover, and S. D. Aust. 1992. Production of hydroxyl radical by lignin peroxidase from *Phanerochaete chrysosporium*. Arch. Biochem. Biophys. 298:480-485.

7. Bell, B. A. and W. D. Burrows. 1989. Removal and degradation of TNT in a semicontinuous activated sludge system, p. 344-356. *In* C.A Cole and D. A. Long (eds.) Hazardous and industrial wastes: proceedings of the 21st Mid-Atlantic industrial waste conference. Technomic Publishing Co., Inc., Lancaster.
8. Boopathy, R. and C. F. Kulpa. 1993. Nitroaromatic compounds serve as nitrogen source for *Desulfovibrio* sp. (B strain). Can. J. Microbiol. 34:430-433.
9. Boopathy, R., C. F. Kulpa and M. Wilson. 1993. Metabolism of 2,4,6-trinitrotoluene by *Desulfovibrio* sp. (B strain). Appl. Microbiol. Biotechnol. 39:270-275.
10. Boopathy, R., J. Manning, C. Montemagno and C. Kulpa. 1994. Metabolism of 2,4,6-trinitrotoluene by a *Pseudomonas* consortium under aerobic conditions. Curr. Microbiol. 28:131-137.
11. Boopathy, R., M. Wilson and C. F. Kulpa. 1993. Anaerobic removal of 2,4,6-trinitrotoluene (TNT) under different electron accepting conditions: laboratory study. Water Environ. Res. 65:271-275.
12. Boopathy, R., M. Wilson, C. D. Montemagno, J. F. Manning, Jr., and C. F. Kulpa. 1994. Biological transformation of 2,4,6-trinitrotoluene (TNT) by soil bacteria isolated from TNT-contaminated soil. Bioresource Technol. 47:19-24.
13. Bueding, E. and N. Jolliffe. 1946. Metabolism of trinitrotoluene (TNT) in vitro. J. Pharmacol. Exp. Ther. 88:300-312.
14. Bumpus, J. A. and S. D. Aust. 1986. Biodegradation of environmental pollutants by the white rot fungus *Phanerochaete chrysosporium*: involvement of the lignin degrading system. BioEssays 6:166-170.
15. Bumpus, J. A. and S. D. Aust. 1987. Biodegradation of DDT [1,1,1-trichoro-2,2-bis(4-chlorophenyl)ethane] by the white rot fungus *Phanerochaete chrysosporium*. Appl. Environ. Microbiol. 53:2001-2008.
16. Bumpus, J. A. and M. Tatarko. 1994. Biodegradation of 2,4,6-trinitrotoluene by *Phanerochaete chrysosporium*: Identification of initial degradation products and the discovery of a TNT metabolite that inhibits lignin peroxidases. Curr. Microbiol. 28:185-190.
17. Bumpus, J. A., M. Tien, D. Wright, and S. D. Aust. 1985. Oxidation of persistent environmental pollutants by a white rot fungus. Science 228:1434-1436.
18. Carpenter, D. F., N. G. McCormick, J. H. Cornell, and A. M. Kaplan. 1978. Microbial transformation of ^{14}C-labeled 2,4,6-trinitrotoluene in an activated-sludge system. Appl. Environ. Microbiol. 35:949-954.
19. Channon, H. J., G. T. Mills, and R. T. Williams. 1944. The metabolism of 2,4,6-trinitrotoluene (alpha-TNT). Biochem. J. 38:70-85.
20. Chung, N., M. M. Shah, T. A. Grover, and S. D. Aust. 1993. Reductive activity of a manganese-dependent peroxidase from *Phanerochaete chrysosporium*. Arch. Biochem. Biophys. 306:70-75.
21. Clark, M. G., E. J. Partick, G. S. Patten, F. L. Crane, H. Low, and C. Grebing. 1981. Evidence for the extracellular reduction of ferricyanide by rat liver. Biochem. J. 200:565-572.
22. Crane, F. L., H. Roberts, A. W. Linnane, and H. Low. 1982. Transmembrane ferricyanide reduction by cells of the yeast *Saccharomyces cerevisae*. J. Bioeng. Biomembranes 14:191-205.
23. Crawford, R. L. 1981. The chemical structure of lignin and the use of various lignin preparations for microbial studies, p. 7-19. *In* R. L. Crawford (ed.) Lignin biodegradation and transformation. John Wiley and Sons. New York.
24. Cripps, C., J. A. Bumpus, and S. D. Aust. 1990. Biodegradation of azo- and heterocyclic dyes by *Phanerochaete chrysosporium*. Appl. Environ. Microbiol. 56:1114-1118.
25. Dagley, S. 1975. Microbial degradation of organic compounds in the biosphere. Am. Sci. 63:681-689.
26. Davenport, R., L. R. Johnson, D. J. Shaeffer, and H. Balbach. 1994. Phototoxicology. 1. Light-enhanced toxicity of TNT and some related compounds to *Daphnia magna* and *Lytechinus variagatus*. Ecotoxicity Environ. Safety 27:14-22.
27. Duque, E., A. Haidour, F. Godoy and J. L. Ramos. 1993. Construction of a *Pseudomonas* hybrid strain that mineralizes 2,4,6-trinitrotoluene. J. Bacteriol. 175:2278-2283.
28. Eaton, D. C. 1984. Mineralization of polychlorinated biphenyls by *Phanerochaete chrysosporium*: a ligninolytic fungus. Enzyme Microb. Technol. 7:194-196.
29. Fernando, T. and S. D. Aust. 1991. Biodegradation of munition waste, TNT (2,4,6-trinitrotoluene), and RDX (hexahydro-1,3,5-trinitro-1,3,5-triazine) by *Phanerochaete chrysosporium*, p. 214-232. *In* D.W. Tedder and F.G. Pohland (eds.), Emerging technologies in hazardous waste management II. American Chemistry Society, Washington, D.C.
30. Fernando, T., J. A. Bumpus, and S. D. Aust. 1990. Biodegradation of TNT (2,4,6-trinitrotoluene) by *Phanerochaete chrysosporium*. Appl. Environ. Microbiol. 56: 1667-1671.
31. Fewson, C.A. 1981. Biodegradation of aromatics with industrial relevance, p. 141-170. *In* T. Leisinger, A. M. Cook, R. Huttler, and J. Neusch (eds.) Microbial degradation of xenobiotics and recalcitrant compounds. Academic Press, London.

32. Funk, S. B., D. J. Roberts, D. L. Crawford and R. L. Crawford. 1993. Initial-phase optimization for bioremediation of munition compound-contaminated soils. Appl. Environ. Microbiol. 59:2171-2177.
33. Gold, M. H., M. Kuwahara, A. A. Chiu, and J. K. Glenn. 1984. Purification and characterization of an extracellular H_2O_2-requiring diaryl propane oxygenase from the white rot basidiomycete *Phanerochaete chrysosporium*. Arch. Biochem. Biophys. 234:353-362.
34. Hoffsommer, J. C., L. A. Kaplan, D. J. Glover, D. A. Kubose, C. Dickinson, H. Goya, E. G. Kayser, C. L. Groves, and M. E. Sitzmann. 1978. Final report. Contract WOL-TR-77-136. Biodegradability of TNT: a three year pilot plant study. Naval Surface Weapons Center.
35. Hudock, G. A. and D. M. Gring. 1970. Biological effects of trinitrotoluene. Contract No. 0164-69-CO822. Naval Environ. Health Center, Norfolk.
36. Isbister, J. D., G. L. Anspach, J. F. Kitchens, and R. C. Doyle. 1984. Composting for decontamination of soils containing explosives. Microbiologica 7:47-73.
37. Johnson, L. R., R. Davenport, H. Balbach, and D. J. Shaeffer. 1994. Phototoxicology. 2. Near-ultraviolet light enhancement of Microtox assays of trinitrotoluene and aminodinitrotoluenes. Ecotoxicity Environ. Safety 27:23-33.
38. Johnson, L. R., R. Davenport, H. Balbach, and D. J. Shaeffer. 1994. Phototoxicology. 3. Comparative toxicity of trinitrotoluene and aminodinitrotoluenes to *Daphnia magna, Dugesia dorotocephala*, and sheep erythrocytes. Ecotoxicity Environ. Safety 27:34-44.
39. Jolliffe, L. K., R. J. Doyle, and U. N. Streips. 1981. The energized membrane and cellular autolysis in *Bacillus subtilis*. Cell 25:753-763.
40. Kaplan, D. L. 1992. Biological degradation of explosives and chemical agents. Curr. Opin. Biotechnol 3:253 260.
41. Kaplan, D. L. and A. M. Kaplan. 1982. Thermophilic biotransformations of 2,4,6-trinitrotoluene under simulated composting conditions. Appl. Environ. Microbiol. 44:757-760.
42. Kaplan, D. L. and A. M. Kaplan. 1982. 2,4,6-Trinitrotoluene surfactant complexes: decomposition, mutagenicity, and soil leaching studies. Environ. Sci. Technol. 16:566-571.
43. Keither, L. H. and W. A. Telliard. 1979. Priority pollutants. I. A perspective view. Environ. Sci. Technol. 13:416-423.
44. Kelley, R. L. and C. A. Reddy. 1986. Identification of glucose oxidase activity as the primary source of hydrogen peroxide production in ligninolytic cultures of *Phanerochaete chrysosporium*. Arch. Microbiol. 144:248-253.
45. Kelley, R. L. and C. A. Reddy. 1986. Purification and characterization of glucose oxidase from ligninolytic cultures of *Phanerochaete chrysosporium*. J. Bacteriol. 166:269-274.
46. Kersten, P. J. 1990. Glyoxal oxidase of *Phanerochaete chrysosporium*: Its characterization and activation by lignin peroxidase. Proc. Natl. Acad. Sci. USA 87:2936-2940.
47. Kersten, P. J. and T. K. Kirk. 1987. Involvement of a new enzyme, glyoxal oxidase, in extracellular H_2O_2 production by *Phanerochaete chrysosporium*. J. Bacteriol. 169:2195-2201.
48. Kersten, P. J., M. Tien, B. Kalyanaraman, and T. K. Kirk. 1985. The ligninase of *Phanerochaete chrysosporium* generates cation radicals for methoxybenzenes. J. Biol. Chem. 260:2609-2612.
49. Kirk, T. K. and R. L. Farrell. 1987. Enzymatic combustion: the microbial degradation of lignin. Ann. Rev. Microbiol. 41:465-505.
50. Kirk, T. K., E. Schultz, W. J. Connors, C. F. Lorenz, and J. G. Zeikus. 1978. Influence of culture parameters on lignin metabolism by *Phanerochaete chrysosporium*. Arch. Microbiol. 117:277-285.
51. Klausmeier, R. E., J. L. Osmon and D. R. Walls. 1973. The effect of trinitrotoluene on microorganisms. Dev. Ind. Microbiol. 15:309-317.
52. Lemberg, R. and J. P. Callaghan. 1944. Metabolism of symmetrical trinitrotoluene. Nature 154:768-769.
53. Low, H., F. L. Crane, E. J. Partick, G. S. Patten, and M. G. Clark. 1984. Properties and regulation of a trans-plasma membrane redox system of perfused rat heart. Biochim. Biophys. Acta 804:253-260.
54. Mason, R. P. and P. D. Josephy. 1985. Free radical mechanism of nitroreductases, p. 121-140. *In* D. Rickert (ed.) Toxicity of nitroaromatic compounds. Hemisphere Publishing Corporation, New York.
55. McCormick, N. G., F. E. Feeherry and H. S. Levinson. 1976. Microbial transformation of 2,4,6-trinitro-toluene and other nitroaromatic compounds. Appl. Environ. Microbiol. 31:949-958.
56. Michels, J. and G. Gottschalk. 1994. Inhibition of the lignin peroxidase of *Phanerochaete chrysosporium* by hydroxylamino-dinitrotoluene, an early intermediate in the degradation of 2,4,6-trinitrotoluene. Appl. Environ. Microbiol. 60:187-194.
57. Mitchell, P. 1961. Conduction of protons through the membranes of mitochondria and bacteria by uncouplers of oxidative phosphorylation. Biochem. J. 81:24.

58. Montemagno, C. D. 1991. Evaluation of the feasibility of biodegrading explosives-contaminated soils and groundwater at the Newport Army Ammunition Plant (NAAP). USATHAMA Report CETHA-TS-CR-9200. Argonne National Laboratory, Chicago.
59. Moore, B. 1918. The causation and prevention of trinitrotoluene (T.N.T.) poisoning. J. Amer. Med. Assoc. 70:412-413.
60. Moreno, S. N. J. and R. Docampo. 1985. Mechanism of toxicity of nitro compounds used in the chemotherapy of trichomoniasis. Environ. Health Perspectives 64:199-208.
61. Morton, A. R., D. O. Manmohan, U. Ranadive, and J. A. Hathaway. 1976. Biological effects of trinitrotoluene from exposure below the threshold limit value. Amer. Ind. Hyg. Assoc. J. 37:56-60.
62. Osmon, J. L., C. C. Andrews, and A. Tatyrek. 1978. The biodegradation of TNT in enhanced soil and compost systems. Report ARLCD-TR-77032. Large Caliber Weapon Systems Laboratory, United States Army Armament Research and Development Command, Dover.
63. Osmon, J. L. and R. E. Klausmeier. 1972. The microbial degradation of explosives. Dev. Ind. Microbiol. 14:247-252.
64. Parrish, F. W. 1977. Fungal transformation of 2,4-dinitrotoluene and 2,4,6-trinitrotoluene. Appl. Environ. Microbiol. 34:232-233.
65. Pearse, A. G. E. 1972. Principles of oxidoreductase histochemistry, p. 881-920. Histochemistry: theoretical and applied, 3rd ed., vol. 2. The Williams and Wilkins Co., Baltimore.
66. Pederson, G. L. 1970. Evaluation of toxicity of selected TNT wastes on fish. Phase I - acute toxicity of alpha-TNT to bluegills. Sanitary Engineering Sp. Study No. 24-007-70/71, United States Army Environmental Hygiene Agency, Edgewood Arsenal.
67. Pereira, W. E., D. L. Short, D. B. Manigold and P. K. Roscio. 1979. Isolation and characterization of TNT and its metabolites in groundwater by gas chromatograph-mass spectrometer-computer techniques. Bull. Environ. Contam. Toxicol. 21:554-562.
68. Priby Council, Medical Research Council. 1921. T.N.T. poisoning and the fate of T.N.T. in the animal body. His Majesty's Stationary Office, U.K.
69. Rosenblatt, D. H. and R. B. Turkeltaub. 1987. Army installation restoration program: contaminated soil cleanup objectives for an inactive army ammunition plant, p. 523-527. *In* Proceedings of the NWWA FOCUS Conference on Midwestern Groundwater Issues. National Water Well Association, Dublin.
70. Ross, R. H. and W. R. Hartley. 1990. Comparison of water quality criteria and health advisories for 2,4,6-trinitrotoluene (TNT). Regul. Toxicol. Pharmacol. 11:114-117.
71. Ruchoft, C. C., M. LeBosquet, Jr., and W. G. Meckler. 1945. TNT wastes from shell-loading plants. Ind. Eng. Chem. 37:937-943.
72. Ryon, M. G. 1987. Water quality criteria for 2,4,6-trinitrotoluene (TNT). United States Army Project 84PP4845. Report No. AD-ORNL-6304.
73. Ryon, M. G. and R. H. Ross. 1990. Water quality criteria for 2,4,6-trinitrotoluene. Regul. Toxicol. Pharmacol. 11:104-113.
74. Schackmann, A. and R. Muller. 1991. Reduction of nitroaromatic compounds by different *Pseudomonas* species under aerobic conditions. Appl. Microbiol. Biotechnol. 34:809-813.
75. Schott, S., C. C. Ruchoft, and S. Migregian. 1943. TNT wastes. Ind. Eng. Chem. 35:1122-1132.
76. Servent, D., C. Ducrocq, Y. Henry, A. Guissani and M. Lenfant. 1991. Nitroglyerin metabolism by *Phanerochaete chrysosporium*: evidence for nitric oxide and nitrite formation. Biochim. Biophys. Acta. 1074:320-325.
77. Shah, M. M., T. A. Grover, D. P. Barr, and S. D. Aust. 1992. On the mechanism of inhibition of the veratryl alcohol oxidase activity of lignin peroxidase H2 by EDTA. J. Biol. Chem. 267:21564-21569.
78. Smock, L. A., D. L. Stonburner, and J. R. Clark. 1976. The toxic effects of trinitrotoluene (TNT) and its primary degradation products on two species of algae and the fathead minnow. Water Res. 10:537-543.
79. Spiker, J. K., D. L. Crawford and R. L. Crawford. 1992. Influence of 2,4,6-trinitrotoluene (TNT) concentration on the degradation of TNT in explosive-contaminated soils by the white rot fungus *Phanerochaete chrysosporium*. Appl. Environ. Microbiol. 58:3199-3202.
80. Stahl, J. D. and S. D. Aust. 1993. Metabolism and detoxification of TNT by *Phanerochaete chrysosporium*. Biochem. Biophys. Res. Comm. 192:477-482.
81. Stahl, J. D. and S. D. Aust. 1993. Plasma membrane dependent reduction of 2,4,6-trinitrotoluene by *Phanerochaete chrysosporium*. Biochem. Biophys. Res. Comm. 192:471-476.
82. Stahl, J. D. and S. D. Aust. Concepts for the bioremediation of trinitrotoluene-contaminated wastes by the white rot fungus *Phanerochaete chrysosporium*, p.172-185. *In* Proceedings of the 9th Annual Conference on Hazardous Waste Remediation. Elseuier Scientific, Netherlands.
83. Sublette, K. L., E. V. Ganapathy, and S. Schwartz. 1992. Degradation of munition wastes by *Phanerochaete chrysosporium*. Appl. Biochem. Biotechnol. 34/35:709-723.

84. Tan, E. L., C. H. Ho, W. H. Griest and R. L. Tyndall. 1992. Mutagenicity of trinitrotoluene and its metabolites formed during composting. J. Toxicol. Environ. Health 36:165-175.
85. Tien, M. 1987. Properties of ligninase from *Phanerochaete chrysosporium* and their possible applications. CRC Crit. Rev. Microbiol. 15:141-168.
86. Tien, M. and T. K. Kirk. 1984. Lignin-degrading enzyme from *Phanerochaete chrysosporium*. Purification, characterization, and catalytic properties of a unique H_2O_2-requiring oxygenase. Proc. Natl. Acad. Sci. USA 81:2280-2284.
87. Traxler, R. W., E. Wood and J. M. Delaney. 1974. Bacterial degradation of alpha-TNT. Dev. Ind. Microbiol. 16:71-76.
88. Valli, K., B. J. Brock, D. K. Joshi, and M. H. Gold. 1992. Degradation of 2,4-dinitrotoluene by the lignin-degrading fungus *Phanerochaete chrysosporium*. Appl. Environ. Microbiol. 58:221-228.
89. Valli, K., H. Wariishi, and M. H. Gold. 1992. Degradation of 2,7-dichlorobenzo-*p*-dioxin by the lignin-degrading basidiomycete *Phanerochaete chrysosporium*. J. Bacteriol. 174:2131-2137.
90. Venkatadri, R. and R. L. Irvine. 1993. Cultivation of *Phanerochaete chrysosporium* and production of lignin peroxidase in novel biofilm reactor systems: hollow fiber and silicone membrane reactor. Water Res. 27:591-596.
91. Walsh, J. T., R. C. Chalk and C. Merritt, Jr. 1973. Application of liquid chromatography to pollution abatement studies on munition wastes. Anal. Chem. 45:1215-1220.
92. Westermark, V. and K.-E. Eriksson. 1974. Cellobiose:quinone oxidoreductase, a new wood-degrading enzyme from white rot fungi. Acta Chemica Scandinavica B 28:209-214.
93. Westermark, V. and K.-E. Eriksson. 1975. Purification and properties of cellobiose:quinone oxidoreductase from *Sporotrichum pulverulentum*. Acta Chemica Scandinavica B 29: 419-424.
94. Williams, R. T., P. S. Ziegenfuss, and P. J. Marks. 1988. Field demonstration composting of explosives-contaminated sediments at the Louisiana Army Ammunition Plant (LAAP). Final report. Report #AMXTH-IR-TE 88242. Contract DAAK-11-85-D-C07.
95. Williams, R. T., P. S. Ziegenfuss, and W. E. Sisk. 1992. Composting of explosives and propellant contaminated soils under thermophilic and mesophilic conditions. J. Ind. Microbiol. 9:137-144.
96. Won, W. D., L. H. DiSalvo and J. Ng. 1976. Toxicity and mutagenicity of 2,4,6-trinitrotoluene and its microbial metabolites. Appl. Environ. Microbiol. 31:576-580.
97. Won, W. D., R. J. Heckley, D. J. Glover and J. C. Hoffsommer. 1974. Metabolic disposition of 2,4,6-trinitroluene. Appl. Microbiol. 27:513-516.
98. Woodward, R. E. 1990. Evaluation of composting implementation: a literature review. Report TCN 89363. United States Army Toxic and Hazardous Materials Agency, Aberdeen Proving Ground.
99. Yamashima, I., S. Shikata and F. Egami. 1954. Enzymatic reduction of aromatic nitro, nitroso, and hydroxyl amino compounds. Chem. Soc. Jap. Bull. 27:42-45.

9

PATHWAY OF 2,4,6-TRINITROTOLUENE (TNT) DEGRADATION BY *PHANEROCHAETE CHRYSOSPORIUM*

Jochen Michels and Gerhard Gottschalk

Institut für Mikrobiologie
Georg August-Universität Göttingen
Grisebach-straße 8
37077 Göttingen, Germany

INTRODUCTION

2,4,6-Trinitrotoluene (TNT) was the most important military explosive during World War II. It was released into the environment by contaminated wastewaters during production and by destruction of ammunition plants. It has become the predominant contaminant of soil and ground water at many munition production and processing sites in Germany and other countries. TNT is toxic to microorganisms (25), algae (55), and freshwater fish (34). Chronic exposure to TNT causes anemia (33) and toxic hepatitis to rats and humans (26). Its metabolic transformation products are also mutagenic in the Ames-Test (12).

The white rot fungus *Phanerochaete chrysosporium* is able to mineralize lignin (24) and a wide variety of recalcitrant xenobiotics. The fungus secretes an extracellular lignin degradation system including lignin peroxidases (48), manganese-dependent peroxidases (16), and extracellular hydrogen peroxide generating enzymes (13, 21, 22). The ligninolytic system has been reported to be involved in the degradation of organopollutants (5).

TNT can be very easily transformed by successive metabolic reduction of the three nitro groups. Under aerobic conditions aminodinitrotoluene and diaminonitrotoluene are formed (40). 2,4,6-Triaminotoluene additionally appears under anaerobic conditions and a low redox potential (4, 30, 38). TNT is considered to be resistant to oxidative attack by oxygenases (19).

P. chrysosporium is able to mineralize TNT (15) but seems to be inhibited by higher concentrations of the explosive (> 20 ppm) (44). Recently, the plasma membrane-dependent reduction of TNT by *P. chrysosporium* was reported and aminodinitrotoluene isomers were identified as the reduction products (46). With the exception of these initial reductive steps the pathway of TNT mineralization by the fungus is still unknown. In contrast the metabolism of 2,4-dinitrotoluene (DNT) by *P. chrysosporium* was elucidated recently (53).

Biodegradation of Nitroaromatic Compounds
Edited by J. Spain, Plenum Press, New York, 1995

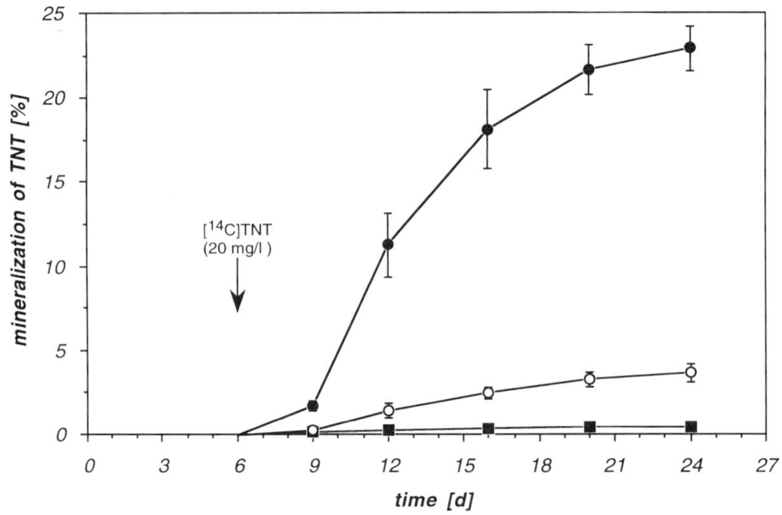

Figure 1. Release of $^{14}CO_2$ from ring-labeled [^{14}C]TNT by ligninolytic or non-ligninolytic cultures of *P. chrysosporium* (32). Shallow stationary cultures (7 ml) were incubated in 100 ml serum bottles for 6 days at 39°C according to (49). Then 91 nCi of ring-labeled [^{14}C]TNT was added (↓) to each culture to give a final concentration of 20 mg/liter. Bottles were flushed with O_2 and the $^{14}CO_2$ evolved was trapped. Data points represent the means ± standard deviation (n=5). Ligninolytic culture, ●; non-ligninolytic culture, O; autoclaved control, ■.

We have also studied the mineralization of TNT by white rot fungi (32). We shall focus here on two topics from this work, the pathway of TNT degradation by *P. chrysosporium* and the toxicity of TNT to the fungus.

PATHWAY OF TNT DEGRADATION BY *P. CHRYSOSPORIUM*

The production of extracellular peroxidases by *P. chrysosporium* is suppressed in media containing nitrogen and carbon sources in excess (49). The rate of TNT mineralization by cultures grown under these conditions is very low (Fig. 1). In defined low-nitrogen medium, manganese-dependent peroxidases and lignin peroxidases are produced and secreted; the activity of the latter usually peaks on the sixth day. It is apparent from Figure 1 that [^{14}C]TNT added on day 6 was mineralized significantly faster in ligninolytic cultures. We conclude from these results that ligninolytic enzymes are involved in the degradation of TNT. The mineralization of [^{14}C]TNT under nitrogen-insufficient (15, 45) and nitrogen-sufficient conditions (44) was reported previously.

The formation of early intermediates from TNT following its addition to 6-day-old agitated ligninolytic cultures is depicted in Figure 2. A few hours after the addition of TNT hydroxylaminodinitrotoluenes, and subsequently, aminodinitrotoluenes accumulated in the culture fluid. The identities of the hydroxylaminodinitrotoluenes and aminodinitrotoluenes were confirmed by mass spectroscopy. This initial reduction sequence did not require ligninolytic conditions (45), because hydroxylaminodinitrotoluenes and aminodinitrotoluenes could also be detected after addition of TNT to non-ligninolytic cultures. The isomers of hydroxylaminodinitrotoluenes could not be resolved by HPLC (32), but the aminodinitrotoluenes were resolved, and it could be demonstrated that both reduction products, 2-amino-4,6-dinitrotoluene and 4-amino-2,6-dinitrotoluene, were formed (data not shown).

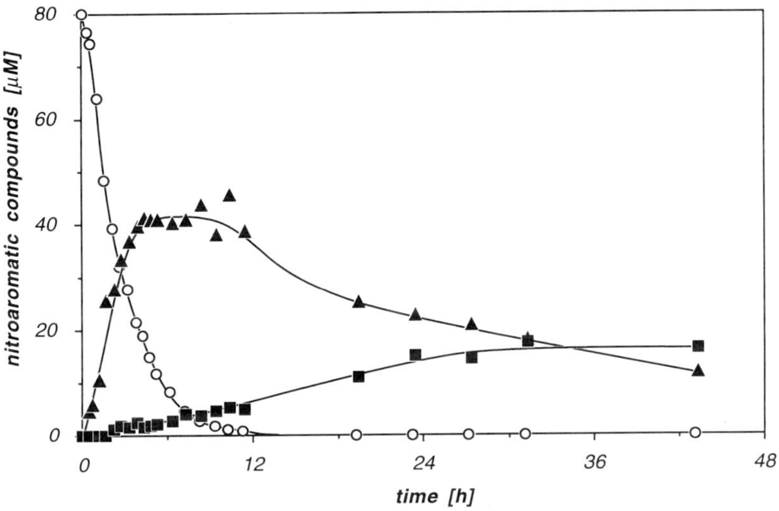

Figure 2. Transient accumulation of hydroxylaminodinitrotoluenes after the addition of TNT to ligninolytic cultures of *P. chrysosporium* (32). Agitated ligninolytic cultures (40 ml), grown as described previously (3), were incubated for 6 days at 37°C in 125-ml Erlenmeyer flasks before the addition of TNT. Product formation was followed by HPLC. TNT, O; hydroxylaminodinitrotoluene, both isomers, ▲; aminodinitrotoluene, both isomers, ■.

Under ligninolytic conditions a darkening of the mycelial pellets occurred at high manganese content (30 ppm) of the medium. This darkening resulted from the precipitation of MnO_2 (17, 36), a strong oxidant (51). When hydroxylaminodinitrotoluenes were added to darkened mycelia with MnO_2 precipites, oxidation products, azoxytetranitrotoluenes (11, 30), were immediately formed from hydroxylaminodinitrotoluenes (Table 1). They arose from the condensation of hydroxylaminodinitrotoluenes and nitrosodinitrotoluenes in a reaction analogous to the formation of azoxybenzene from phenylhydroxylamine and nitrosobenzene (20). After addition of hydroxylaminodinitrotoluene the mycelial mat decolorized rapidly, probably due to MnO_2 reduction. The same phenomenon could be observed following the addition of TNT to ligninolytic cultures with darkened mycelial mats, because hydroxylaminodinitrotoluenes are the first detectable reduction products of TNT. Therefore, it can be assumed that the extent of azoxytetranitrotoluene appearance during TNT degradation is dependent on the availability of MnO_2 in ligninolytic cultures of *P. chrysosporium*.

Following the identification of aminodinitrotoluenes as products formed from TNT by *P. chrysosporium*, we tested whether these compounds were further metabolized by the fungus. For that purpose [^{14}C]aminodinitrotoluenes were added to ligninolytic cultures, and the mineralization was compared with that of similar amounts of [^{14}C]TNT (Fig. 3). The results suggest that the aminodinitrotoluenes are not only products of TNT reduction but also intermediates in the TNT mineralization pathway. Aminodinitrotoluenes were mineralized by the fungus and the rates of mineralization were quite similar to those of TNT. This suggests also that azoxytetranitrotoluenes are not dead-end products if formed from TNT, and their appearance and disappearance in cultures could indeed be observed (data not shown). The degradation of azoxytetranitrotoluenes by *P. chrysosporium* is not unexpected, because the reduction of azoxy compounds to azo compounds is chemically easy and the mineralization of various azo-dyes by *P. chrysosporium* has been demonstrated (35, 43).

Table 1. Spontaneous oxidation of hydroxylaminodinitrotoluenes by MnO_2 precipitated on the mycelial mat of ligninolytic cultures of *P. chrysosporium*

culture conditions	hydroxylamino dinitrotoluenes	nitroso dinitrotoluenes	azoxy tetranitrotoluenes
ligninolytic (with MnO_2 precipitation)	12[a]	4	13
non ligninolytic (without MnO_2 precip.)	23	0	0

Hydroxylaminodinitrotoluene isomers were added to 6 days old shallow stationary cultures to a final concentration of 20 mg/l. Samples were taken one min after the disappearance of the brownish precipitate of MnO_2 from the mycelial mat of the ligninolytic culture.
[a]Integration unit of the peak area from the HPLC-chromatogram (no calibration was made for nitrosodinitrotoluene and azoxytetranitrotoluene).

These results led to the conclusion that the aminodinitrotoluenes are intermediates in TNT degradation and that the ligninolytic system was not required for this reduction as was also observed by Stahl and Aust (45). The reduction of TNT by *P. chrysosporium* proceeds stepwise (56) and hydroxylaminodinitrotoluenes are intermediates in addition to aminodinitrotoluenes (6, 32). Enzyme activity for the reduction of 2,4-dinitrotoluene by *P. chrysosporium* was mentioned by Valli et al. (53). According to our experiments with cell-free extracts of non-ligninolytic cultures of *P. chrysosporium* the activity is intracellular

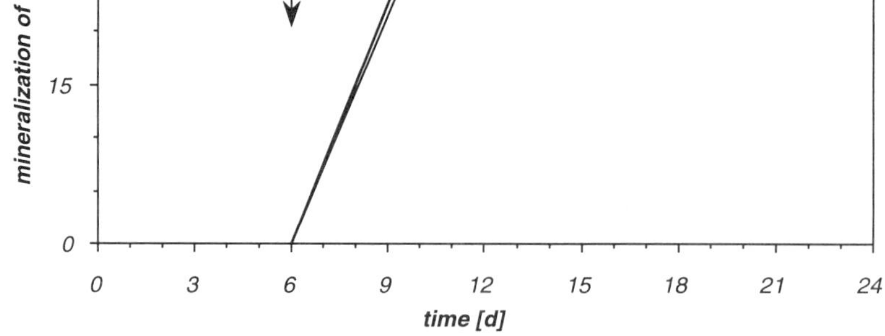

Figure 3. Comparison of the mineralization of ring-labeled [^{14}C]TNT and [^{14}C]aminodinitrotoluenes by ligninolytic cultures of *P. chrysosporium* (32). Shallow stationary cultures (7 ml) were incubated for 6 days at 39°C in 100 ml serum bottles. Then 0.36 mg/liter of [^{14}C]TNT or 0.31 mg/liter of [^{14}C]aminodinitrotoluene was added to each culture (↓). Bottles were flushed with O_2 and $^{14}CO_2$ evolved was trapped every three to four days. Data points represent the means ± standard deviation (n=3). [^{14}C]TNT, □; [^{14}C]aminodinitrotoluenes, ■.

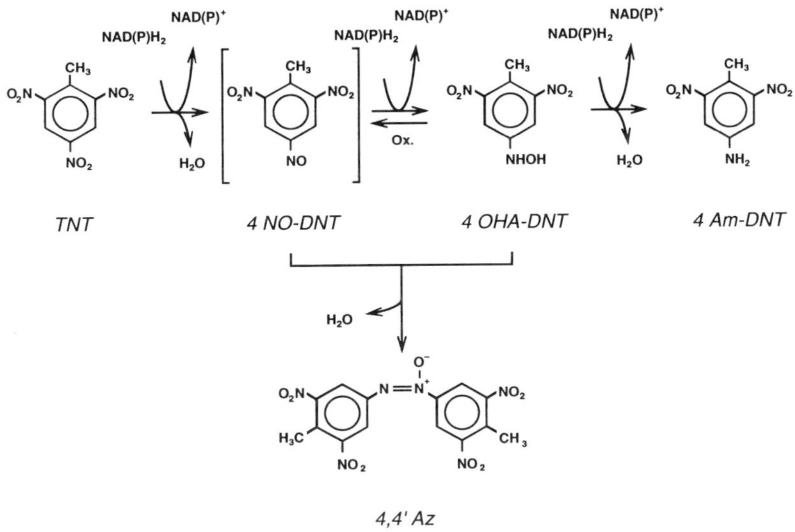

Figure 4. Reduction of TNT to 4-amino-2,6-dinitrotoluene (4Am-DNT). 4-Nitroso-2,6-dinitrotoluene (4NO-DNT) could not be detected in the reduction of TNT. 4-Hydroxylamino-2,6-dinitrotoluene (4OHA-DNT) is the first detectable intermediate in the reduction course of TNT by *P. chrysosporium*. 4,4'-Azoxy-2,2',4,4'-tetranitrotoluene is the spontaneous condensation product of 4OHA-DNT and 4NO-DNT. It is assumed that the latter is generated from 4OHA-DNT by an oxidative force such as manganese dioxide in the culture fluid.

and dependent on NADH and NADPH (unpublished results)(Fig. 4). Recently, Stahl and Aust (46) reported that reduction of TNT was dependent on intact mycelia of *P. chrysosporium*. They proposed the involvement of a membrane-bound redox system that reduces TNT to aminodinitrotoluenes in a six electron step. Since hydroxylaminodinitrotoluenes could be found to accumulate in the supernatant, the stepwise reduction of TNT seems to be predominant.

Since ligninolytic activity was required for extensive mineralization of TNT, the accumulation of substrates of the ligninolytic enzymes originating from aminodinitrotoluenes was expected under non-ligninolytic conditions. Nitrogen-sufficient (12 mM ammonium tartrate) agitated cultures were incubated with 4-amino-2,6-dinitrotoluene for 48 h, and a transformation product was extracted, purified by preparative thin layer chromatography and analyzed by mass spectroscopy. By comparison with chemically synthesized reference compounds it was identified as 4-formamido-2,6-dinitrotoluene.

The experiment was then repeated with chemically synthesized 4-formamido-2,6-dinitrotoluene instead of 4-amino-2,6-dinitrotoluene. The reduction product isolated was identified as 2-amino-4-formamido-6-nitrotoluene. Using this compound as substrate, very slow formation of 2,4-diamino-6-nitrotoluene (detectable only after 96 h of incubation) could be observed. The sequence of reactions that follows 4-amino-2,6-dinitrotoluene is depicted in Figure 5. We could not obtain any evidence for a direct reduction of 4-amino-2,6-dinitrotoluene to 2,4-diamino-6-nitrotoluene.

Our next goal was to find out which compound serves as the substrate for the ligninolytic system. It is apparent from Figure 6B that 2-amino-4-formamido-6-nitrotoluene disappeared rapidly in ligninolytic cultures, but not in non-ligninolytic cultures (Fig. 6A).

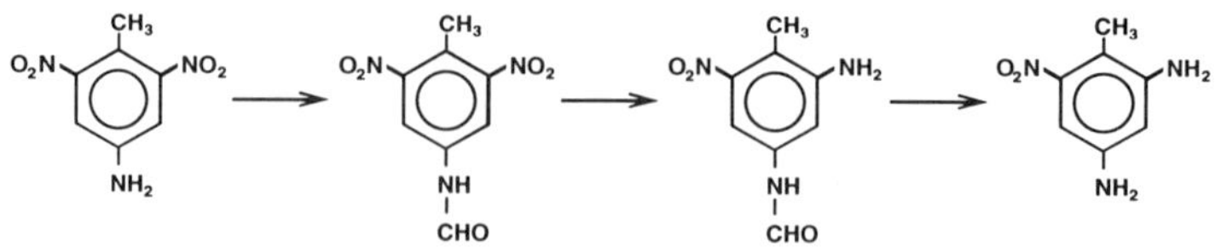

Figure 5. Reduction of 4-amino-2,6-dinitrotoluene (4Am-DNT) to 2,4-diamino-6-nitrotoluene (2,4DAm-NT) by non-ligninolytic cultures of *P. chrysosporium*. The formylation of the amino group of 4Am-DNT led to 4-formamido-2,6-dinitrotoluene (4FAm-DNT). The latter is reduced to 2-amino-4-formamido-6-nitrotoluene (2Am4FAm-NT) which is subsequently deformylated.

Therefore, we assume that Stahl and Aust (45) detected the diamino-derivative only in non-ligninolytic cultures because its precursor 2-amino-4-formamido-6-nitrotoluene is preferentially oxidized by the ligninolytic enzymes. The purified lignin peroxidase isozyme H8 was used to confirm that 2-amino-4-formamido-6-nitrotoluene serves as a substrate for this enzyme. In contrast to the oxidation of hydroxylaminodinitrotoluene by the lignin peroxidase (6, 32) veratryl alcohol is not involved in this reaction.

These results provide some insight into the pathway of TNT degradation by *P. chrysosporium*. To our knowledge this is the first description of the formylation of an aromatic amino group by *P. chrysosporium*. Acylation of amino groups by white rot fungi has been reported previously. For example, aniline is acetylated and formylated by *Trametes versicolor* (41) presumably in a kind of detoxification reaction (52). We could not find the acetylated derivative in cultures of *P. chrysosporium* incubated with 4-amino-2,6-dinitrotoluene even though an authentic standard was available for comparison.

The mineralization of 2,4-dinitrotoluene by *P. chrysosporium* also started with the reduction of one nitro group leading to either 2-amino-4-nitrotoluene or 4-amino-2-nitrotoluene (53), which was analogous to the initial reduction of TNT. 2-Amino-4-nitrotoluene was a substrate for the manganese-dependent peroxidase and was converted to 4-nitro-1,2-benzoquinone. The authors did not investigate the metabolic fate of 4-amino-2-nitrotoluene but assumed that the reactions involved in its degradation are similar. No formylated products were described (53). But it is possible that the transformation of aminonitrotoluenes by the manganese-dependent peroxidase occurs faster than a possible formylation.

Acetylation of 2-nitro-4-aminotoluene by *Mucrosporium* sp. was observed in the course of 2,4-dinitrotoluene degradation (31). The isomer 2-acetamido-4-nitrotoluene could not be detected as a transformation product. This indicates that acetylation of an aromatic amino group is affected by its position on the ring. In this case the methyl residue *ortho* to the amino group could hinder the acetylation of 2-amino-4-nitrotoluene. Since our attempts to synthesize an authentic standard of 2-formamido-4,6-dinitrotoluene have failed so far, we could not investigate a formylation reaction of 2-amino-4,6-nitrotoluene by the fungus.

The degradation of another aromatic amine, 4-chloroaniline, by *P. chrysosporium* was recently investigated by Chang and Bumpus (10). Although no steric hindrance is expected, acylation of the amino group of 4-chloroaniline was not observed. Formylation and acetylation of 4-chloroaniline is possible as was shown with *Streptomyces* sp. (39). Like 2-amino-4-nitrotoluene, 4-chloroaniline is a substrate for the ligninolytic system. Lignin peroxidase isozyme H2 is able to oxidize 4-chloroaniline to various

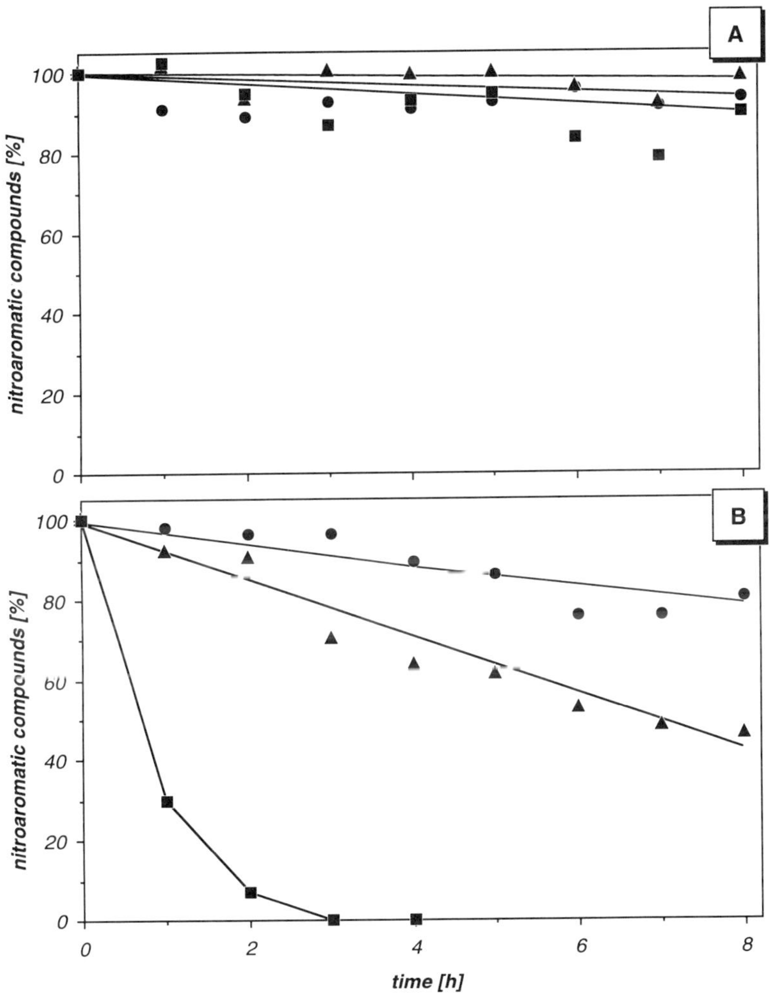

Figure 6. Further degradation of the formylated intermediates in TNT breakdown by non-ligninolytic (A) or ligninolytic cultures (B) of *P. chrysosporium*. Shallow stationary cultures (7 ml) were incubated for 6 days at 39°C in 100 ml serum bottles. The intermediates were added to separate cultures to a final concentration of 20 μg/ml. 4-Amino-2,6-dinitrotoluene, ●; 4-formamido-2,6-dinitrotoluene, ▲; 2-amino-4-formamido-6-nitrotoluene, ■.

products (10). Presumably, oxidation of 4-chloroaniline rather than acylation is the dominant reaction.

The identification of 2-amino-4-formamido-6-nitrotoluene as a substrate of the lignin peroxidase raises questions about the role of 2,4-diamino-6-nitrotoluene. Its formation from 2-amino-4-formamido-6-nitrotoluene under non-ligninolytic conditions can be expected, since acylation reactions (formylation and acetylation) of aromatic amino goups are clearly reversible (29).

Research is now being conducted to identify the oxidation products of 2-amino-4-formamido-6-nitrotoluene. Also the metabolic fate of 2,4-diamino-6-nitrotoluene under non-ligninolytic conditions is of interest, since some non-ligninolytic mineralization of TNT by *P. chrysosporium* was demonstrated (32, 44).

INHIBITION OF THE VERATRYL ALCOHOL OXIDASE ACTIVITY OF LIGNIN PEROXIDASE H8 BY HYDROXYLAMINODINITROTOLUENE

It was an unexpected finding in our work on the degradation of TNT that aminodinitrotoluene, at concentrations above 25 μM, was mineralized by ligninolytic cultures to a higher extent than TNT at comparable concentrations (Fig. 7) (32). While mineralization of aminodinitrotoluene was more or less proportional to the amount added, TNT mineralization stagnated at a constant level. The rate of TNT-degradation was apparently affected by the transient accumulation of hydroxylaminodinitrotoluene which inhibits the veratryl alcohol oxidase activity of the lignin peroxidases, as was independently reported by Bumpus and Tatarko (6) (Fig. 8). Neither TNT itself nor the two isomers of aminodinitrotoluene affected the lignin peroxidase in vitro (6, 32).

Inhibition of lignin peroxidase by hydroxylaminodinitrotoluenes was studied using purified lignin peroxidase H8 (49). In the presence of hydroxylaminodinitrotoluene the oxidation of veratryl alcohol was strongly inhibited for some time. The duration of the lag phase was dependent on the concentration of hydroxylaminodinitrotoluene in the assay mixture (Fig. 9). However, the veratryl alcohol oxidase activity returned, without a pronounced transition, following this lag period. Furthermore, the veratryl alcohol oxidase activity that reappeared was also dependent on the initial concentration of hydroxylaminodinitrotoluene in the test assay. In other words, the inhibition was only partially reversible. After separation of the two isomers 2-hydroxylamino-4,6-dinitrotoluene and 4-hydroxylamino-2,6-dinitrotoluene by preparative TLC using the solvent system ethyl acetate:hexane (1:1, vol/vol), the effect of each isomer on the purified lignin peroxidase H8 was tested separately. Because the inhibition pattern of the isolated isomers are similar, all inhibition experiments were carried out with the mixture of both isomers.

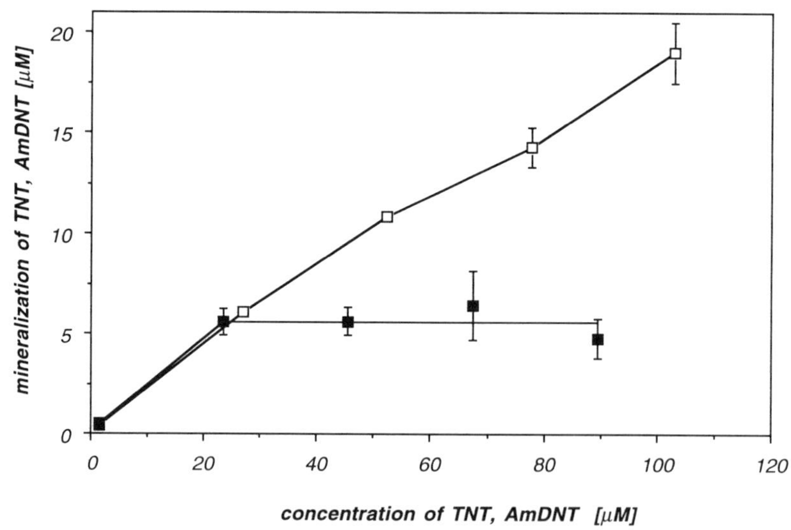

Figure 7. Comparison of the mineralization of increasing concentrations of ring-labeled [^{14}C]TNT and [^{14}C]aminodinitrotoluenes by ligninolytic cultures of *P. chrysosporium*. Shallow stationary cultures (7 ml) were incubated for 6 days at 39°C in 100-ml serum bottles. Then [^{14}C]TNT or [^{14}C]aminodinitrotoluene was added to the cultures. After 72 h bottles were flushed with O_2 and $^{14}CO_2$ evolved was trapped. Data points represent the means ± standard deviation (n=3). [^{14}C]TNT, ■; [^{14}C]aminodinitrotoluene (both isomers), □.

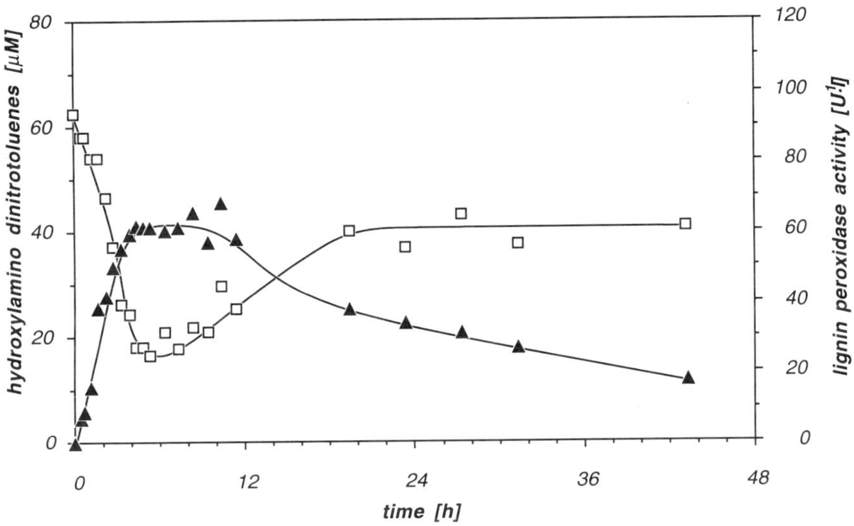

Figure 8. Correlation of the inhibition of the lignin peroxidase with the transient accumulation of hydroxylaminodinitrotoluene following the addition of TNT. The activity of the enzyme was measured according to (49). Lignin peroxidase activity, □; concentration of hydroxylaminodinitrotoluene, ▲.

As previously reported, oxidation products of hydroxylaminodinitrotoluene could be extracted from the reaction mixture; azoxytetranitrotoluene (32) and TNT (6) were identified by chromatographic techniques and UV absorption spectroscopy. Oxidation of hydroxylaminodinitrotoluene occurred only if lignin peroxidase, H_2O_2, and veratryl alcohol were present in the test mixture. Hydroxylaminodinitrotoluene alone is, therefore, a poor substrate for the lignin peroxidase and veratryl alcohol is necessary for its oxidation. But hydroxylamino-

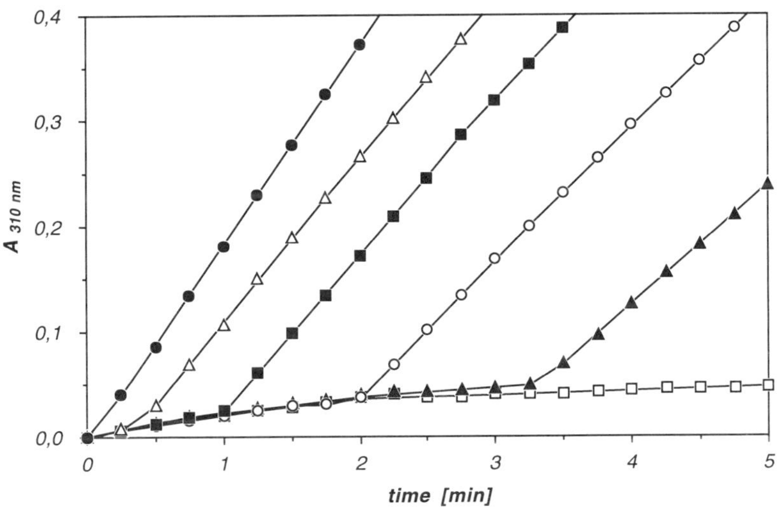

Figure 9. Effect of increasing amounts of hydroxylaminodinitrotoluenes on the veratryl alcohol oxidase activity of lignin peroxidase H8 (32). The reaction mixture contained in a final volume of 600 µl: 0.05 µM lignin peroxidase H8, 40 mM sodium tartrate buffer, pH 3.5, 130 µM H_2O_2, 5.6 mM veratryl alcohol, and various concentrations of hydroxylaminodinitrotoluene (both isomers) [µM]: 0, ●; 5,Δ; 10, ■; 16, O; 21, ▲; 24, □.

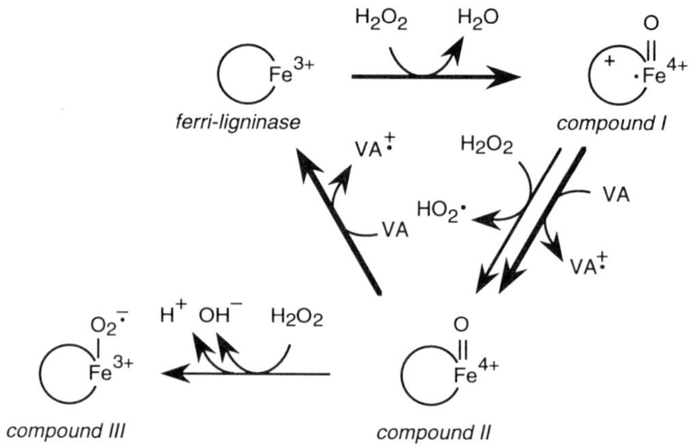

Figure 10. The catalytic cycle of the lignin peroxidase (Wariishi and Gold (54), modified). The thick arrows denote the catalytic cycle in the presence of H_2O_2 and a reducing substrate like veratryl alcohol. $VA^{+\cdot}$ = veratryl alcohol cation radical; $O_2^{\dot{-}}$ = superoxide radical.

dinitrotoluene is able to protect the enzyme from inactivation by H_2O_2 (32). This was also shown for cyanide, another inhibitor of the veratryl alcohol-oxidizing activity of the lignin peroxidase (42). The protection of peroxidases by reducing substrates from inactivation caused by H_2O_2 has been previously reported (2).

The catalytic properties of the lignin peroxidases have been reviewed by Tien (47). The lignin peroxidases of *P. chrysosporium* possess a heme as prosthetic group (27, 48) and the catalytic cycle is comparable to other peroxidases such as horseradish peroxidase (47). In Figure 10 the catalytic cycle is indicated by the thick arrows. The resting enzyme "ferri-ligninase" is activated by H_2O_2 in a two-electron step to yield compound I. A reducing substrate like veratryl alcohol is then oxidized to the veratryl alcohol radical cation ($VA^{+\cdot}$) by compound I yielding compound II. The resting enzyme is regenerated by oxidizing a second substrate molecule. With an excess of H_2O_2 the catalytically inactive compound III will be formed (thin arrows). By releasing a superoxide radical, compound III reverts spontaneously to the native enzyme (8, 54) (not shown in Figure 10). In the absence of a reducing substrate lignin peroxidase is irreversibly inactivated by H_2O_2 (47).

Veratryl alcohol plays an important role as a radical cation mediator in lignin biodegradation (18). The oxidative decarboxylation of oxalate (1) or EDTA (42) by the lignin peroxidase is mediated by the veratryl alcohol cation radical ($VA^{+\cdot}$) leading to an apparent inhibition of the veratryl alcohol oxidase activity of the lignin peroxidase.

The effect of the initial concentration of veratryl alcohol on the duration of the lag period during oxidation of hydroxylaminodinitrotoluene was investigated (Fig. 11). The length of the lag period was directly related to the veratryl alcohol concentration, up to 2.8 mM, thus indicating its role as a radical cation mediator in the oxidation of hydroxylamino-dinitrotoluene. Surprisingly, the activity of the lignin peroxidase that reappeared was also affected by the concentration of veratryl alcohol. The rates were lower when veratryl alcohol was below 1.1 mM (Fig. 11). A K_m value of 167 μM was determined for veratryl alcohol and lignin peroxidase H2 at pH 3.5 (50).

An unexpected result was the effect of H_2O_2 on the lag phase and the lignin peroxidase activity at constant levels of veratryl alcohol and inhibitor (Fig. 12). An increase of the concentration of H_2O_2 from 64 μM to 512 μM in the assay mixture led to an extension of the lag period. At lower H_2O_2-concentrations, down to 13 μM, the veratryl alcohol oxidase

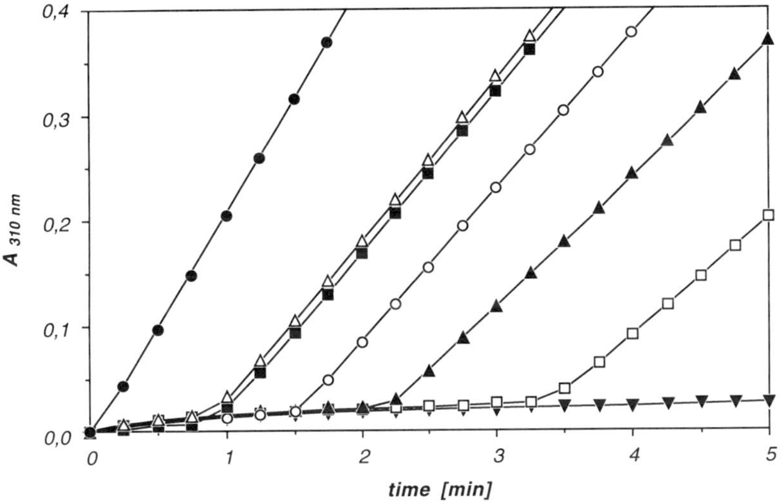

Figure 11. Effect of decreasing concentrations of veratryl alcohol on veratryl alcohol oxidase activity of lignin peroxidase H8 during inhibition by hydroxylaminodinitrotoluenes (32). The reaction mixture contained in a final volume of 600 µl: 0.05 µM lignin peroxidase H8, 40 mM sodium tartrate buffer pH 3.5, 260 µM H_2O_2, 10 µM hydroxylaminodinitrotoluenes (both isomers) and various concentrations of veratryl alcohol [mM]: 11.4, Δ; 2.8, ■; 1.7, O; 1.1, ▲; 0.9, □; 0.6, ▼; control (without hydroxylaminodinitrotoluene) 1.1 mM, ●.

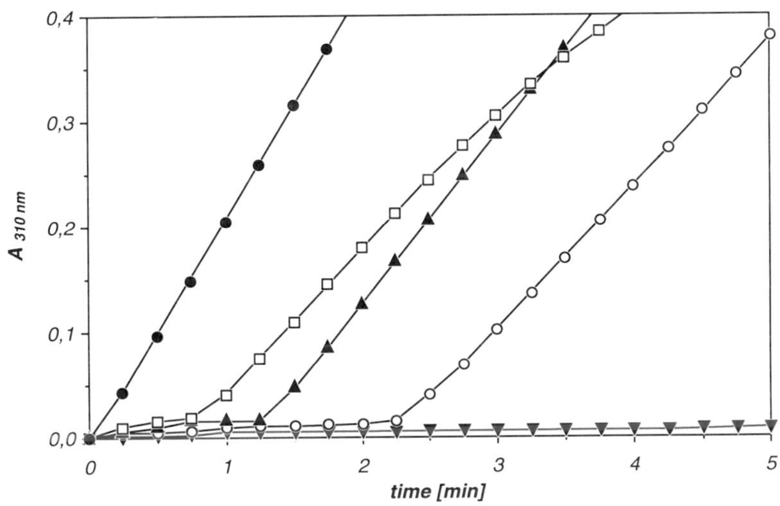

Figure 12. Effect of increasing concentrations of hydrogen peroxide on the veratryl alcohol oxidase activity of lignin peroxidase H8 during inhibition by hydroxylaminodinitrotoluenes. The reaction mixture contained in a final volume of 600 µl: 0.05 µM lignin peroxidase H8, 40 mM sodium tartrate buffer pH 3.5, 1.1 mM veratryl alcohol, 10 µM hydroxylaminodinitrotoluenes (both isomers) and various concentrations of H_2O_2 [µM]: 64, □; 128, ▲; 256, O; 512, ▼; control (without hydroxylaminodinitrotoluene) 256 µM, ●.

activity that reappeared was strongly affected by the limitation of H_2O_2. Although some H_2O_2 is required, because it participates in the catalytic cycle of the lignin peroxidase, the oxidation of hydroxylaminodinitrotoluene is inhibited by increasing concentrations of H_2O_2. Oxidative inactivation of the enzyme by H_2O_2 is not a sufficient explanation for this effect, since veratryl alcohol and hydroxylaminodinitrotoluene can protect the lignin peroxidase from this kind of inactivation (32). Compound II could be partially withdrawn from the catalytic cycle by H_2O_2 (see Figure 10), especially under conditions in which the concentration of veratryl alcohol radical cation is low. The latter has been proposed to convert compound III back to the ferric state (7).

TOXICITY OF HYDROXYLAMINODINITROTOLUENE

Hydroxylaminodinitrotoluene is responsible for the hemotoxic symptoms observed after exposure of ammunition workers to TNT (26). Oxyhemoglobin is oxidized by hydroxylaminodinitrotoluene to form ferrihemoglobin and nitrosodinitrotoluene (23). In red blood cells this reaction becomes a futile cycle, since nitrosodinitrotoluene can be reduced back to hydroxylaminodinitrotoluene which can react again with oxyhemoglobin (23). Furthermore, as demonstrated with radioactive nitrosobenzene, covalent binding to different amino acid residues of the globin moiety is part of the toxic effect of arylhydroxylamines (14, 23, 33). The predominant and best examined reaction is the binding of arylhydroxylamines to SH-groups of the globin moiety forming hemoglobin-cysteine-sulphinanilides (14, 33).

Since compound III is the dioxygen complex of ferrous peroxidase, it has similarities to oxymyoglobin and oxyhemoglobin (9, 54). From that point of view it can be proposed that compound III and hydroxylaminodinitrotoluene co-oxidize to yield ferric enzyme and nitrosodinitrotoluene (Fig. 13). Therefore, two independent mechanisms may exist to oxidize hydroxylaminodinitrotoluene: one via the catalytic cycle mediated by veratryl alcohol, and the second via oxidation by compound III.

An enzymatically inactive complex may be formed by reaction of nitrosodinitrotoluene with certain amino acid residues of the enzyme. A binding of nitrosodinitrotoluene to cysteine residues seems unlikely, since all eight cysteine residues of lignin peroxidase isozyme H2 are linked through disulfide bridges (37). The toxic effect of hydroxylaminodinitrotoluene on living cells of *P. chrysosporium* might be explained by a similar mechanism. In red blood cells nitrosobenzene is covalently bound to hemoglobin, other enzymes, and membrane constituents resulting in a disturbance of the normal cellular function of these cells and finally in hemolysis (14).

CONCLUSIONS

The results reviewed here provide a reasonable explanation for the toxic effect of TNT on *P. chrysosporium*. Toxicity originates from the accumulation of hydroxylaminodinitrotoluene which interferes with veratryl alcohol oxidase activity. Radioactive binding experiments to test this model are underway. It also is clear now that the pathway of TNT degradation involves an initial reduction to aminodinitrotoluene. Uncontrolled oxidation of 4-amino-2,6-dinitrotoluene is prevented by formylation of the first amino group generated on the TNT molecule. Oxidation reactions, which yet have to be elucidated in detail, occur only after a second amino group is formed on the aromatic ring. Furthermore, the fate of 2-amino-4,6-dinitrotoluene must be examined. It may be first reduced to 2,4-diamino-6-nitrotoluene and then formylated. Until now no evidence in support of this has been obtained.

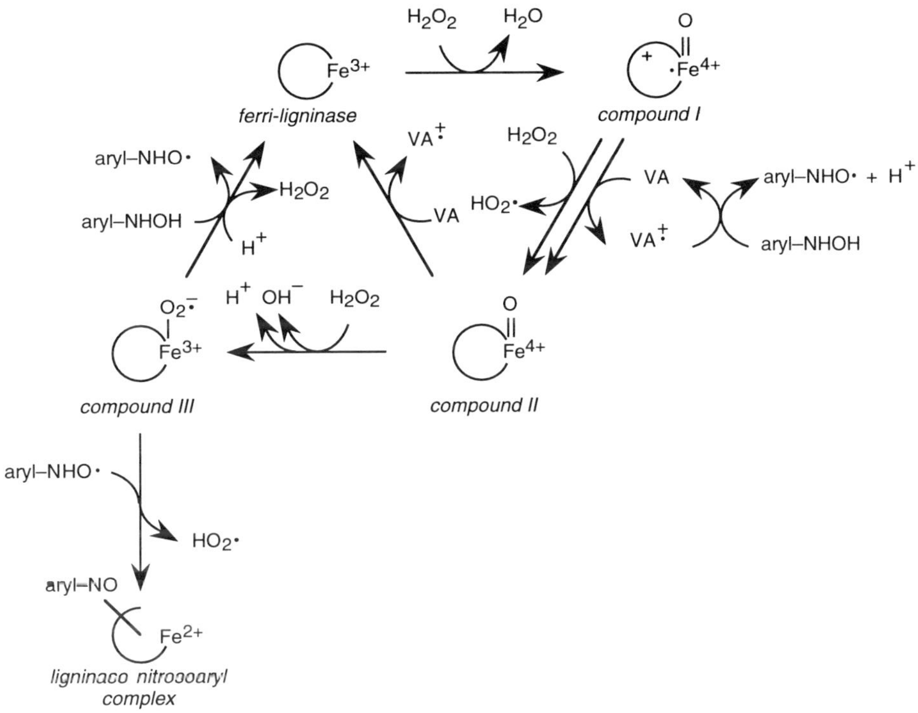

Figure 13. Proposed mechanism of the inhibition of the lignin peroxidase (LiP) by hydroxylaminodinitrotoluenes. Compound III is co-oxidized with hydroxylaminodinitrotoluene to ferri-ligninase and nitrosodinitrotoluene (33). The (catalytically inactive) ligninase-nitrosoaryl-complex results from the reaction of nitrosodinitrotoluene with various amino acid residues of the enzyme. The radical Aryl-NHO· can either disproportionate spontaneously (a) or react with compound III (b).

$$2\ \text{aryl-NHO}^{\cdot} \leftrightarrow \text{aryl-NHOH} + \text{aryl-NO} \rightarrow \text{azoxyaryl} \qquad \text{(a)}$$

$$\text{LiP}^{3+}\text{-O}_2^{-} + \text{aryl-NHO}^{\cdot} \rightarrow \text{LiP}^{2+}\text{-aryl-NO} + \text{O}_2^{-} + \text{H}^{+} \qquad \text{(b)}$$

Since (a) and (b) compete for Aryl-NHO· the extent of reaction (b) is dependent on the concentration of LiP^{3+}-O_2^- (compound III).

The knowledge of some key intermediates of TNT degradation by *P. chrysosporium* will facilitate the investigation of fungal species with are better adapted to soil and to high TNT-concentrations and more suitable for bioremediation.

REFERENCES

1. Akamatsu, Y., D. B. Ma, T. Higuchi, and M. Shimada. 1990. A novel enzymatic decarboxylation of oxalic acid by the lignin peroxidase system of white-rot fungus *Phanerochaete chrysosporium*. FEBS Lett. 269:261-263.
2. Arnao, M. B., M. Acosta, J. A. del Río, and F. García-Cánovas. 1990. Inactivation of peroxidase by hydrogen peroxide and its protection by a reductant agent. Biochim. Biophys. Acta. 1038:85-89.
3. Bonnarme, P. and T. W. Jeffries. 1990. Mn(II) regulation of lignin peroxidases and manganese-dependent peroxidases from lignin-degrading white rot fungi. Appl. Environ. Microbiol. 56:210-217.
4. Boopathy, R., C. F. Kulpa, and M. Wilson. 1993. Metabolism of 2,4,6-trinitrotoluene (TNT) by *Desulfovibrio* sp. (B-strain). Appl. Microbiol. Biotechnol. 39:270-275.

5. Bumpus, J. A., and S. D. Aust. 1986. Biodegradation of environmental pollutants by the white rot fungus *Phanerochaete chrysosporium*: involvement of the lignin degrading system. BioEssays 6:166-170.
6. Bumpus, J. A., and M. Tatarko. 1994. Biodegradation of 2,4,6-trinitrotoluene by *Phanerochaete chrysosporium*: Identification of initial degradation products and the discovery of a TNT metabolite that inhibits lignin peroxidases. Curr. Microbiol. 28:185-190.
7. Cai, D., and M. Tien. 1989. On the reactions of lignin peroxidase compound III (Isozyme H8). Biochem. Biophys. Res. Commun. 162:464-469.
8. Cai, D., and M. Tien. 1990. Characterization of the oxycomplex of lignin peroxidases from *Phanerochaete chrysosporium*: equilibrium and kinetics studies. Biochemistry 29:2085-2091.
9. Cai, D., and M. Tien. 1992. Kinetic studies on the formation and decomposition of compound-II and compound-III - reactions of lignin peroxidase with H_2O_2. J. Biol. Chem. 267:11149-11155.
10. Chang, C.-W., and J. A. Bumpus. 1993. Oligomers of 4-chloroaniline are intermediates formed during its biodegradation by *Phanerochaete chrysosporium*. FEMS Microbiol. Lett. 107:337-342.
11. Channon, H. J., G. T. Mills, and R. T. Williams. 1944. The metabolism of 2,4,6-trinitrotoluene (α-TNT). Biochem. J. 38:70-85.
12. Einistö, P. 1991. Role of bacterial nitroreductase and *O*-acetyltransferase in urine mutagenicity assay of rats exposed to 2,4,6-trinitrotoluene (TNT). Mutat. Res. 262:167-169.
13. Eriksson, K.-E., B. Pettersson, J. Volc, and V. M. Musilek. 1986. Formation and partial characterisation of glucose 2-oxidase, a H_2O_2 producing enzyme in *Phanerochaete chrysosporium*. Appl. Microbiol. Biotechnol. 23:257-262.
14. Eyer, P., and E. Lierheimer. 1980. Biotransformation of nitrosobenzene in the red cell and the role of glutathione. Xenobiotica 10:517-526.
15. Fernando, T., J. A. Bumpus, and S. D. Aust. 1990. Biodegradation of TNT (2,4,6-trinitrotoluene) by *Phanerochaete chrysosporium*. Appl. Environ. Microbiol. 56:1666-1671.
16. Glenn, J. K., and M. H. Gold. 1985. Purification and characterization of an extracellular Mn(II)-dependent peroxidase from lignin-degrading basidiomycete *Phanerochaete chrysosporium*. Arch. Biochem. Biophys. 244:329-341.
17. Glenn, J., L. Akileswaran, and M. Gold. 1986. Mn(II) oxidation is the principal function of the extracellular Mn-peroxidase from *Phanerochaete chrysosporium*. Arch. Biochem. Biophys. 251:688-696.
18. Harvey, P. J., H. E. Schoemaker, and J. M. Palmer. 1986. Veratryl alcohol as a mediator and the role of radical cations in lignin biodegradation by *Phanerochaete chrysosporium*. FEBS Lett. 195:242-246.
19. Higson, F. K. 1992. Microbial degradation of nitroaromatic compounds. Adv. Appl. Microbiol. 37:1-19.
20. Kaufman, D. D., J. R. Plimmer, J. Iwan, and U. I. Klingebiel. 1972. 3,3',4,4',-Tetrachloroazoxybenzene from 3,4-dichloroaniline in microbial culture. J. Agr. Food. Chem. 20:916-919.
21. Kelley, R. L., and C. A. Reddy. 1986. Identification of glucose oxidase activity as the primary source of hydrogen peroxide production in ligninolytic cultures of *Phanerochaete chrysosporium*. Arch. Microbiol. 144:248-253.
22. Kersten, P. J., and T. K. Kirk. 1987. Involvement of a new enzyme, glyoxal oxidase, in extracellular H_2O_2 production by *Phanerochaete chrysosporium*: synthesized in the absence of lignin in response to nitrogen starvation. J. Bacteriol. 135:790-797.
23. Kiese, M., and K. Taeger. 1976. The fate of phenylhydroxylamine in human red cells. Naunyn-Schmiedeberg's Arch. Pharmacol. 292:59-66.
24. Kirk, T. K., E. Schulz, W. J. Connors, L. F. Lorenz, and J. G. Zeikus. 1978. Influence of culture parameters on lignin metabolism by *Phanerchaete chrysosporium*. Arch. Microbiol. 117:277-285.
25. Klausmeier, R. E., J. L. Osmon, and D. R. Walls. 1973. The effect of trinitrotoluene on microorganisms. Dev. Ind. Microbiol. 15:309-317.
26. Koss, G., A. Lommel, I. Ollroge, I. Tesseraux, R. Haas, and A. D. Kappos. 1989. Zur Toxikologie der Nitrotoluole und weiterer Nitroaromaten aus rüstungsbedingten Altlasten. Bundesgesundhbl. 32:527-536.
27. Kuwahara, M., J. K. Glenn, M. A. Morgan, and M. H. Gold. 1984. Separation and characterization of two extracellular H_2O_2-dependent oxidases from ligninolytic cultures of *Phanerochaete chrysosporium*. FEBS Lett. 169:247-250.
28. Lenk, W., and M. Riedl. 1989. N-Hydroxy-N-arylacetamides. V. Differences in the mechanism of haemoglobin oxidation in vitro by N-hydroxy-4-chloroacetanilide and N-hydroxy-4-chloroaniline. Xenobiotica 19:453-475.
29. Lyons, C. D., S. Katz, and R. Bartha. 1984. Mechanisms and pathways of aniline elimination from aquatic environments. Appl. Environ. Microbiol. 48:491-496.
30. McCormick, N. G., F. E. Feeherry, and H. S. Levinson. 1976. Microbial transformation of 2,4,6-trinitrotoluene and other nitroaromatic compounds. Appl. Environ. Microbiol. 31:949-958.

31. McCormick, N. G., J. H. Cornell, and A. M. Kaplan. 1978. Identification of biotransformation products from 2,4-dinitrotoluene. Appl. Environ. Microbiol. 35:945-948.

32. Michels, J., and G. Gottschalk. 1994. Inhibition of the lignin peroxidase of *Phanerochaete chrysosporium* by hydroxylaminodinitrotoluene, an early intermediate in the degradation of 2,4,6-trinitrotoluene. Appl. Environ. Microbiol. 60:187-194.

33. Neumann, H. G. 1988. Biomonitoring of aromatic amines and alkylating agents by measuring hemoglobin adducts. Int. Arch. Occup. Environ. Health 60:151-155.

34. Osmon, J. L., and R. E. Klausmeier. 1972. The microbial degradation of explosives. Dev. Ind. Microbiol. 14:247-252.

35. Pasti-Grigsby, M. B., A. Paszczynski, S. Goszczynski, D. L. Crawford, and R. L. Crawford. 1992. Influence of aromatic substitution patterns on azo dye degradability by *Streptomyces* spp. and *Phanerochaete chrysosporium*. Appl. Environ. Microbiol. 58:3605-3613.

36. Perez, J., and T. W. Jeffries. 1992. Roles of manganese and organic acid chelators in regulating lignin degradation and biosynthesis of peroxidases by *Phanerochaete chrysosporium*. Appl. Environ. Microbiol. 58:2402-2409.

37. Poulos, T. L., S. L. Edwards, H. Wariishi, and H. Gold. 1993. Crystallographic refinement of lignin peroxidase at 2 Å. J. Biol. Chem. 268:4429-4440.

38. Preuss, A., J. Fimpel, and G. Diekert. 1993. Anaerobic transformation of 2,4,6-trinitrotoluene (TNT). Arch. Microbiol. 159:345-353.

39. Russel, S., and Bollag, J. M. 1977. Formylation and acetylation of 4-chloroaniline by a *Streptomyces* sp. Acta Microbiol. Pol. 26:59-64.

40. Schackmann, A., and R. Müller. 1991. Reduction of nitroaromatic compounds by different *Pseudomonas* species under aerobic conditions. Appl. Microbiol. Biotechnol. 34:809-813.

41. Scheibner, K., T. Günther, and W. Fritsche. 1993. Comparison of aniline metabolism by white rot fungi and autochthonous soil fungi. VAAM Frühjahrstagung, Leipzig, Germany, Poster No. P244.

42. Shah, M. M., T. A. Grover, and S. D. Aust. 1991. Metabolism of cyanide by *Phanerochaete chrysosporium*. Arch. Biochem. Biophys. 290:173-178.

43. Spadaro, J. T., M. H. Gold, and V. Renganathan. 1992. Degradation of azo dyes by the lignin-degrading fungus *Phanerochaete chrysosporium*. Appl. Environ. Microbiol. 58:2397-2401.

44. Spiker, J. K., D. L. Crawford, and R. L. Crawford. 1992. Influence of 2,4,6-trinitrotoluene (TNT) concentration on the degradation of TNT in explosive-contaminated soils by the white rot fungus *Phanerochaete chrysosporium*. Appl. Environ. Microbiol. 58:3199-3202.

45. Stahl, J. D., and S. D. Aust. 1993. Metabolism and detoxification of TNT by *Phanerochaete chrysosporium*. Biochem. Biophys. Res. Commun. 192:477-482.

46. Stahl, J. D., and S. D. Aust. 1993. Plasma membrane dependent reduction of 2,4,6-trinitrotoluene by *Phanerochaete chrysosporium*. Biochem. Biophys. Res. Commun. 192:471-476.

47. Tien, M. 1987. Properties of ligninases from *Phanerochaete chrysosporium* and their possible applications. Crit. Rev. Microbiol. 15:141-168.

48. Tien, M., and T. K. Kirk. 1984. Lignin-degrading enzyme from *Phanerochaete chrysosporium*: purification, characterization and catalytic properties of a unique H_2O_2-requiring oxygenase. Proc. Natl. Acad. Sci. USA 81:2280-2284.

49. Tien, M., and T. K. Kirk. 1988. Lignin peroxidase of *Phanerochaete chrysosporium*. Methods Enzymol. 161:238-249.

50. Tuisel, H., R. Sinclair, J. A. Bumpus, W. Ashbaugh, B. J. Brock, and S. D. Aust. 1990. Lignin peroxidase H2 from *Phanerochaete chrysosporium*: purification, characterization and stability to temperature and pH. Arch. Biochem. Biophys. 279:158-166.

51. Turney, T. A. 1965. Oxidation mechanisms, p. 46-48. Butterworth & Co., London.

52. Tweedy, B. G., C. Loeppky, and J. A. Ross. 1970. Metobromuron: acetylation of the aniline moiety as a detoxification mechanism. Science 168:482-483.

53. Valli, K., B. J. Brock, D. K. Joshi, and M. H. Gold. 1992. Degradation of 2,4-dinitrotoluene by the lignin-degrading fungus *Phanerochaete chrysosporium*. Appl. Environ. Microbiol. 58:221-228.

54. Wariishi, H., and M. H. Gold. 1990. Lignin peroxidase compound III. J. Biol. Chem. 265:2070-2077.

55. Won, W. D., L. H. Disalvo, and J. Ng. 1976. Toxicity and mutagenicity of 2,4,6-trinitrotoluene and its microbial metabolites. Appl. Environ. Microbiol. 31:576-580.

56. Yamashina, I., S. Shikata, and F. Egami. 1954. Enzymatic reduction of aromatic nitro, nitroso and hydroxyl amino compounds. Bull. Chem. Soc. Jpn. 27:42-45.

10

BIOORGANIC CHEMISTRY OF THE ARYLHYDROXYLAMINE AND NITROSOARENE FUNCTIONAL GROUPS

Michael D. Corbett and Bernadette R. Corbett

Eppley Institute for Cancer Research
University of Nebraska Medical Center
600 S. 42nd Street
Omaha, NE 68198-6805

INTRODUCTION

Arylamine and nitroaromatic chemicals make up one of the largest groups of man-made chemicals that find their way into our environment, including food, water and air. The vast majority of herbicides, the mainstay of no-till agriculture, are derivatives of these classes of compounds. Even cooking high protein foods leads to the formation of arylamine chemicals, some of which are potent mutagens and probable human carcinogens. Arylamine and nitroaromatic compounds are at either end of an oxidation/reduction equilibrium and are both relatively unreactive. Both arylamine and nitroaromatic compounds can be metabolized into the more reactive intermediates between the two, that is, the arylhydroxylamine and nitrosoarene compounds (Fig. 1). This review will focus attention on the chemistry of these arylhydroxylamine and nitrosoarene intermediates under biological conditions.

The first report on the production of an hydroxylamine metabolite from a nitroaromatic or other xenobiotic compound was in 1944, when 2,6-dinitro-4-hydroxylaminotoluene was obtained in crystalline form from the urine of rabbits treated with 2,4,6-trinitrotoluene (TNT) (7). The well-known ability of TNT to induce methemoglobinemia was rationalized by the ability of this metabolite to react chemically with oxyhemoglobin and cause oxidation of the hemoglobin iron. TNT was a fortuitous choice because its hydroxylamine metabolite is among the most stable known. Most nitroso and hydroxylamine metabolites defy direct isolation from *in vivo* studies because of their chemical reactivity. In addition, they are rarely found as metabolites in bile or urine, except for some conjugated forms of the arylhydroxylamine compounds.

Nearly 20 years later the metabolic production of an arylhydroxylamine from the opposite direction, i.e. by N-oxidation, led to the proposal that arylamine genotoxicity was the result of the production of hydroxylamine metabolites (29). Considerable research has expanded on this seminal discovery, to create an extensive literature dealing with the role of

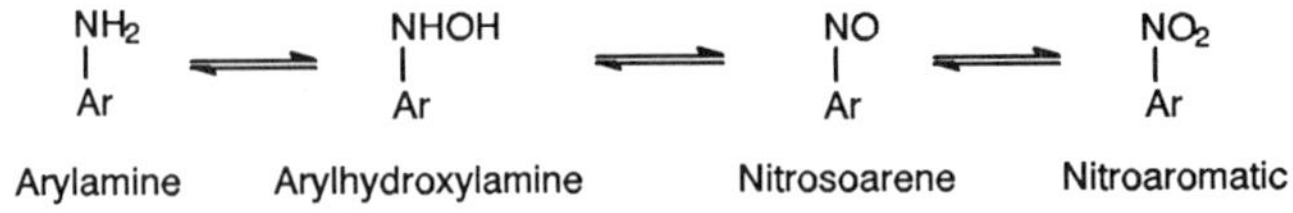

Figure 1. Redox interconversions of arylamine and nitroaromatic compounds via intermediate arylhydroxylamine and nitrosoarene compounds.

hydroxylamine metabolites in chemical mutagenesis and carcinogenesis (43, 47, 54, 89). In this review, their role in chemical carcinogenesis will be presented only to illustrate the chemical behavior of these intermediate compounds under physiological conditions.

This review will describe what is currently known about the often novel metabolic pathways available to hydroxylamine and nitroso metabolites under certain physiological conditions. The basic chemistry of these functional groups is the key to understanding their roles in the metabolic pathways of nitroaromatic chemicals and how structural changes are likely to affect these pathways. Much of the current knowledge on this subject has been obtained from *in vitro* studies with mammalian systems; nevertheless, its general applicability to microbial systems is to be expected.

FORMATION AND INTERCONVERSIONS OF ARYLHYDROXYLAMINES AND NITROSOARENES

Only the two-, four- and six-electron reduction products of nitroaromatic chemicals are potentially stable because the odd-electron reduction products are organic radicals. Stability or reactivity is relative; thus, the stability of the two- and four-electron reduction products will be highly dependent upon the chemical structure of the parent aromatic system. The sequential two-electron reduction of nitroaromatic compounds is shown as the sum of three reversible interconversions in Figure 1. Thermodynamically, the amine oxidation state is generally of lowest energy and the nitro oxidation state highest within a given aromatic ring system; thus the sequential reduction of the nitro group proceeds with the release of energy for each two-electron reduction step. Although they are thermodynamically more stable than the nitro group, the nitroso and hydroxylamine groups are certainly much more chemically or kinetically reactive than the nitro group because of the reaction pathways that are uniquely possible for each of these intermediary oxidation states. The thermodynamic tendency to release the "excess energy" present in the nitroso and hydroxylamine groups results in some of this energy being converted to tissue damage as the nitroso and hydroxylamine groups find kinetically viable pathways to attain the more stable amine oxidation state. The energy rich nitro functional group does not have alternative pathways to release this energy; thus, only by a catalyzed reduction process can the nitro group attain the more thermodynamically stable amine oxidation state.

If, during the chemical or enzymatic reduction of the nitro group, the nitroso or hydroxylamine metabolites are produced as discrete intermediates, then numerous reactions other than continued reduction to the amine becomes possible. These new reactions include many that can occur under normal physiological conditions which has led biomedical scientists to propose that the toxic actions of nitroaromatic and arylamine compounds are mediated largely by the production of their nitroso and hydroxylamine metabolites. The same fundamental chemical reactions are likely to contribute to microbial toxicity. On the other hand, several microbes appear to have taken advantage of a major reaction available to arylhydroxylamine compounds, and to have developed enzymes to direct hydroxylamine metabolism towards a pathway that can result in mineralization.

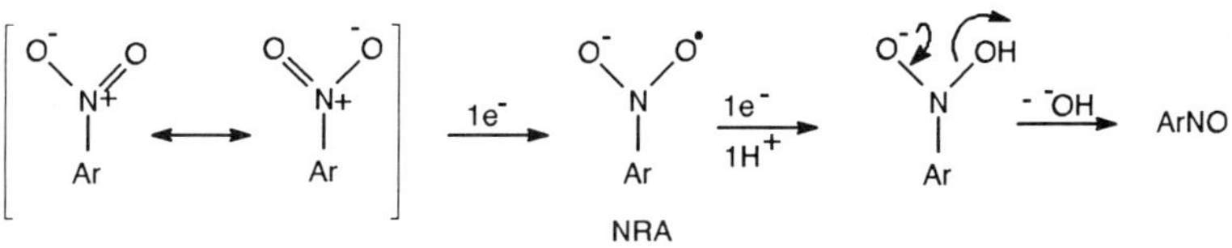

Figure 2. Conversion of nitroaromatic to nitrosoarene compounds by a two-electron addition reaction.

The one-electron reduction product of a nitroaromatic compound is the nitro radical anion (NRA) which is rarely more than a fleeting intermediate as in the shuttling of electrons from NAD(P)H to oxygen or as a putative intermediate in the reduction of a nitro group to a nitroso group (73). Because of the generally favorable reaction with the ground state (triplet) of oxygen, the NRA is capable of producing superoxide anion and reformation of the nitroaromatic compound. If a facile one-electron reduction of a nitroaromatic compound is coupled to this reaction, then such cycling can contribute to oxidative stress in a cell (49). On the other hand, the NRA is an unlikely candidate to mediate the toxic effects of nitroaromatic compounds that arise from the formation of covalent adducts to certain biomolecules.

Both the nitroso and hydroxylamine groups are electrophiles, a property absent in the amine group and not readily expressed by the nitro group. This electrophilicity is consistent with the thermodynamic tendency of the intermediate oxidation states, the nitroso and hydroxylamine species, to undergo reduction to the amine state. The nitro group is kinetically unreactive as an electrophile because both the one- and two-electron addition products disrupt the resonance stabilization of the ground state nitro group (Fig. 2).

The nitroso functional group is readily reduced in a thermodynamically favored and kinetically facile reaction to the hydroxylamine state (1). The relatively rapid kinetics of nitroso reduction can be reasonably explained on the basis of probable stabilities of transition state intermediates produced via electron addition. Unlike the nitro group, for which addition of an electron disrupts the major resonance of the ground state, addition of an electron to the nitroso group yields transitional intermediates with structures quite similar to the major resonance forms of the ground state nitroso group (Fig. 3).

A compilation of reduction potentials for a series of substituted nitrosobenzenes is often cited (69). Unfortunately, these determinations were conducted at pH 1 in aqueous acetone, and while they illustrate the general tendency of electron-withdrawing substituents to increase the reduction potential of nitrosoarene compounds, their quantitative values should be viewed with skepticism for reactions occurring at physiological pH.

The reduction of the hydroxylamine group involves an energy barrier. The generally sluggish reduction of an hydroxylamine group compared to reduction of its nitroso analog is reflected by the ease of reduction of many nitrosoarene compounds to the corresponding arylhydroxylamines by reduced pyridine and flavin coenzymes (1, 2, 65). The lack of

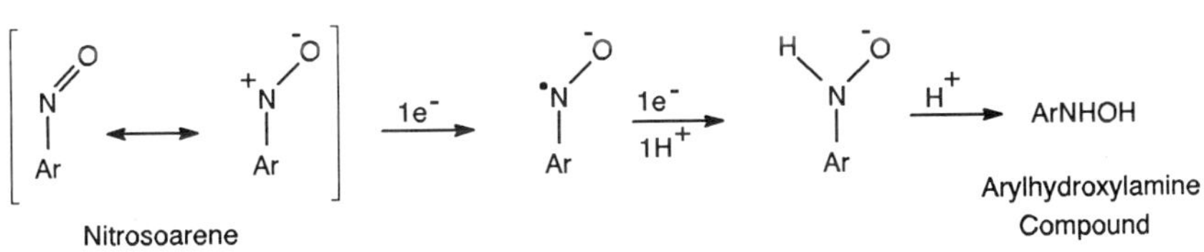

Figure 3. Reduction of nitrosoarene compounds to arylhydroxylamine compounds by a two-electron addition reaction.

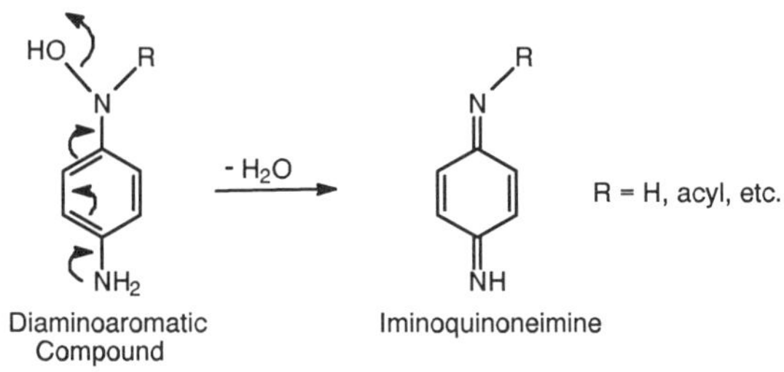

Figure 4. Dehydration reaction of a diaminoaromatic compound to form an iminoquinoneimine.

reactivity between most arylhydroxylamine compounds and NADPH is due to a kinetic barrier since the reaction is thermodynamically favorable. The notable exceptions are the special situations in which the presence of strong resonance interactions with *ortho-* and *para-*amino or hydroxy substituents can facilitate the N-O bond cleavage that is required to reduce the hydroxylamine to the corresponding amine (14, 16, 61). In such special substrates, N-O cleavage might be viewed as the result of a "push-pull" type of bond formation and cleavage to produce the relatively stable iminoquinone or iminoquinoneimine (Fig. 4), either of which is readily reduced to the hydroquinone state (i.e. aminophenols and diaminoaromatics) (65).

It is the combination of these kinetic barriers at the extremes of the redox sequence in Figure 1 with the kinetic ease of interconversion of the hydroxylamine and nitroso groups that allows the discrete formation of the nitroso and hydroxylamine groups during the reduction of the nitro group or oxidation of the amine group. Unfortunately, the analytical detection of these intermediates is often difficult, and requires special experimental design (19, 54). Therefore, reports of the absence of these intermediates during metabolic processes may be a result of improper methodology, particularly in those cases where authentic samples of the arylhydroxylamine and nitrosoarene compounds have not been available.

Reductive conditions favor the hydroxylamine side of what can be considered an equilibrium reaction between the hydroxylamine and nitroso functional groups trapped between the arylamine and nitroaromatic oxidative extremes. This is why all the standard methods for the partial reduction of nitroaromatic compounds inevitably yield the hydroxylamine product rather than the nitroso product. For example, a common technique to accomplish such partial reduction is the use of zinc powder and ammonium chloride in hydroxylic solvents (19). Stronger reducing conditions, such as zinc powder in acidified hydroxylic solvents, result in the complete reduction of the nitro group to the amine. This is because the acidic conditions provide a catalyst for the N-O bond breaking reaction.

Isolation of purified arylhydroxylamine products is often complicated by their tendency to undergo further reactions, particularly condensation with nitrosoarene impurities to produce azoxy products (Fig. 5) (60). Arylhydroxylamine compounds with high electron density in the aromatic ring system tend to undergo complex oxidative reactions and polymerization. Purified arylhydroxylamines tend to have a short shelf life, even when stored refrigerated in the dark under an inert gas.

Oxidative conditions favor the nitroso side of the equilibrium between arylhydroxylamine and nitrosoarene compounds. This property is illustrated by our observation that dilute solutions of mono-substituted arylhydroxylamines in seawater underwent spontaneous and nearly quantitative conversions to the respective nitrosoarene derivative (Corbett, unpublished). The rates of these oxidations are surprisingly fast, as demonstrated by the apparent first-order rate constant of 0.71 min^{-1} for 4-methylphenylhydroxylamine in sterile-filtered seawater at 25°C. Sternson

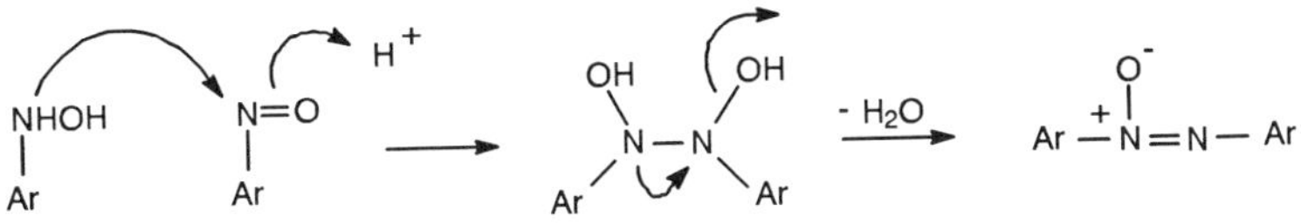

Figure 5. Condensation reaction between nitrosoarene and arylhydroxylamine compounds to form azoxy products.

reported that the oxidation of phenylhydroxylamine in metal-free phosphate buffers gave nitrosobenzene as the major product with about 10% production of nitrobenzene and that the apparent first-order rate constant for consumption of starting material was about 0.06 min^{-1} for O_2-saturated buffer (57). The origin of the nitroaromatic product was not due to subsequent oxidation of the nitrosoarene product but appeared to be a more complex oxidative reaction of the arylhydroxylamine compound that occurred in competition with the major reaction of nitrosoarene formation.

Biochemical Oxidations

In studies with chloroperoxidase (CPX), a peroxidative enzyme that is amazingly similar in certain properties to cytochrome P-450, we found the nitrosoarene metabolite to be the terminal product of arylamine oxidation (Fig. 6) (23, 31). CPX has been used to prepare nitrosoarene chemicals on a micro scale (17) because few chemical techniques are available for the direct conversion of arylamines to nitrosoarene compounds (28). It is probable that this enzymatic oxidation produces an intermediary hydroxylamine compound which, under the reaction conditions, is rapidly converted to the nitroso level. An apparent kinetic block in the oxidation of nitrosoarene to nitroaromatic compounds allows for the fairly selective production of the former by mild oxidants, particularly for those arylamines with electron-withdrawing substituents.

A notable exception to the rule of thumb that the nitrosoarene compound is favored under oxidative conditions is the observation of arylhydroxylamine rather than nitrosoarene metabolite in microsomal oxidase studies of arylamine metabolism (43). *In vitro* studies of N-oxidation generally contain the biological reductant NADPH, which can, by a mechanism independent of cytochrome P-450 catalysis, reduce any nitroso product back to the hydroxylamine state. Our studies on CPX-catalyzed oxidation of arylamines by H_2O_2 did not require a reductant such as NADPH, and thus, may better reflect the actual enzymatic or equilibrium product of oxidative reactions, including microsomal oxidations. Microsomal oxidations in which substrates such as cumene hydroperoxide are used in place of O_2/NADPH should yield the nitrosoarene rather than arylhydroxylamine metabolite. In the case of 4-chloroaniline, the nitroso metabolite rather than the hydroxylamine metabolite was produced; however, the major product was the nitroaromatic compound, indicative of further oxidation (55). This raises the hypothetical question as to whether

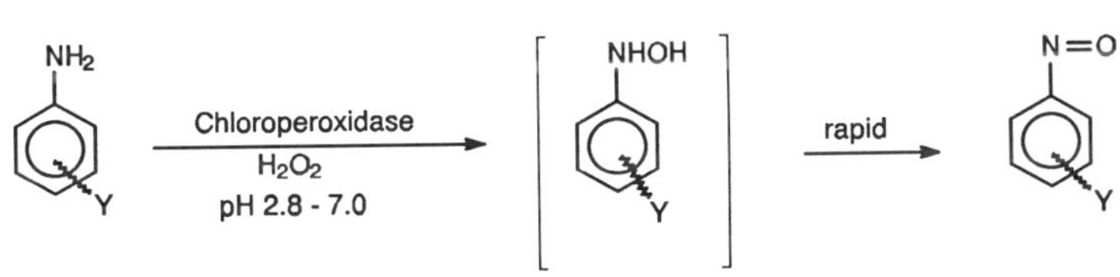

Figure 6. Oxidation of arylamine compounds by the fungal enzyme chloroperoxidase to form nitrosoarene products.

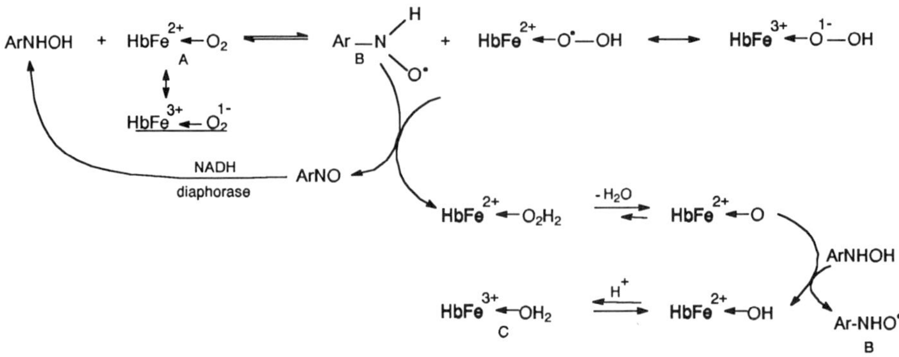

Figure 7. Mechanism for methemoglobin production during the redox cycling of arylhydroxylamine (ArN-HOH) and nitrosoarene (ArNO) compounds. A, oxyhemoglobin; B, arylhydronitroxide radical; C, methemoglobin.

the arylhydroxylamine or the nitrosoarene compound is the actual metabolite that dissociates from the cytochrome P-450 enzyme under normal conditions.

In general, arylhydroxylamine compounds are much more readily oxidized to the nitrosoarene compounds than are the corresponding arylamine compounds. All reasonable chemical mechanisms for these oxidations suggest the existence of major mechanistic differences between arylamine and arylhydroxylamine oxidation since the former requires N-O bond formation, while the latter is a dehydrogenation (31). The enzymatic conversion of arylhydroxylamine to nitrosoarene compounds is generally much more facile than the conversion of arylamine compounds to arylhydroxylamine compounds which suggests that the latter metabolic intermediate may be rapidly oxidized to the nitrosoarene compound before it can escape the immediate area of the enzyme active site. No study has yet determined the relative kinetic parameters for arylamine and arylhydroxylamine oxidations by an enzyme, such as CPX, capable of oxidizing both.

Methemoglobinemia frequently follows acute exposure of humans to most arylamine and nitroaromatic compounds. The probable mechanism for its induction involves a redox cycling between the arylhydroxylamine and nitrosoarene metabolites as illustrated in Figure 7 (35). The reaction of oxyhemoglobin (A in Fig. 7) with arylhydroxylamine compounds results in the overall oxidation of both the hemoglobin Fe^{2+} and the arylhydroxylamine compound, with molecular oxygen serving as the electron acceptor (71). An initial one-electron transfer from the arylhydroxylamine compound to oxyhemoglobin yields an arylhydronitroxide radical (B in Fig. 7) and a species suggestive of superoxide-coordinated hemoglobin. The arylhydronitroxide radical has been proposed as a distinct transient intermediate which can rapidly transfer another one-electron or hydrogen atom either to its coproduct or another molecule of oxyhemoglobin, and thus, be converted to a nitrosoarene compound (71). The driving force in this overall oxidation of arylhydroxylamine to the corresponding nitrosoarene compound is its coupling with the reduction of O_2 to H_2O. Methemoglobin (C in Fig. 7) can be viewed as the three-electron reduction product of oxyhemoglobin, or as the one-electron oxidation product of hemoglobin. The latter view is conceptually simpler (i.e. Fe^{2+} to Fe^{3+}), while the former view is mechanistically more accurate.

Nucleophilic and Electrophilic Reactivity

The valence-bond structures of arylhydroxylamine and nitrosoarene compounds are generalized in Figure 8. Each has structural features that endow it with both nucleophilic

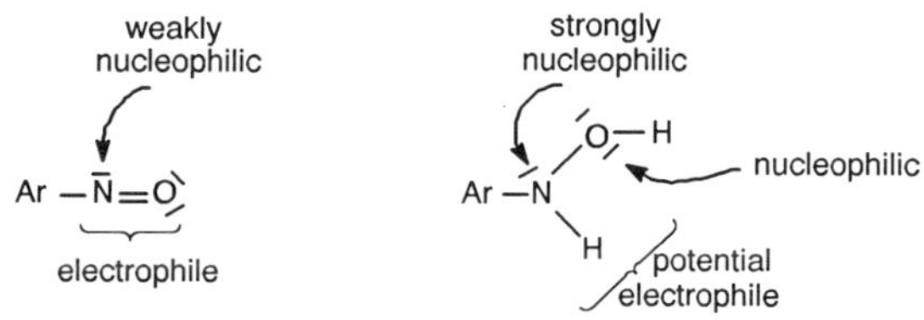

Figure 8. The valence bond structures of arylhydroxylamine and nitrosoarene compounds indicating the potential for both compounds to react as either nucleophiles or electrophiles.

and electrophilic properties. The presence of three unshared pairs of electrons is also common to both; however, the bond hybridization in the hydroxylamine group is sp^3 for the nitrogen, while it is sp^2 in the nitroso group. The nucleophilic properties of both groups originate from the unshared pair of electrons on the nitrogen atom. The sp^3 orbital containing the non-bonded electron pair of the nitrogen in arylhydroxylamine compounds is generally the site of highest available electron density and thus, the principal site of nucleophilicity for most arylhydroxylamine compounds. However, the nucleophilicity of the oxygen in the hydroxylamine group is significant, and for some arylhydroxylamine compounds, it is the primary nucleophilic site. The site of acylation in arylhydroxylamine compounds is the primary factor controlling the formation of reactive electrophiles in most situations. Arylhydroxylamine compounds are generally about an order of magnitude more nucleophilic than the analogous arylamine compounds because of the "alpha-effect" (56); however, the basicities of arylamine compounds generally exceed those of arylhydroxylamine analogs. Phenylhydroxylamine, with a pK_a reported to be 3.3 (56) or 1.9 (62, 87) is a weaker Bronsted base than aniline with a pK_a of 4.6 (56). These combined data suggest that the nitrogen atom of arylhydroxylamine compound is a soft base, at least softer than the analogous arylamine compound (83).

The electrophilic properties of the hydroxylamine and nitroso functional groups are quite different. The nitroso group is a modest electrophile, similar in reactivity to the aldehyde carbonyl group (44). The 1,2-addition of a nucleophile to the polarized double bond of both the nitroso and carbonyl groups is often a favorable reaction and is the reason for the electrophilicity of nitrosoarenes.

The electrophilic reactivity of the hydroxylamine group results from the heterolytic cleavage of the N-OH bond, either by spontaneous ionization or nucleophilic displacement at the nitrogen atom. This is expected to be a thermodynamically favorable bond-breaking reaction, and it often yields a nitrenium ion intermediate which is a highly electrophilic species that can react with many nucleophilic sites. As in the reduction of the hydroxylamine to amine functional group, there is often a kinetic barrier to this bond breaking reaction. Such a barrier allows biological systems to control the pathway of N-O cleavage. That N-O bond cleavage produces a strong electrophile is evidenced by the nature of the nucleophiles with which it reacts.

Although the geno- and cytotoxic properties of arylhydroxylamine and nitrosoarene chemicals are most likely due to their electrophilic behavior, those reactions in which they behave as nucleophiles are very important to their overall fate, and may even determine the specific metabolic pathway by which their toxicity is expressed. The chemistry of these functional groups under physiological conditions is a mix of chemical and enzymatic reactions, not all of which will necessarily occur in a specific biochemical system. Thus, the fate of such groups in a particular cell or subcellular environment will depend upon the availability of coreactants and enzymes. The relative kinetics of the various potential reactions must be considered because of the potential redox interconversion of nitrosoarene and arylhydroxylamine compounds. In a defined biochemical system, the prevalence of one reaction pathway over others may change as concentrations of reactants change or as enzymes are inactivated. This latter process is commonly observed in studies with arylhydroxylamine and nitrosoarene chemicals and appears to be the result of the electrophilic properties of these functional groups and/or their metabolites.

The electrophilic properties of arylhydroxylamine and nitrosoarene compounds arise from very different bond-forming and bond-breaking reactions and lack a common electrophilic intermediate. The differences in electrophilic intermediates between arylhydroxylamine and nitrosoarene compounds are sufficient to exert a strong effect on the preference of target nucleophile. In turn, the arylhydroxylamine and nitrosoarene compounds are expected to produce somewhat different products, including protein binding by two chemically distinct electrophilic mechanisms.

THE BAMBERGER REARRANGEMENT

Electrophilicity of Arylhydroxylamine Compounds and Their Reactivity with Biological Nucleophiles

N-Phenylhydroxylamine is a weak organic base; however, it has two linked heteroatoms, and thus, the potential for two sites of basicity (Fig. 9). Spectroscopy has established that the most basic site is the nitrogen atom, with a $pK_a = 1.9$, while protonation at oxygen proceeds with a pK_a of -6 (62, 87). Even though the ratio of nitrogen to oxygen protonation is 10^8, protonation at oxygen allows for a competing reaction that is largely irreversible and thus causes the consumption of the arylhydroxylamine compound. This competing reaction is the heterolytic cleavage of the N-O bond, generally with the intermediate formation of a nitrenium ion. This is the basis for the acid catalyzed rearrangement of arylhydroxylamine compounds to aminophenols known as the Bamberger rearrangement (87). The Bamberger rearrangement can be broadly defined to include all N-O bond heterolyses for structures Ar-N-O-Y that result in the formation of a new bond to the aromatic ring and the simultaneous reduction of the hydroxylamine group to the amine functional group.

Acid catalysis of N-O bond cleavage is best explained in terms of nucleofugacity (defined as the ability of a leaving group to carry away an electron pair) because this explanation is readily extrapolated to other pathways in which the breaking of the N-O bond

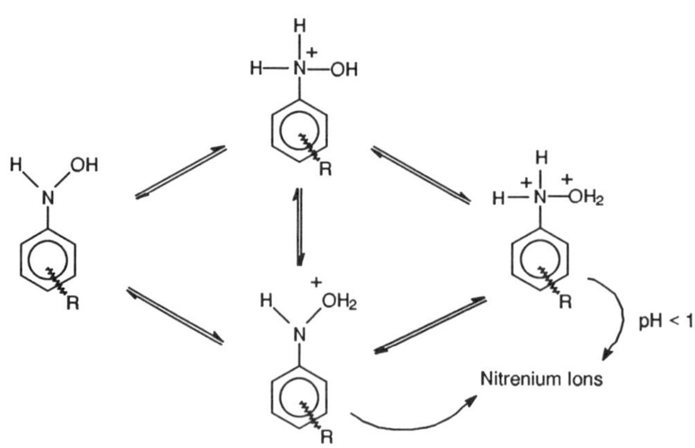

Figure 9. Protonation equilibria of arylhydroxylamine compounds and potential bond cleavage reactions.

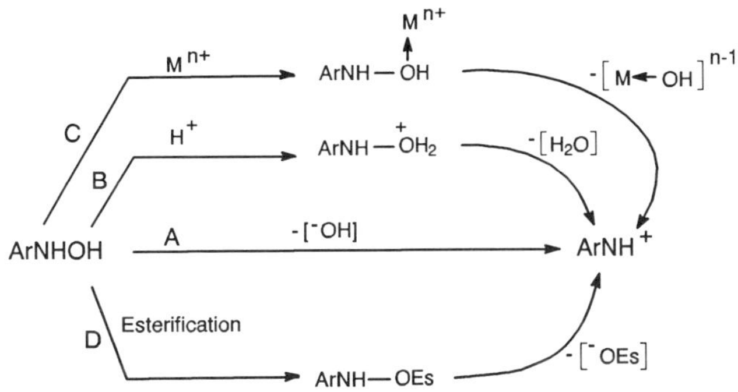

Figure 10. Heterolytic cleavage of arylhydroxylamine compounds: A, uncatalyzed; B, Bronsted acid catalysis; C, Lewis acid catalysis; D, facilitated mechanism.

is facilitated. Simple heterolysis of the hydroxylamine group with loss of the hydroxide ion (path A in Fig. 10) is generally a very slow reaction in most solvents including water which, as a highly polar solvent, is expected to favor such bond-breaking. Under sufficiently acidic conditions, protonation of the arylhydroxylamine compound occurs with a minor contribution of the O-protonated form (Fig. 9). Nevertheless, the protonated form can undergo N-O heterolysis (path B in Fig. 10) much faster than the unprotonated form (path A) because H_2O is a better leaving group than the OH anion. That a Lewis acid might replace H^+ as a catalyst (path C) has been proposed for certain enzyme catalyzed rearrangements. Conjugation of the hydroxylamine hydroxyl group with an organic or inorganic acid produces the O-ester. Heterolysis of the N-O bond in such O-esters (path D in Fig. 10) is generally quite facile, consistent with the high nucleofugacity of most acid anions. In fact, most N,O-esters are much too reactive to allow for their isolation in pure form, or even for their manipulation in solutions at room temperature (52, 75), being, for the most part, fleeting intermediates in certain chemical and enzymatic reactions (5, 88).

Nitrenium Ion Production

Formation of an intermediate nitrenium ion is often used to explain the products of these N-O heterolyses, although a nitrenium/carbenium ion species is a better representation (Fig. 11). It has been proposed that the charge delocalization into the aromatic ring is so extensive that the nitrogen atom has little charge, and might even exist as an intermediate imine-like structure (62). Nitrenium ions that form a close ionic association with their anionic leaving groups are considered to be ion pairs which are further classified into tight ion pairs and solvent separated ion pairs, depending upon their susceptibility to trapping by external nucleophiles (80). As electron withdrawal from the ring increases with substituent changes, the stability of potential nitrenium ions decreases and this leads to an increased contribution of the tight ion pair to the nitrenium ion structure (39).

The tight ion pair is often used to explain Bamberger products arising from an apparent intramolecular rearrangement; however, products from an intermolecular rear-

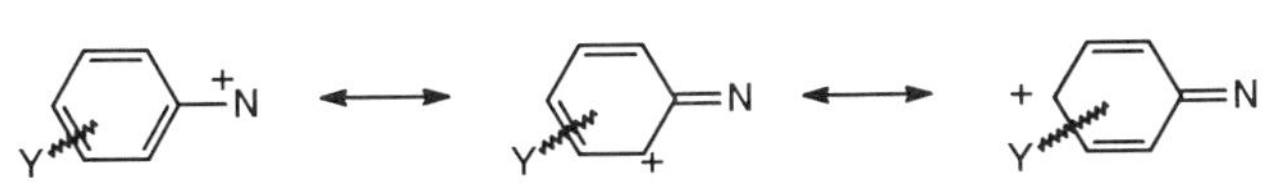

Figure 11. Resonance stabilization of the nitrenium ion.

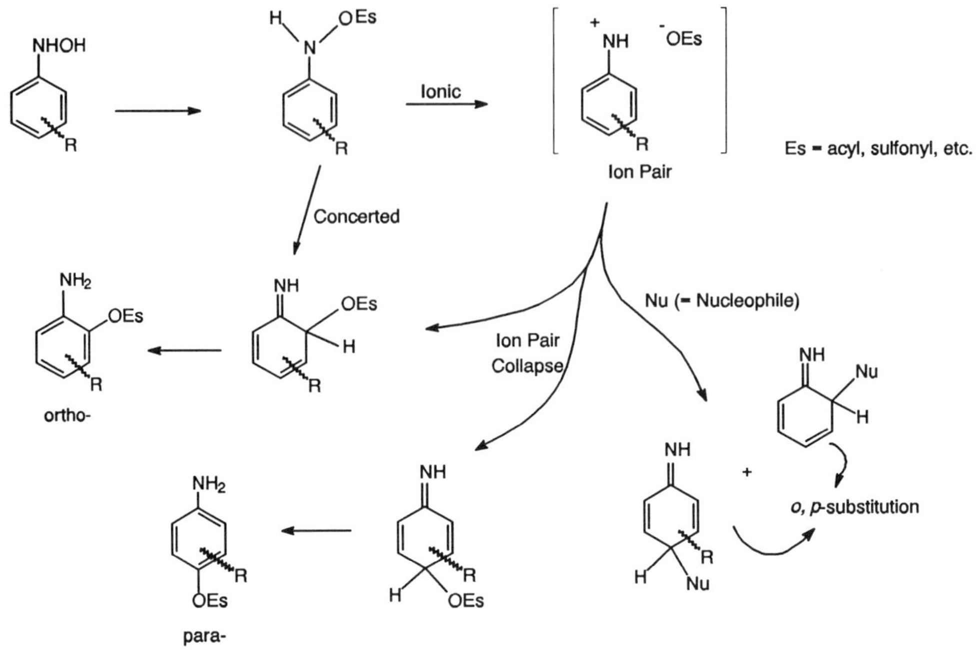

Figure 12. Mechanistic considerations of the Bamberger rearrangement shown for the general case of O-esterified arylhydroxylamine compounds.

rangement arise from a relatively free nitrenium ion species. Support for the intermediacy of a nitrenium ion in both the intramolecular and intermolecular rearrangements arises from observations of the very strong dependency of reaction rates (typically displaying ρ values in the range of -5 to -9) upon σ^+ for a series of ring substituents (40, 52). In this context σ is the substituent constant based on the electron withdrawing/electron donating properties of the ring substituent and σ^+ is the substituent constant in aromatic systems where a positive charge is delocalized between the substituent and reaction center through resonance. Reactions proceeding through either a free or ion paired nitrenium ion comprise the ionic mechanisms of N-O bond cleavages (Fig. 12), and display first-order reaction kinetics (S_N1), as contrasted to nucleophilic displacement at the nitrogen atom which is a bimolecular or S_N2-like mechanism. Bamberger rearrangement products can be readily explained by these cationic species; in special cases, however, a concerted mechanism is more likely (41, 80), and even a homolytic cleavage process to produce radical species has been reported in non-polar solvents (52, 81).

The reformation of a covalent bond between the charged members of an ion pair usually proceeds in such a way that the anionic member bonds to a different position on the cation. Most often these are the positions in the aromatic ring that are *ortho* and *para* to the nitrenium nitrogen. In general, an *ortho* carbon position is the major site for such new bond formation, with ratios of *ortho/para* products ranging from 2 (38) to near infinity (40), with ratios of 5 to 10 being typical of such intramolecular rearrangements (80). In contrast, intermolecular rearrangements occur when an outside anion or nucleophile successfully competes with the daughter anion for bond formation to the nitrenium species. Typically these occur without preference for an *ortho* position and are expected for nitrenium ions that have little or no tight ion pair associations.

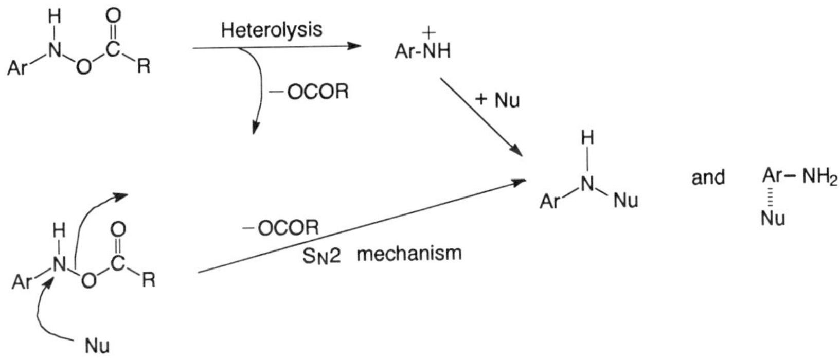

Figure 13. S_N1 vs S_N2 mechanism for the reaction of an arylhydroxylamine-O-ester. (Nu = nucleophile).

The concept that nitrenium ion production is fundamental to Bamberger rearrangement of arylamine compounds has resulted in some misconceptions concerning the actual chemical mechanisms behind certain reactions. Although it is not unreasonable to propose N-O bond heterolysis as proceeding via formation of a discrete nitrenium ion, actual studies with acetyl esters of arylhydroxylamine compounds have revealed a second mechanism that is quite distinct from simple nitrenium ion formation. O-Acetate esters of arylhydroxylamine compounds are about 10^6 times more reactive than those of the corresponding hydroxamic acids yet produce similar Bamberger products, with the provision that the hydroxamic acid compounds give N-acetylated products. Such information led to an over generalization of the reaction mechanisms for apparent Bamberger rearrangements and for DNA adduct formation by arylhydroxylamine-O-esters. A study of the kinetics of the reactions of such esters with exogenous nucleophiles revealed a dependency on the concentration of the nucleophile in addition to the ester (52, 53, 88). This suggested that an S_N2 or direct nucleophilic displacement at the nitrogen atom of the hydroxylamine O-ester could occur as the rate-limiting step, at least in competition with rate-limiting spontaneous N-O ionization, which is an S_N1 process (Fig. 13) . Prior to the kinetic studies, Boche's lab had demonstrated that certain O-esters of arylhydroxylamine compounds reacted with amines by an obvious displacement reaction to yield hydrazines (3, 4). These products were not normally expected from nitrenium ion reactions; thus, the S_N2 pathway provided a seemingly easy explanation for what were considered to be unusual Bamberger rearrangement products (88). There was a tendency to explain the formation of N-centered products as being exclusively the result of the S_N2 pathway which contributed to the erroneous concept that nitrenium ions are not attacked at the nitrogen by nucleophiles. The S_N2 mechanism explained some of the same products expected for the S_N1 process and appeared to be an even better explanation for the structures of some nucleic acid adducts. The possibility that the S_N2 pathway might contribute to the formation of covalent adducts in some cases is of major interest.

The strongest evidence for the S_N2 process was the observation of bimolecular kinetics for nucleophile trapping, which, in turn, was found to be considerably less sensitive to ring effects (ρ= -3) than is expected for an S_N1 process (ρ= -8) (77). Because of solubility limitations, many of the earlier studies were conducted in methanol or less polar solvents. Such solvents strongly suppress the S_N1 mechanism, thus favoring the change to an S_N2 mechanism. In aqueous systems, the S_N1 path was found to be favored over the S_N2 displacement in the case of the biphenyl ring system (77). This constitutes a complete reversal from reactions run in methanol (79). Thus the O-esters of arylhydroxylamine and arylhydroxamic acid compounds both yield nitrenium ion intermediates even though the former react 10^6 times faster. Only the O-esters of arylhydroxylamine compounds appear to react via S_N2 displacement, and the extent to which this competes with the S_N1 pathway is dependent on the stability of the nitrenium ion and the solvent employed.

Unlike the S_N1 mechanism, the S_N2 mechanism is not highly dependent on aromatic rings that can readily stabilize a positive charge in a transition state intermediate. Thus, the S_N2 mechanism might be a path for formation of DNA adducts by an arylhydroxylamine-O-ester unable to form a sufficiently stable nitrenium ion which is conducive to the S_N1 mechanism. To the extent that the S_N1 process becomes less favorable with a change in the aromatic ring, it is expected that the degree of O-acyl hydrolysis will also increase since hydrolysis is also very minimally dependent on ring substituents.

A further consideration of nitrenium ion intermediates is useful for recognizing the potential significance of non-nitrenium pathways (i.e. the S_N2 pathway) for forming the adducts between arylamine residues and biomolecules. Kinetic studies have led to estimates of the longevities of nitrenium ions derived from various arylamines (77). Phenylnitrenium has an estimated lifetime in aqueous solutions of 2×10^{-10} seconds (s), that for the more stabilized 2,6-dimethylphenylnitrenium ion is 1.4×10^{-9} s, while the putative ultimate carcinogenic species, 4-biphenylnitrenium ion has a much longer life of 5.9×10^{-7} s. The N-acetyl analog of the latter is about three times more reactive with a lifetime of 2×10^{-7} s. A major question in arylamine toxicology is whether those arylhydroxylamine-O-ester metabolites which cannot give sufficiently stable nitrenium ions might effect DNA adduct formation by the S_N2 process.

Role of the Nucleophile

In theory, any nucleophile present in the reaction medium might react with the intermediate nitrenium ion species. The original Bamberger rearrangement described the products as those resulting from water or chloride ion as the nucleophiles. In a broader definition, the nucleophiles include many that yield products probably never perceived by Bamberger including the nucleophilic sites of nucleic acid bases. The production of DNA adducts (Fig. 14) with arylhydroxylamine compounds is the result of DNA bases competing with other biological nucleophiles for the nitrenium/carbenium ion intermediate of the Bamberger rearrangement. That this is a minor reaction relative to the overall fate of the reactive species is not surprising in view of the great abundance of strong nucleophiles in biochemical systems. The relative distribution of products should be explained by kinetic factors such as relative nucleophilicity, particularly the hard and soft acid and base (HSAB) theory (83). The distribution can also be affected by steric factors and concentration effects

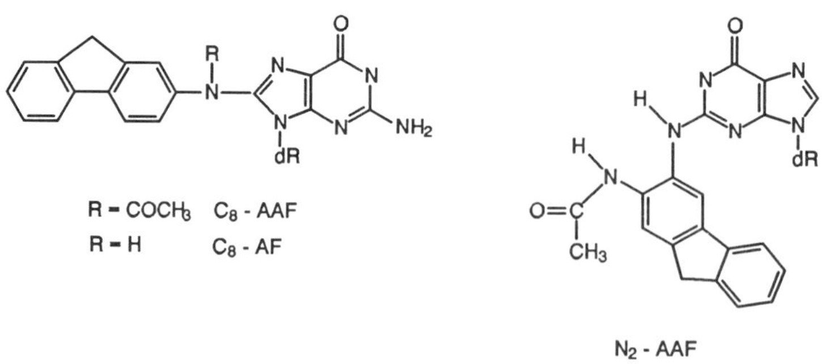

Figure 14. Major known adducts of 2-aminofluorene to DNA. C_8-AF is the adduct to ring position 8 of a guanine base; C_8-AAF is identical to C_8-AF except for an N-acetyl group.

which are pronounced and complex in physiological systems. Also of importance is the ability of a potential nucleophile to compete with the anionic half of an ion pair for the nitrenium/carbenium ion.

The nature of the nucleophile that reacts to form a covalent bond with the delocalized positive charge is likely to have a significant effect upon the site of bond formation. In general, several ring carbon sites and the nitrogen atom of the nitrenium/carbenium ion species are the sites for bond formation with the competing nucleophiles. Carbon- and nitrogen-centered nucleophiles can form stable covalent bonds with the nitrenium nitrogen, as illustrated by the structures of the known purine adducts with 2-aminofluorene (Fig. 14). The adducts in which the nucleophilic attack occurred on nitrogen (i.e. C_8-2-acetylaminofluorene and C_8-2-aminofluorene) have been thought by some to be abnormal for an S_N1 mechanism, even though they are the major adducts quantitatively.

The formation of a covalent bond to the nitrenium nitrogen atom by an oxygen-centered nucleophile is not a favorable reaction on the basis of HSAB and would in many cases be a futile reaction in that the arylhydroxylamine (or arylhydroxamic acid) compound or its O-ester is regenerated. Hydration of a primary arylnitrenium ion at the nitrogen results in an arylhydroxylamine (or arylhydroxamic acid) compound. However, experimental observations suggest that this is a minor reaction in many cases (53, 82). The formation of arylhydroxylamine or arylhydroxamic acid compounds in reactions of analogous O-esters is most often the result of solvolysis of the ester and not the result of hydration of the nitrenium ion at nitrogen.

Enzymatic Esterification and the Facilitated Bamberger Rearrangement

Most of what is now known about the heterolysis of arylhydroxylamine-O-esters and arylhydroxamic acid-O-esters has resulted from studies of the mechanism(s) by which arylhydroxylamine and arylhydroxamic acid compounds form covalent bonds with nucleic acid bases. The literature prior to 1985 has been reviewed by Hanna and Banks (50); however, since then, some major new advances have been made, particularly concerning enzymatic O-esterification of arylhydroxylamine compounds.

The nucleophilic properties of the hydroxylamine functional group are such that one would expect acylation reactions to occur predominantly on the nitrogen atom (Fig. 15). Until recently, this was a rather general rule, particularly when dealing with organic synthetic reagents such as acid anhydrides and acyl halides. The discovery of O-acylation of arylhydroxylamine compounds via enzyme catalysis (50) was not considered to represent a major violation of this rule, since enzymes often catalyze reactions that synthetic organic chemists cannot. The enzyme, O-acetyltransferase, probably achieves this feat in a single step by first binding the arylhydroxylamine compound so that the nucleophilicity of the nitrogen atom is effectively blocked. Subsequently, acetyl-CoA reacts with the hydroxyl group of the bound arylhydroxylamine compound to produce the O-acetate ester which is released from the enzyme (Fig. 16). The enzyme probably blocks the nitrogen atom by binding the arylhydroxylamine compound to an acidic site close to the active-site.

In recent years, several organic reagents have been found which rather selectively carry out O-acylation of arylhydroxylamine compounds (e.g. Fig. 17). This apparent violation of the general rule was first reported for acetyl cyanide (36, 68, 78). Recently, aspirin was reported to facilitate the covalent binding of carcinogenic arylhydroxylamine compounds to DNA, and it was proposed that an intermolecular transfer of the acetyl group from the phenolic ester of aspirin to the -OH of the arylhydroxylamine compound occurred to produce the arylhydroxylamine-O-acetyl ester (75). Unfortunately, the direct detection of this intermediate was not accomplished. Explanations of these unusual and specific O-acylations are not yet available.

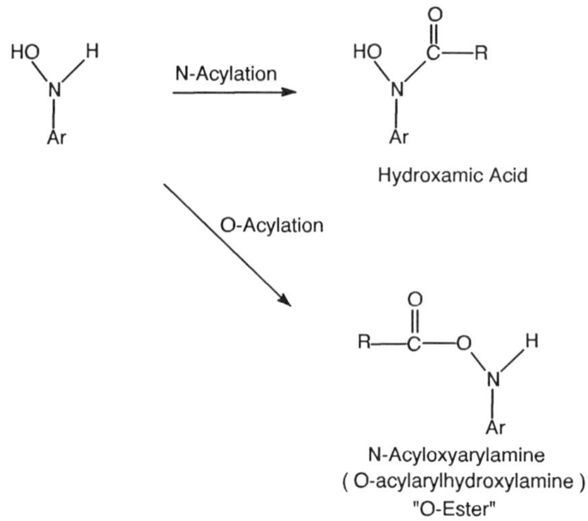

Figure 15. N- vs O-Acylation of arylhydroxylamines.

An unusual enzymatic action has been described in which arylhydroxamic acids were converted to arylhydroxylamine-O-ester (Fig. 16). This is thermodynamically unfavorable in most cases and is probably achieved as the result of the further rapid reactions of the latter which serve to displace the enzyme-mediated equilibrium between the arylhydroxamic acid compound and arylhydroxylamine-O-ester. The enzyme that catalyzes this reaction is called arylhydroxamic acid-N,O-acyl transferase (50, 51) and is either closely associated with or identical to the enzyme which catalyzes O-acetylation of arylhydroxylamine compounds (64). In turn, it appears that the enzyme which catalyzes these two reactions may be the same as N-acetyltransferase, which is a rather non-specific enzyme that catalyzes reversible N-acetylations of arylamine and arylhydroxylamine compounds. These functions are depicted in Figure 16 for arylhydroxylamine metabolism, and their contributions to the bioactivation of arylhydroxylamine and aryl hydroxamic acid compounds are a major research area in arylamine toxicity (51).

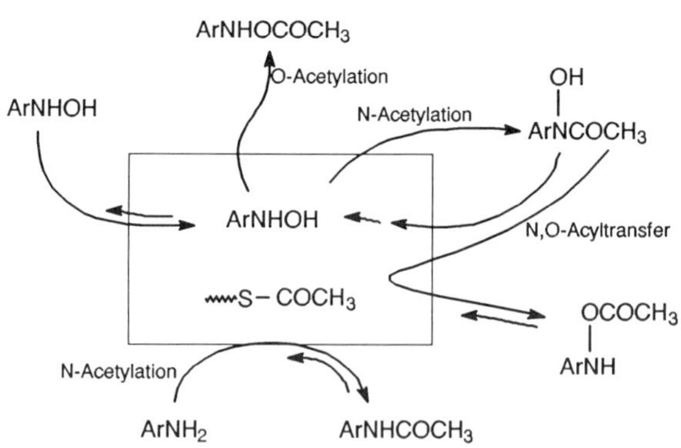

Figure 16. Schematic presentation of reversible enzymatic acetylation reactions of arylamines (Ar-NH_2) and arylhydroxylamines (Ar-NHOH) and the N,O-acyltransfer reaction of arylhydroxamic acids ($ArNOHCOCH_3$).

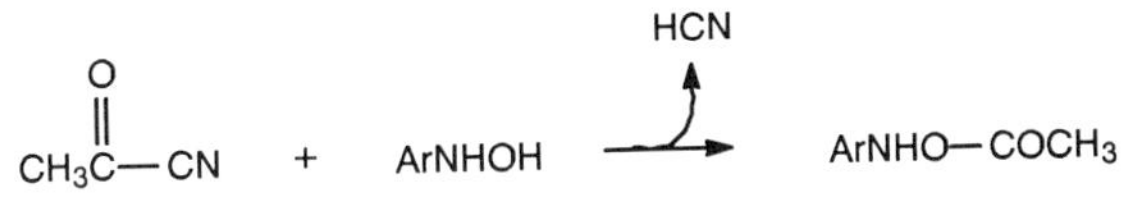

Figure 17. Specific O-acetylation of arylhydroxylamine compounds by acetyl cyanide.

Another type of metabolic esterification is the O-sulfation of both arylhydroxylamine and arylhydroxamic acid compounds, which is catalyzed by various sulfotransferases and requires phosphoadenosine-phosphosulfate (PAPS) as a cosubstrate (Fig. 18). These O-sulfate esters are quite reactive by virtue of the high nucleofugacity of the sulfate anion and are even more unstable than the analogous O-acetyl and O-pivaloyl esters (42). For the most part arylhydroxylamine-O-sulfate esters are fleeting intermediates, and their existence was proposed on the basis of observed chemical reactions. As expected, the O-sulfate esters of arylhydroxamic acid compounds are more stable than the O-sulfate esters of the corresponding arylhydroxylamine compounds. Examples of such "stable" hydroxamic acid O-sulfate esters include those of two closely-related ring systems. The half-life in aqueous solution for the sulfate ester of N-OH-2-acetylaminofluorene is estimated to be less than 1 min whereas the sulfate ester of N-OH-4-acetylaminobiphenyl is 180 min (42). These relative values reflect the significantly greater stabilization of nitrenium ions by the 2-aminofluorene system compared to the 4-aminobiphenyl system, thus, the O-sulfate ester of the former is more reactive.

Enzymatic conversion of arylhydroxylamine and arylhydroxamic acid compounds to highly reactive O-esters is often accompanied by progressive inactivation of the responsible enzyme (66). This "suicide inhibition" has been extensively investigated, particularly in the case of N,O-acyltransferase (6) and more recently for sulfotransferase (70,85). The probable cause is covalent binding of some of the activated metabolite to the enzyme. Such adduction is to be expected when a highly reactive metabolite is produced in close proximity to a biomolecule that possesses nucleophilic groups. The responsible enzyme often becomes the first biomolecule to be damaged by the electrophilic species that it produced.

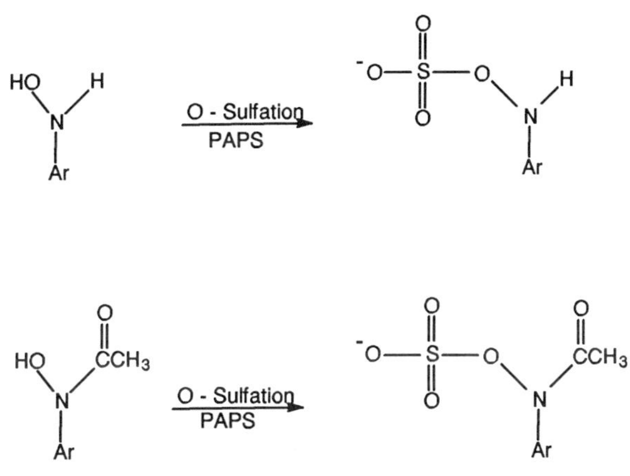

Figure 18. Enzymatic O-sulfation of arylhydroxylamines and arylhydroxamic acids.

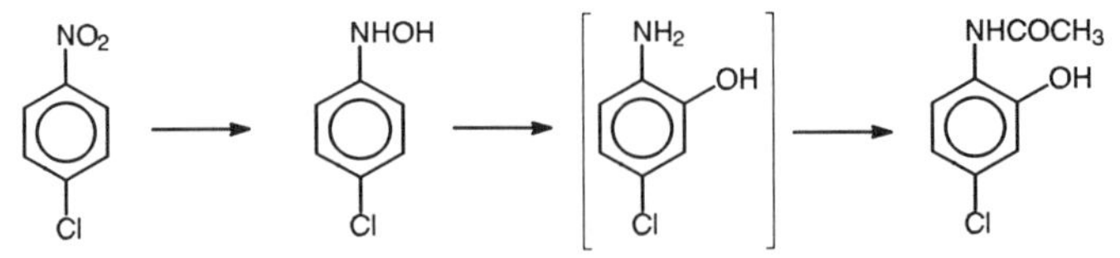

Figure 19. Metabolism of 4-chloronitrobenzene in a yeast.

The Bamberger Rearrangement at Enzyme Active Sites

During a study of the metabolism of 4-chloronitrobenzene by a yeast, we observed the production of 2-hydroxy-4-chloroacetanilide along with the nitroso and hydroxylamine intermediary metabolites, and the expected metabolites 4-chloroacetanilide and 4-chloroaniline (15). The probable metabolic pathway to the rearrangement product is depicted in Figure 19. The most interesting feature is that the phenolic products were not the result of microsomal oxidations, as is the normal pathway for production of such as metabolites. Rather, we found that the phenolic metabolites arose directly from the arylhydroxylamine compound and that yeast cells were responsible for the conversion. Incubation in growth media alone or in filter-sterilized spent growth media did not cause this reaction. Further studies to identify possible enzymes that might catalyze this microbial Bamberger rearrangement were not conducted; however, a few years later we again observed a similar rearrangement but with a purified mammalian enzyme (24).

Products indicative of the Bamberger rearrangement were unexpectedly observed during a study of the interaction of nitrosoarene compounds and α-ketoglutarate dehydrogenase (α-KGD) (24). This study was part of our research concerning the action of thiamine-dependent enzymes on nitrosoarene compounds. As will be explained later, we had reason to expect that a nitrosoarene compound could readily be reduced to the corresponding arylhydroxylamine compound at the active sites of such enzymes. The conversion of the arylhydroxylamine compound to the *o*-aminophenol and its N-succinyl amide (Fig. 20) was the first observation that thiamine-dependent enzymes could cause such product formation and led us to speculate on the mechanism by which an enzyme could cause such a rearrangement (24, 26, 30).

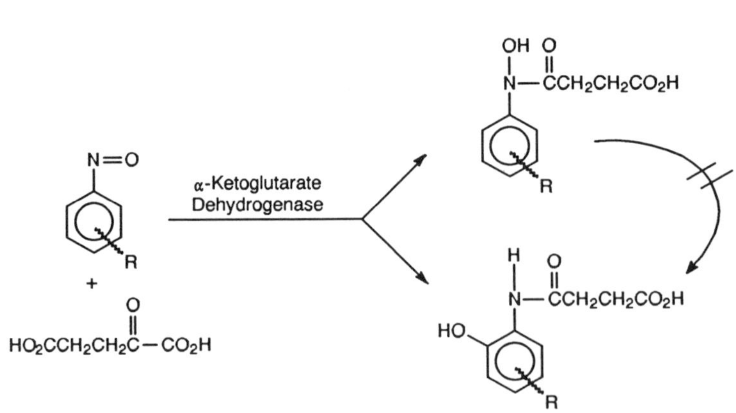

Figure 20. Major products resulting from the cometabolism of nitrosoarene compounds by α-ketoglutarate dehydrogenase.

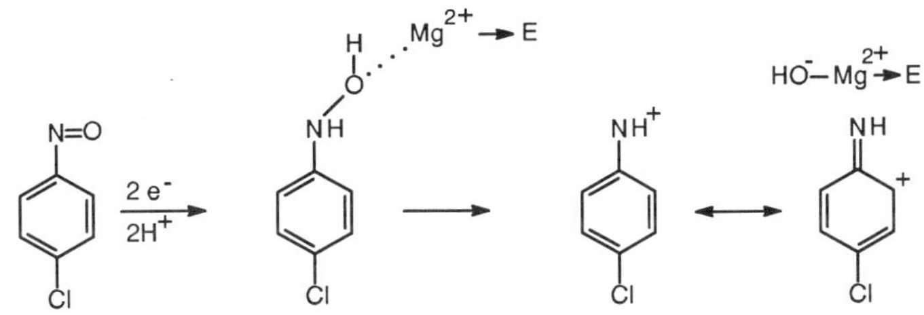

Figure 21. Proposed actions of enzyme-bound Mg^{2+} in effecting the Bamberger rearrangement at the enzyme's active site.

One possibility that might also explain a similar reaction by certain non-thiamine enzymes is that some metal ion cofactors (e.g. Mg^{2+} for α-KGD) might behave as Lewis acids, particularly if present at a hydrophobic site on the enzyme. As shown in Figure 21, the interaction of an hydroxylamine group with such a site might serve to catalyze N-O bond heterolysis and thus account for the observed Bamberger rearrangement type products (30). This is similar to the proposed catalytic function of enzyme-bound Zn^{2+} in effecting N-O bond cleavage of an arylhydroxylamine compound by liver alcohol dehydrogenase (61). More recently, Winston's lab described the Bamberger rearrangement of N-hydroxy-2-aminofluorene during the reduction of 2-nitrosofluorene by alcohol dehydrogenase (72). Further reduction of a portion of the arylhydroxylamine to arylamine compound also occurred. This process is very similar to that observed for α-KGD, in that both involve an initial two-electron reduction of a nitrosoarene, followed by the rearrangement of a portion of the arylhydroxylamine product. Winston (72) proposed that active-site Zn^{2+} served as the catalyst for the alcohol dehydrogenase-catalyzed Bamberger rearrangement but did not consider the possibility that this same Lewis acid complex might also be involved in the reduction reaction. It is reasonable that the two reactions proceed via a common intermediate. Although it is likely that some enzymes with bound metal cations can at times interact with the hydroxylamine group and cause N-O bond cleavage, another mechanistic possibility exists to explain this Bamberger chemistry in the case of α-KGD.

The interaction of nitrosoarene compounds with α-KGD while metabolizing its normal substrate, α-ketoglutaric acid, produces the arylhydroxylamine compound as a discrete mechanistic intermediate (Fig. 22). This reduction is effected by the simultaneous production of enzyme-bound "active succinate", a reactive acyl intermediate that is converted in several steps to succinyl-CoA during the normal enzyme turnover. If the arylhydroxylamine compound is near this reactive acyl group, subsequent transfer of the succinyl group to either the nitrogen or oxygen atom of the arylhydroxylamine compound is possible (Fig. 22). According to the rule of thumb concerning the site of arylhydroxylamine nucleophilicity, N-acylation should be favored, and indeed this is the major reaction for most nitrosoarene substrates. If, however, O-acylation occurs in competition with N-acylation, one might expect the Bamberger rearrangement to proceed as indicated in Figure 22 (24). The final stable product would be the amide as the result of a favorable O-N acyl transfer. A similar phenomenon has been observed with another thiamine-dependent enzyme, pyruvate dehydrogenase (Corbett, unpublished).

In research soon to be published, we have found that one or more of the potential electrophilic species generated from nitrosoarene compounds by both pyruvate dehydrogenase and α-KGD can effect extensive covalent adduction of nitrosoarene compounds to DNA. The arylhydroxylamine-O-esters are likely candidates for these phenomena.

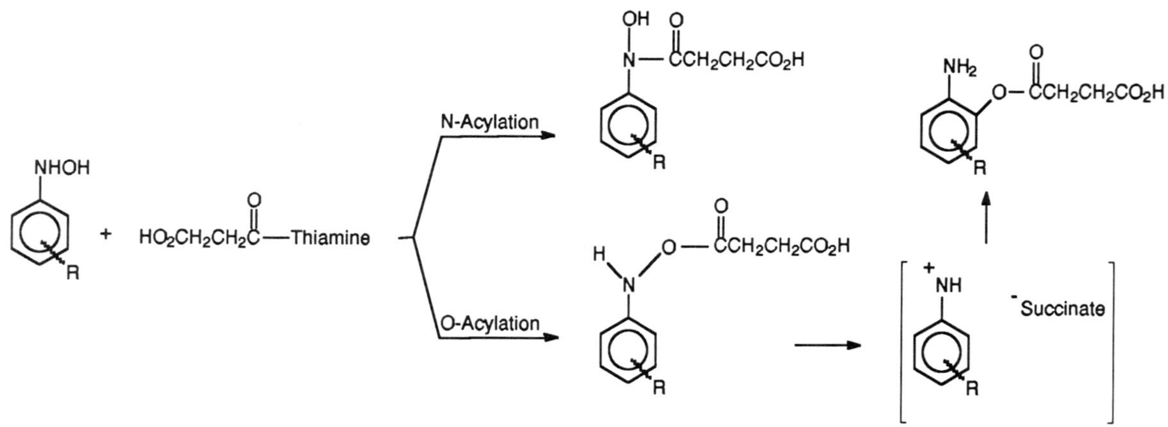

Figure 22. Proposed mechanism in which facilitated N,O-bond heterolysis arises from the O-acylation of an arylhydroxylamine compound bound at the active site of α-KGD.

Gutmann's lab has reported an interesting action of bovine serum albumin (BSA) in effecting the Bamberger rearrangement of the O-sulfate of N-OH-2-acetylaminofluorene to the isomeric and O-sulfated phenolic metabolites (63). How this common protein family can catalyze this rearrangement remains unknown. In the absence of BSA, this arylhydroxamic acid-O-sulfate is extensively adducted to nucleophiles including guanosine, nucleic acids and sulfhydryl (-SH) nucleophiles (63).

The above considerations of the possible production of a nitrenium ion species also suggest a likely general mechanism for enzymatic reduction of arylhydroxylamine and arylhydroxamic acid compounds to the corresponding arylamine and arylamide compounds, respectively. During the studies on α-KGD, we observed the production of small amounts of arylamine products which increased upon addition of KI. The presence of KBr or KCl diverted much of the rearrangement product into ring-halogenated products. This suggested that reduction of an intermediate nitrene, or O-succinoylated-arylhydroxylamine can also occur (26). A similar reductive process on O-sulfate esters of some arylhydroxamic acid compounds was subsequently reported (84), and similar explanations were proposed. Enzymatic conversion of arylhydroxylamine to arylamine compounds is not uncommon and can conceivably be effected by many enzymes. On the other hand, the direct reduction of an arylhydroxamic acid to the corresponding amide analog ($ArNHCOCH_3$) probably requires a higher activation energy or utilization of coupled reactions.

Microbiological Bamberger Rearrangements Revisited

In recent years, studies on the microbial degradation of nitroaromatic chemicals have led to additional reports on the possible occurrence of the Bamberger rearrangement. These include results on 4-nitrobenzoic acid (46), 4-nitrotoluene (48), and nitrobenzene (76), all of which have been found to undergo mineralization via a novel reductive route (Fig. 23). In these cases, the conversion of the arylhydroxylamine compound to the catechol analogs allowed subsequent ring cleavage and mineralization. The corresponding arylamine compounds were not converted to the catechols by these microbial systems, and the arylhydroxylamine nitrogen atom was released as ammonia. The detection of the putative aminophenol intermediates was not reported in the first two cases; however, the more recent study from Spain's lab did establish the intermediacy of the aminophenol, at least in the case of nitrobenzene mineralization (76).

It is easy to envision several possible mechanisms for the conversion of the aminophenol to a catechol, or perhaps the former as well as the latter might be a substrate for the subsequent enzymatic ring cleavage. Even an enzyme such as quinone reductase might indirectly effect the conversion of the aminophenol to the catechol via an intermediary quinoneimine. On the other hand, ring-monooxygenation *ortho-* to the hydroxylamine functional group would likely be followed rapidly by the loss of H_2O to give the quinoneimine (Fig. 24). This is another extreme case of the facilitation of N-O bond heterolysis by a ring substituent ("push-pull" mechanism). The electron pair of the acidic phenol group might be seen as displacing the N-O bond via direct resonance interaction. Because of the rather stable electronic structure of the quinoneimine product relative to a nitrenium ion and

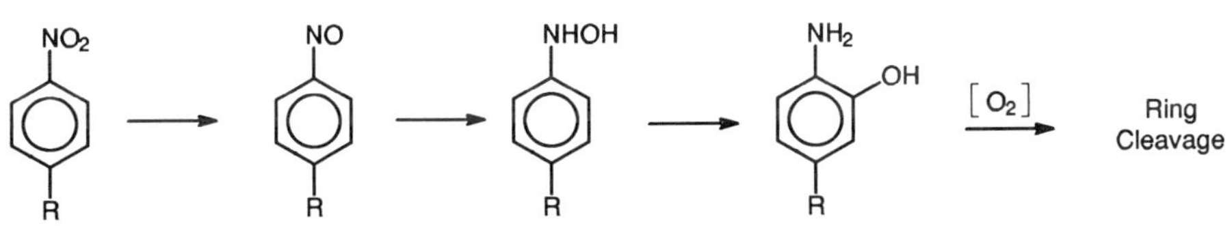

Figure 23. A reductive pathway for microbial nitroaromatic metabolism leading to ring cleavage.

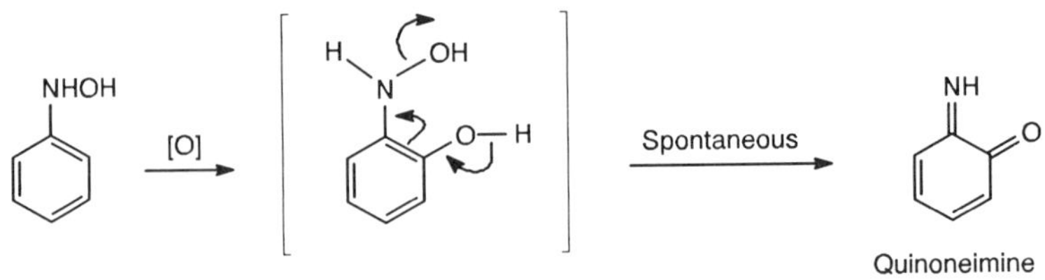

Figure 24. Speculation on a mechanism for the conversion of arylhydroxylamine compounds to the *ortho*-aminophenols.

to the extent that the transitional intermediate shares in this stability, the elimination of H_2O to form the quinoneimine is a very favorable reaction (80) in contrast to what is required for the vast majority of N-O bond cleavages.

Quinoneimines are not innocuous in terms of potential toxicity (25). Although they are rightfully viewed as less electrophilic than a nitrenium ion, it is possible for a Michael-type addition reaction with a protein -SH group to occur causing toxicity. The stability of quinoneimines could actually strengthen their role as direct-acting cytotoxic agents, in that they might reach a broader range of targets than will more reactive electrophiles. The quinoneimine formed by any route is prone to hydrolysis, which would produce the analogous quinone and NH_3. Because they can also serve as electrophiles, especially to protein -SH groups, the quinone group can also be potentially cytotoxic.

It should prove most interesting as more is discovered about how the, as yet unknown, enzymes effect N-O bond cleavage. Precedents certainly exist for several potential mechanisms ranging from Lewis acid catalysis to O-esterification reactions as a basis for enzymatic catalysis. That rearrangement of the arylhydroxylamine compound to the *ortho*-aminophenol is within the realm of microbial transformations was previously reported (15); however, only recently has the potential significance of this process been recognized. If the Bamberger rearrangement does prove to be a vital process in the mineralization of nitroaromatic pollutants, then this reaction might be viewed in a more positive connotation. Previously, the Bamberger rearrangement was associated with mediating the toxic effects of nitroaromatic compounds.

There may be a price to pay for the microorganism that dabbles in this chemistry. Those microbial strains that have an incomplete nitroreductase or a leaky enzyme that allows relatively high concentrations of nitrosoarene and arylhydroxylamine compounds to develop are subjecting the microbial population to toxic metabolites. On the other hand, perhaps the potential mutagenicity of these intermediates will contribute to the "natural" development of newer microbial strains that will more efficiently process these toxins in order to minimize autotoxicity.

SOME ORGANIC AND BIOCHEMICAL REACTIONS OF NITROSOARENE COMPOUNDS

The principal reactions of nitrosoarene compounds under physiological conditions are those in which the nitroso group behaves as an electrophile. Unlike the electrophilic reactions of arylhydroxylamine compounds, these reactions occur spontaneously without the need for further activation. On the other hand, the level of reactivity of the nitroso group is significantly less than that of the electrophilic species derived from arylhydroxylamine compounds.

Reaction with the -SH Group

Aside from the redox reactions presented above, the major single reaction of nitrosoarene compounds in biochemical systems is with the sulfhydryl group of amino acids and proteins (Fig. 25). This reaction was first described by Eyer in 1979, although earlier Kiese had demonstrated the binding of phenylhydroxylamine to protein (33). The facile oxidation of arylhydroxylamine compounds to nitrosoarene compounds in aqueous solutions was responsible for Kiese's observation and was the reason Eyer chose to investigate nitrosobenzene. Since then considerable interest in the potential for nitrosoarene compounds to inactivate protein function has led to the belief that this general reaction might be responsible for a variety of cytotoxic effects (34). Neumann and coworkers noted the production of arylamine compounds during the reaction of a series of nitrosoarene compounds with reduced glutathione (32) and modified Eyer's original mechanism.

The first step is the rapid and reversible nucleophilic attack of -SH on the nitrosoarene group to produce a semimercaptal (59), which can then follow several competitive paths to the products (Fig. 25). The major product is usually the sulfinanilide, depending upon reaction conditions. Recently, McClelland proposed a modification of the mechanism for the rearrangement of the semimercaptal to the sulfinanilide, based on kinetic and ^{18}O tracer studies (58). He proposed that the rate limiting step is an N-O heterolysis, in which the semimercaptal gives the sulfenium ion species via elimination of ^{-}OH. This is facilitated by the adjacent sulfur atom, which is estimated to cause a rate enhancement of 10^6 relative to the Bamberger rearrangement of the analogous arylhydroxylamine compound. Rehydration of the sulfenium species at the sulfur atom yields another putative sulfenium species which, upon loss of H^+, results in the sulfinamide product.

In our laboratory, we utilized this reaction with glutathione to convert a radiolabelled sample of the nitroso analog of chloramphenicol (CAP) to amino-CAP, without reductive loss of chlorine atoms from the molecule, which is difficult to achieve by most reductive methods (21). The intermediate formation of the stable sulfinamide-CAP was equally valuable as a method to remove trace impurities of the nitro analog, which is CAP itself,

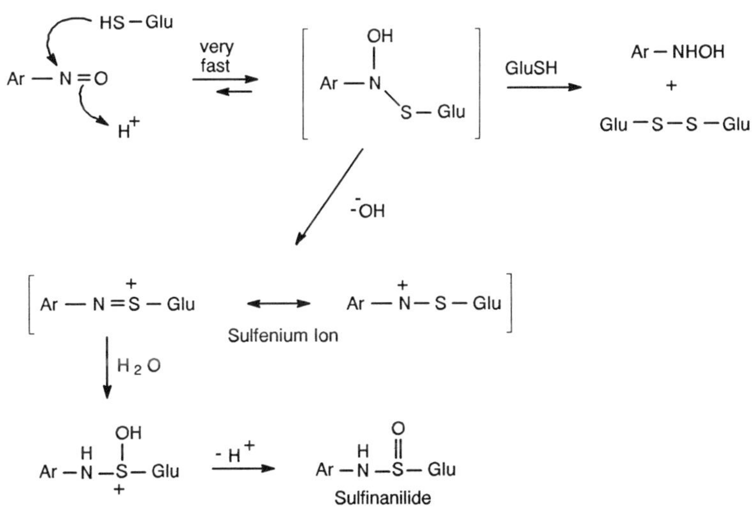

Figure 25. Proposed mechanism for the reaction of nitrosoarene compounds with sulfhydryl groups. (Glu-SH is reduced glutathione; Glu-S-S-Glu is oxidized glutathione).

from the desired amino-CAP. Chromatographic attempts to achieve this were unsuccessful which is not the normal situation for most nitrosoarene/nitroaromatic mixtures.

The "nitroso-thiol reaction" is generally considered to be a major detoxification reaction, in that it produces a stable, relatively H_2O-soluble metabolite in the form of the sulfinamide. On the other hand, this reaction with protein -SH can cause protein inactivation and cytotoxicity which is the basis for the original interest in this reaction. The determining factor is the source of the -SH group, with glutathione and "unimportant" proteins being detoxification processes, at least until the glutathione status of the cell is adversely affected. The cytotoxicity of arylamine and nitroaromatic compounds is likely the result of the nitrosoarene metabolite interacting with proteins in a non-specific manner, and such binding has even been observed for TNT, which led to the proposal that the hepatotoxicity of this chemical is the result of such reactions (67).

Reaction with Thiamine-Dependent Enzymes

The purpose of thiamine as an enzyme cofactor is to provide a chemical mechanism for the stabilization of acyl anions at the active sites of the enzymes. These enzymes include the α-oxoacid decarboxylases, such as pyruvate decarboxylase and the pyruvate and α-ketoglutarate dehydrogenases, and one of the principal transferase enzymes of sugar metabolism, transketolase. The spontaneous decarboxylation of pyruvic acid does not occur because the transitional intermediate of this reaction is unstable. Enzymatic decarboxylation is achieved by a mechanistic change in which thiamine serves as an electron sink (Fig. 26). The electron pair from the heterolysis of the C-CO_2H bond can be stabilized through resonance interaction with the heterocyclic ring of thiamine. This enzyme-bound intermediate is called "active acetaldehyde" to reflect its value as a biochemical synthon equivalent to the acyl anion of acetaldehyde. The organic chemist's equivalent is the cyanohydrin of acetaldehyde, a derivative in which the former non-acidic aldehyde C-H bond is now modestly acidic and whose carbanion is a good nucleophile. In a similar manner, some enzymes are able to stabilize certain acyl anions by substitution of enzyme-bound thiamine for cyanide ion. These relatively stable enzyme intermediates are still highly nucleophilic, yet the enzyme appears to control the subsequent reactions of these nucleophiles depending on the overall function of the enzyme. In the case of pyruvate dehydrogenase, the anion of "active acetaldehyde" is thought to cause the nucleophilic displacement of thiolate anion from the disulfide bond (-S-S-) of enzyme bound lipoic acid. This generates the S-acetyl bond which, in a transesterification reaction with coenzyme

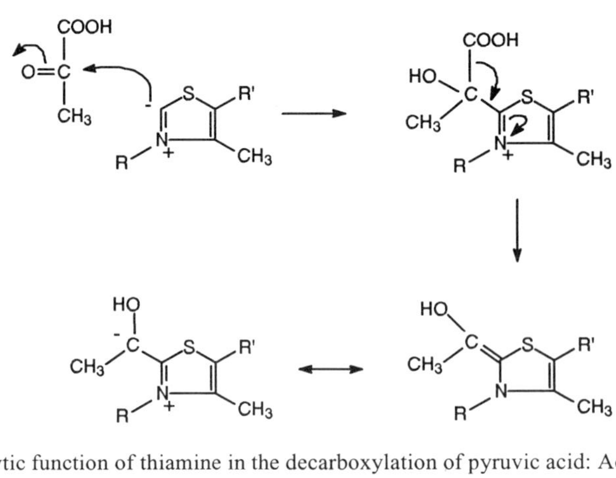

Figure 26. Catalytic function of thiamine in the decarboxylation of pyruvic acid: Active Acetaldehyde.

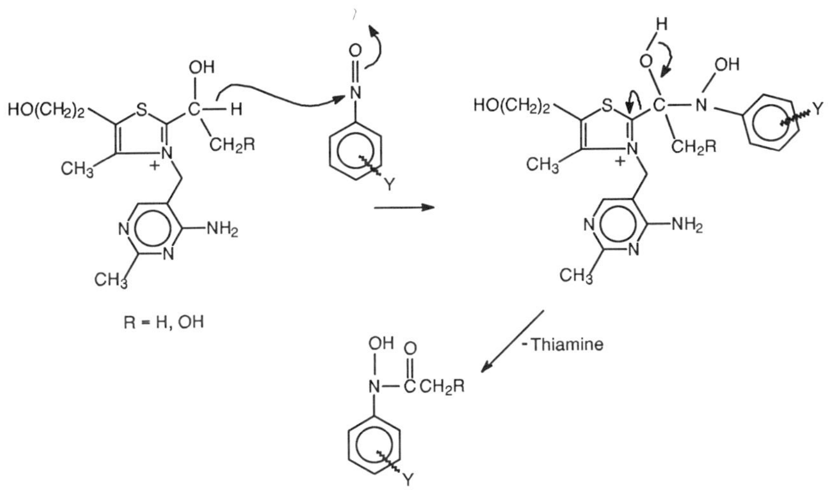

Figure 27. Proposed nucleophilic mechanism for the reaction of nitrosoarene compounds with thiamine-dependent enzymes.

A, becomes acetyl-CoA. This reaction is often depicted as an electron-pair displacement, but two sequential electron transfer processes might be possible. The question arose as to what might happen if the nucleophilic intermediates of thiamine catalysis were to react with nitrosoarene compounds since a theoretical assessment suggested that the analogous arylhydroxamic acid would be produced (8).

A test of the enzymatic reactions based on the mechanism proposed in Figure 27 revealed that the reactions indeed occurred, but that the mechanism might not be the result of nucleophilic addition of the enzyme-bound anion to the nitrosoarene group. In the case of pyruvate decarboxylase, nitrosoarene compounds were converted to a mixture of the expected arylhydroxamic acid compound with the analogous arylhydroxylamine compound (12). Transketolase showed much less tendency to produce the arylhydroxylamine compound as a coproduct of the arylhydroxamic acid (10, 12, 19). On the other hand, α-KGD not only produced the arylhydroxylamine product but also appeared to catalyze a Bamberger rearrangement of this coproduct. The formation of significant amounts of arylhydroxylamine product relative to the arylhydroxamic acid product has prompted a modification of our proposed mechanism to account for this observation. The possibility that the putative intermediate (A in Fig. 28) might also eliminate the arylhydroxylamine product with formation of enzyme-bound "active acetate" (B in Fig. 28) is a slight modification of the initially proposed mechanism (12, 18) but can explain our observations with both pyruvate decarboxylase and transketolase. A more extreme alternative is a mechanism in which the interaction of the nitroso group with the "active aldehyde" was either a two-electron plus H^+ transfer or a hydride ion transfer (12, 18). This is a modification of the mechanism in Figure 28 and also predicts the formation of the arylhydroxylamine as a coproduct with the arylhydroxamic acid. The mechanism in Figure 28 can also explain the strange occurrence of the Bamberger rearrangement during the interaction of nitrosoarene compounds with α-KGD as previously presented.

Certain of the arylhydroxamic acid products from these thiamine-enzyme dependent conversions are quite unusual, in that no other biochemical system has been found to produce arylhydroxamic acids with acyl groups derived from glycolic or succinic acid (Fig. 29). Similar conversions have been observed in more complex biochemical systems which

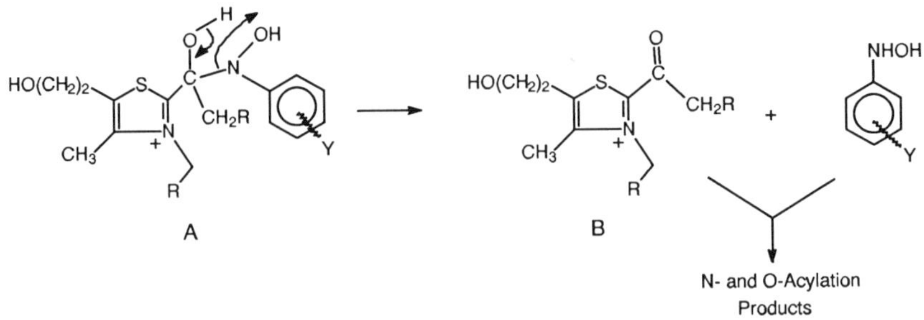

Figure 28. Modification of the nucleophile mechanism for the reaction of nitrosoarene compounds with thiamine-dependent enzymes. Structure A is the putative reaction intermediate at the enzyme active-site. B is "active acetate" when R=H.

suggest that these reactions probably occur in various organisms *in vivo* (9, 11, 90, 91). The significance of these unusual structural types of arylhydroxamic acids is that their subsequent metabolism might be considerably different from that of the N-acetyl type of arylhydroxamic acid for which the metabolism has been well characterized. In the few systems tested to date, the glycolyl type of arylhydroxamic acid has been observed to cause more DNA binding (20) and different DNA adducts (27, 66) than does the analogous acetyl arylhydroxamic acid.

The Nitroso-Glyoxylate Reaction

There is no obvious reason to expect that a nitrosoarene compound would react chemically with glyoxylic acid. That pyruvic acid does not react with any nitrosoarene compound had been repeatedly established from the control experiments of pyruvate dehydrogenase studies; and the structural difference between pyruvate and glyoxylate provided no basis on which to expect these acids to behave so differently. Nevertheless, in 1980 we observed a facile reaction between glyoxylate and 4-chloronitrosobenzene that gave the product we had thought would arise only through enzyme catalysis (13). Radioisotope studies established that decarboxylation of glyoxylate accompanied the formation of the N-formyl hydroxamic acid (Fig. 30). The generality of the reaction extends to most nitrosoarene compounds studied to date, and thus, was called the "nitroso-glyoxylate" reaction (18).

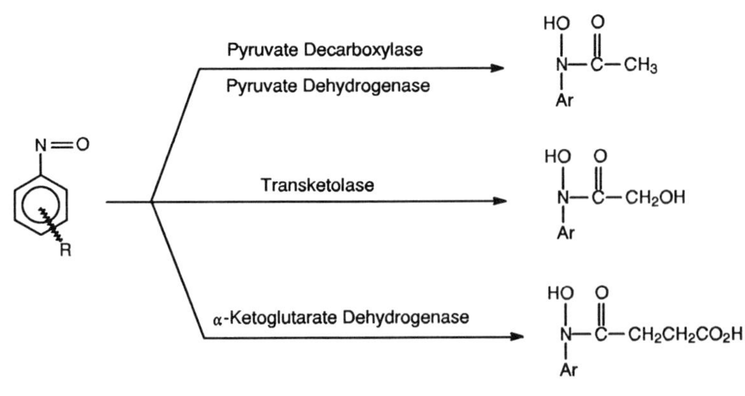

Figure 29. General pathways for hydroxamic acid formation by thiamine-dependent enzymes.

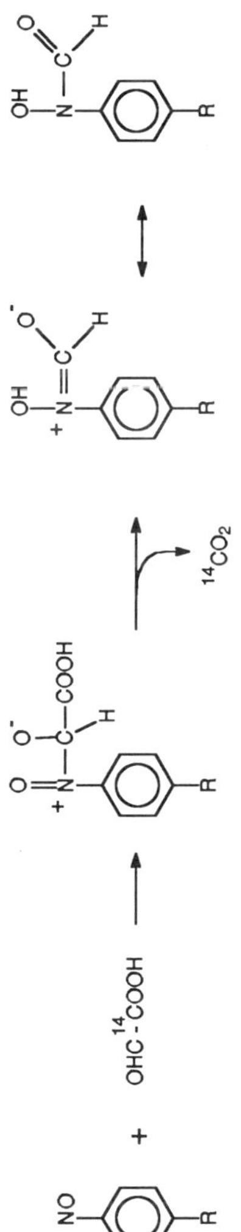

Figure 30. The Nitroso-Glyoxylate Reaction. Initial observations and earliest proposed mechanism.

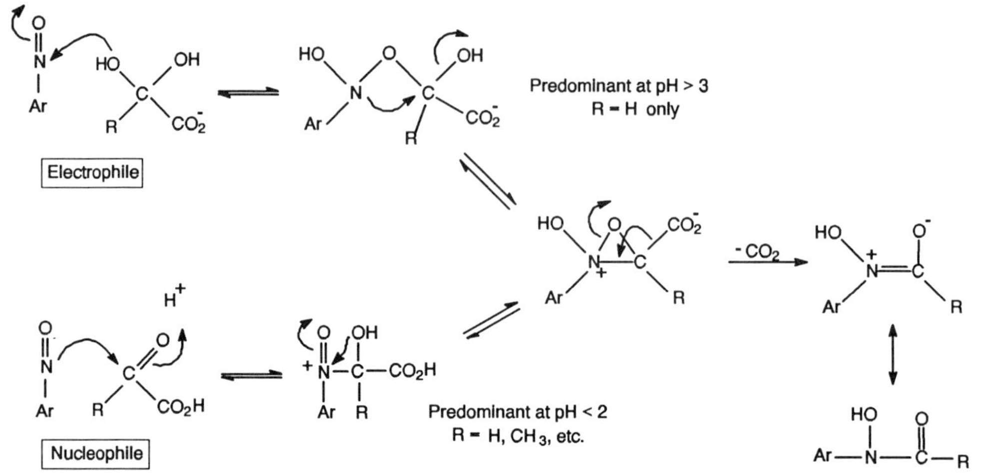

Figure 31. Electrophilic and nucleophilic roles of the nitroso functional group in reactions of nitrosoarene compounds with α-oxoacids.

Kinetic studies showed the reaction to follow a second order rate equation. An unusual observation at the time, but which later proved to be a key point in mechanistic considerations, was the strong solvent dependency of the reaction rate. The nitroso-glyoxylate reaction appears to be restricted to aqueous solvents. The reaction rate is largely independent of pH above 5 and the rate decreased markedly as the pH was decreased from 5 to about 2.

A Hammett analysis of the effect of ring substituents on reaction rates indicated that ρ^0 for this reaction was -1.7 at pH 6.0 when σ^0 were employed; but σ^+ did not give a good linear correlation. This suggests that the reaction is favored by an increased electron density at the nitrogen atom of the nitrosoarene compound but that charge stabilization via resonance interaction is not important. These results led to our proposed mechanism for the nitroso glyoxylate reaction (Fig. 31, bottom), in which the nitroso nitrogen atom behaves as a nucleophile towards the aldehyde group of glyoxylate (13). This is a very unusual mechanism in that the weak nucleophilic properties of the nitroso group are seldom employed to explain the reactions of nitrosoarene compounds.

One weakness of the mechanism in Figure 30 and Figure 31, bottom, is that it offers no obvious explanation as to why the same reaction doesn't occur with pyruvate and other α-oxoacids. Subsequent studies led to observations that pyruvate and certain other α-oxoacids can be forced to convert nitrosoarene compounds to the corresponding arylhydroxamic acid under acidic conditions (22, 86). An expanded study of the effect of pH on these reactions, with particular interest in the more acidic range below pH 2, produced a valuable observation that led to the recognition of two distinct reaction pathways for glyoxylate, but only a single possibility for pyruvate and higher analogs (22). The pH rate profiles of the reactions of nitrosoarene compounds with glyoxylate and pyruvate were found to be markedly different. Only for glyoxylate was the rate profile curve for the reaction suggestive of a pH-dependent change in reaction mechanism. Furthermore, both glyoxylate and pyruvate react with nitrosoarene compounds in non-aqueous systems under acidic conditions (86), but the reaction specific to glyoxylate appeared to be restricted to aqueous solvents. This immediately suggested that the form of glyoxylate that is reacting with the nitroso group (pH> 2) must be the hydrate of the aldehyde group of glyoxylate. Pyruvate is poorly hydrated relative to glyoxylate; therefore, its reactivity with nitrosoarene compounds at higher pH becomes negligible.

The proposal shown in Figure 31 incorporates this new information into a mechanism that can account for the unusual behavior of glyoxylate with nitrosoarene compounds in aqueous solution (22). In this mechanism (Fig. 31, top) the nitroso group behaves as an electrophile in the first step, with the -OH groups of glyoxylate hydrate acting as nucleophiles. This is nearly the opposite of the first step of the mechanism illustrated at the bottom of Figure 31. The second step in the electrophile mechanism of Figure 31 proposes an intramolecular S_N2 displacement of ^-OH by the nitrogen atom of the first intermediate to produce the oxaziridine-N-oxide intermediate. This step is likely to be a reversible equilibrium and precedes the probable rate determining step, which is decarboxylation. The first intermediate in the electrophile mechanism is expected to have a highly nucleophilic nitrogen, since its molecular structure is analogous to an arylhydroxylamine compound. The nucleophilicities of either the nitroso or the hydroxylamine groups are expected to increase with increased electron donation by ring substituents; thus, the ρ of -1.7 determined with σ^0 is consistent with either of these mechanisms.

The nucleophilic mechanism that we originally proposed for the nitroso-glyoxylate reaction (13) actually describes the H^+ catalyzed reaction (Fig. 31) (22). A more detailed Hammett analysis supports this proposal, in that the nucleophilic mechanism of Figure 31 is consistent with a correlation of reaction rates with σ^+. On the other hand, the electrophile mechanism is expected to give a better correlation with σ than with σ^+. This is consistent with our observations from a study of substituent group effects on the reaction rate at near neutral pH. Thus, the electrophile mechanism of Figure 31 is the probable mechanism of the nitroso-glyoxylate reaction while the nucleophile mechanism of Figure 31 is thought to describe the general H^+-dependent reactions of α-oxoacids (22). Only the true nitroso-glyoxylate reaction is potentially significant under normal physiological conditions. The H^+ catalyzed chemical reaction of nitrosoarene compounds with pyruvate and higher α-oxoacids is highly improbable, except possibly in gastric fluid.

The pathways of Figure 31 suggest that the nitrosoarene group is unique, in that it can react by two different mechanisms to produce the same terminal functional group. The two mechanisms merge at the putative oxaziridine-N-oxide intermediate. This intermediate point can be attained as the result of the nitrosoarene group serving as either a nucleophile or electrophile. Additional research is needed to confirm this dual role of the nitrosoarene group, particularly its role as a nucleophile, since this property has been rarely invoked to explain a chemical reaction of the nitrosoarene functional group. In one of its reactions, the nitrosoarene group was initially thought to be a nucleophile, but is now thought to proceed via a free radical process (37). The dual role of the nitrosoarene group both as a nucleophile and electrophile is most often invoked to describe the reversible dimerization reaction of nitrosoarene compounds (45). In solution most nitrosoarene compounds exist as monomers, while in the crystalline state, the dimer is the usual molecular form.

PERSPECTIVE

The bioreduction of nitroaromatic compounds can proceed with the formation of the intermediary nitroso and hydroxylamine metabolites. That the fates of these intermediates in a microbial milieu are potentially numerous should be evident from this presentation of their chemical properties. Much of this information originated from research on the mechanisms by which arylhydroxylamine and nitrosoarene compounds cause toxicity in higher organisms. Covalent binding of these intermediates to cellular macromolecules can occur either directly in the case of nitrosoarene compounds and proteins, or indirectly in the case of arylhydroxylamine compounds and both proteins and nucleic acids. Because of these possibilities, the potential for toxicity (74) to a microbial population might be a limiting factor. As attempts are made to expand microbial degradation processes to other nitroaromatic pollutants, it will be valuable to understand how chemical structure can affect the rates of reactions of the nitroso and hydroxylamine intermediates. The proclivity of a specific arylhydroxylamine compound to produce aminophenols via the Bamberger rearrangement is likely to affect the ability of a microbial arylhydroxylamine mutase to catalyze such a rearrangement. Similarly, the relative rates of competing reactions could compromise the potential for a microorganism to efficiently direct the metabolism of a nitroaromatic compound through a pathway that results in mineralization.

REFERENCES

1. Becker, A. R., and L. A. Sternson. 1980. Nonenzymatic reduction of nitrosobenzene to phenylhydroxylamine by NAD(P)H. Bioorg. Chem. 9:305-312.
2. Bernheim, M. L. C. 1972. The non-enzymic oxidation of NADH by nitrosobenzene. Biochem. Biophys. Res. Commun. 46:1598-1602.
3. Boche, G., F. Bosold and S. Schröder. 1988. N-Aryl-O-acylhydroxylamines; preparation by O-acylation or N→O transacylation and reaction with amines; model reactions for key steps connected with the carcinogenicity of aromatic amines. Angew. Chem. Int. Ed. Engl. 27: 973-974.
4. Boche, G., R. H. Sommerlade and F. Bosold. 1986. N-Aryl-O-(diphenylphosphinoyl)hydroxylamines: electrophilic amination of amines to hydrazines; a model reaction for the carcinogenicity of aromatic amines. Angew. Chem. Int. Ed. Engl. 25: 562-563.
5. Bosold, F. and G. Boche. 1990. The ultimate carcinogen, O-acetyl-N-2-(fluorenyl)hydroxylamine ("N-acetoxy-2-aminofluorene"), and its reaction in vitro to form 2-[N-(deoxyguanosin-8-yl)amino]fluorene. Angew. Chem. Int. Ed. Engl. 29: 63-64.
6. Boteju, L. W. and P. E. Hanna. 1993. Bioactivation of N-hydroxyaminofluorenes by N,O-acyltransferase: substituent effects on covalent binding to DNA. Carcinogenesis 14:1651-1657.
7. Channon, H. J., G. T. Mills, and R. T. Williams. 1944. The metabolism of 2,4,6-trinitrotoluene. Biochem. J. 38: 70-85.
8. Corbett, M. D. 1974. Hydroxamic acids from the reaction of active acetaldehyde with aromatic nitroso compounds. Bioorg. Chem. 3:361-365.
9. Corbett, M. D., D. G. Baden and B. R. Chipko. 1979. The nonmicrosomal production of N-(4-chlorophenyl)-glycolhydroxamic acid from 4-chloronitrosobenzene by rat liver homogenates. Bioorg. Chem. 8:227-235.
10. Corbett, M. D. and B. R. Chipko. 1977. N-Phenylglycolylhydroxamate production by the action of transketolase on nitrosobenzene. Biochem. J. 165:263-267.
11. Corbett, M. D., B. R. Chipko and J. H. Paul. 1978. The production of hydroxamic acid metabolites of nitrosobenzene by *Chlorella pyrenoidosa*. J. Environ. Path. Toxicol. 1:259-266.
12. Corbett, M. D. and B. R. Chipko. 1980. Comparative aspects of hydroxamic acid production by thiamine-dependent enzymes. Bioorg. Chem. 9:273-287.
13. Corbett, M. D. and B. R. Corbett. 1980. The reaction of nitroso aromatics with glyoxylic acid: a new path to hydroxamic acids. J. Org. Chem. 45:2834-2838.
14. Corbett, M. D. and B. R. Corbett. 1981. Reductive formylation of N,N-dimethyl-*p*-nitrosoaniline by glyoxylic acid. Evidence for a hydroxamic acid intermediate. J. Org. Chem. 46:466-468.
15. Corbett, M. D. and B. R. Corbett. 1981. Metabolism of 4-chloronitrobenzene by the yeast *Rhodosporidium* sp. Appl. Environ. Microbiol. 41:942-949.
16. Corbett, M. D. and B. R. Corbett. 1982. Enzymic generation of N-[4-(dimethylamino)phenyl]acetohydroxamic acid by the action of pyruvate decarboxylase on 4-(dimethylamino)nitrosobenzene. Bioorg. Chem. 11:328-337.
17. Corbett, M. D. and B. R. Corbett. 1983. Enzymatic preparation of [U-^{14}C]-4-chloronitrosobenzene. Experientia 39:487-488.
18. Corbett, M. D. and B. R. Corbett. 1985. The reactions of C-nitroso aromatics with α-oxo acids, p. 400-408. *In* J.W. Gorrod and L.A. Damani (ed.), Biological oxidation of nitrogen in organic molecules. Ellis Horwood, Ltd., Chichester.

19. Corbett, M. D. and B. R. Corbett. 1986. Effect of ring substituents on the transketolase catalyzed conversion of nitroso aromatics to hydroxamic acids. Biochem. Pharmacol. 35:3613-3621.
20. Corbett, M. D. and B. R. Corbett. 1987. HRP-catalyzed bioactivation of carcinogenic hydroxamic acids. The greater reactivity of glycolyl- versus acetyl-derived hydroxamic acids. Chem.-Biol. Interact. 63:249-264.
21. Corbett, M. D. and B. R. Corbett. 1990. Biochemical studies on the putative nitroso metabolite of chloramphenicol: a new model for the cause of aplastic anemia, p. 245-255. *In* P.C. Howard, S.S. Hecht and F.A. Beland (ed.), Nitroarenes: occurrence, metabolism and biological impact. Plenum Press, New York.
22. Corbett, M. D. and B. R. Corbett. 1993. Studies on the nitroso-glyoxylate reaction. Relative hydroxamic acid production by glyoxylate, pyruvate, and formaldehyde in reactions with 4-nitrosobiphenyl. Chem. Res. Toxicol. 6:82-90.
23. Corbett, M. D., B. R. Corbett and A. O. Batchelor. 1980. The action of chloride peroxidase on 4-chloroaniline. Biochem. J. 187:893-903.
24. Corbett, M. D., B. R. Corbett and D. R. Doerge. 1982. Hydroxamic acid production and active-site induced Bamberger rearrangement from the action of α-ketoglutarate dehydrogenase on 4-chloronitrosobenzene. J. Chem. Soc. Perkin I 1982:345-350.
25. Corbett, M. D., B. R. Corbett, M.-H. Hannothiaux and S. J. Quintana. 1992. The covalent binding of acetaminophen to cellular nucleic acids as the result of the respiratory burst of neutrophils derived from the HL-60 cell line. Toxicol. Appl. Pharmacol. 113:80-86.
26. Corbett, M. D., D. R. Doerge and B. R. Corbett. 1983. Hydroxamic acid production by α-ketoglutarate dehydrogenase. Part 2. Evidence for an electrophilic reaction intermediate at the enzyme active site. J. Chem. Soc. Perkin I 1983:765-769.
27. Corbett, M. D., L. O. Lim, B. R. Corbett, J. J. Johnston and P. Wiebkin. 1988. Covalent binding of N-hydroxy-2-acetylaminofluorene and N-hydroxy-2-glycolylaminofluorene to rat hepatocyte DNA: In vitro and cell-suspension studies. Chem. Res. Toxicol. 1:41-46.
28. Corbett, M. D., C. Wei and B. R. Corbett. 1985. Nitroreductase-dependent mutagenicity of *p*-nitrophenylhydroxylamine and its N-acetyl and N-formyl hydroxamic acids. Carcinogenesis 6:727-732.
29. Cramer, J. W., J. A. Miller and E. C. Miller. 1960. A new metabolic reaction observed in the rat with the carcinogen 2-AAF. J. Biol. Chem. 235:885-888.
30. Doerge, D. R. and M. D. Corbett. 1985. The action of α-ketoglutarate dehydrogenase on 4-chloronitrosobenzene: evidence for species-dependent differences in active site properties. Comp. Biochem. Physiol. 80C:161-165.
31. Doerge, D. R. and M. D. Corbett. 1991. Peroxygenation mechanism for chloroperoxidase-catalyzed N-oxidation of arylamines. Chem. Res. Toxicol. 4:556-560.
32. Dölle, B., W. Töpner and H.-G. Neumann. 1980. Reaction of arylnitroso compounds with mercaptans. Xenobiotica 10:527-536.
33. Eyer, P. 1979. Reactions of nitrosobenzene with reduced glutathione. Chem.-Biol. Interact. 24:227-239.
34. Eyer, P. 1985. Reactions of nitrosoarenes with sulphydryl groups: reaction mechanism and biological significance, p. 386-399. *In* J.W. Gorrod and L.A. Damani (ed.), Biological oxidation of nitrogen in organic molecules. Ellis Horwood, Ltd., Chichester.
35. Eyer, P. 1988. Detoxication of N-oxygenated arylamines in erythrocytes. Xenobiotica 18:1327-1333.
36. Famulok, M., F. Bosold and G. Boche. 1989. Synthesis of N-acetoxy-2-aminonaphthalene, an ultimate carcinogen of the carcinogenic 2-naphthylamine, and its *in vitro* reactions with (bio)nucleophiles. Tetrahedron Lett. 30:321-324.
37. Forrester, A. R. and R. H. Thomson. 1985. Reaction of quinones with nitrosoarenes. Z. Naturforsch. 40B:1515-1518.
38. Gassman, P. G. and G. A. Campbell. 1971. The mechanism of the chlorination of anilines and related aromatic amines. The Involvement of nitrenium ions. J. Amer. Chem. Soc. 93:2567-2569.
39. Gassman, P. G., G. A. Campbell and R. C. Frederick. 1972. Nucleophilic aromatic substitution of anilines *via* aryl nitrenium ions (Anilenium Ions). J. Amer. Chem. Soc. 94:3884-3890.
40. Gassman, P. G. and J. E. Granrud. 1984. Synthesis and rearrangement of methanesulfonate esters of N-hydroxyacetanilides. A model for a penultimate carcinogen. J. Amer. Chem. Soc. 106:1498-1499.
41. Gassman, P. G. and J. E. Granrud. 1984. Isolation, characterization, and rearrangement of *cis*- and *trans*-N-acetyl-2-amino-5,6-dimethoxy-5-methylcyclohexa-1,3-diene. Models for the proposed precursors of meta-substituted products from carcinogenic aromatic amines. J. Amer. Chem. Soc. 106:2448-2449.

42. Gilissen, R. A. H. J., D. P. Ringer, H. J. F. C. Stavenuiter, G. J. Mulder and J. H. N. Meerman. 1992. Sulfation of hydroxylamines and hydroxamic acids in liver cytosol from male and female rats and purified aryl sulfotransferase IV. Carcinogenesis 13:1699-1703.
43. Gorrod, J. W. and D. Manson. 1986. The metabolism of aromatic amines. Xenobiotica 16:933-955.
44. Gowenlock, B. G. and W. Lüttke. 1958. Structure and properties of C-nitroso-compounds. Quarterly Rev. 12:321-340.
45. Gowenlock, B. G. and K. J. McCullough. 1989. Structure of the *trans*-dimer of 2,6-di-isopropylnitrosobenzene. J. Chem. Soc. Perkin Trans. II 1989:551-553.
46. Groenewegen, P. E. J., P. Breeuwer, J. M. L. M. van Helvoort, A. A. M. Langenhoff, F. P. deVries and J. A. M. deBont. 1992. Novel degradative pathway of 4-nitrobenzoate in *Comamonas acidovorans* NBA-10. J. Gen. Microbiol. 138:1599-1605.
47. Guengerich, F. P. and T. Shimada. 1991. Oxidation of toxic and carcinogenic chemicals by human cytochrome P-450 enzymes. Chem. Res. Toxicol. 4:391-407.
48. Haigler, B. E. and J. C. Spain. 1993. Biodegradation of 4-nitrotoluene by *Pseudomonas* sp. strain 4NT. Appl. Environ. Microbiol. 59:2239-2243.
49. Halliwell, B., J. M. C. Gutteridge, and C. E. Cross. 1992. Free radicals, antioxidants and human disease: where are we now. J. Lab. Clin. Med. 119:598-620.
50. Hanna, P. E. and R. B. Banks. 1985. Arylhydroxylamines and arylhydroxamic acids: conjugation reactions, p. 375-402. *In* M.W. Anders (ed), Bioactivation of foreign compounds. Academic Press, Orlando.
51. Hein, D. W. 1988. Acetylator genotype and arylamine-induced carcinogenesis. Biochim. Biophys. Acta 948:37-66.
52. Helmick, J. S., K. A. Martin, J. L. Heinrich and M. Novak. 1991. Mechanism of the reaction of carbon and nitrogen nucleophiles with the model carcinogens O-pivaloyl-N-arylhydroxylamines: competing S_N2 substitution and S_N1 solvolysis. J. Amer. Chem. Soc. 113:3459-3466.
53. Helmick, J. S. and M. Novak. 1991. Nucleophilic substitution at nitrogen and carboxyl carbon of N-aryl-O-pivaloylhydroxylamines in aqueous solution: competition with S_N1 solvolysis of model carcinogens. J. Org. Chem. 56:2925-2927.
54. Hlavica, P. 1982. Biological oxidation of nitrogen in organic compounds and disposition of N-oxidized products. CRC Crit. Rev. Biochem. 12:39-101.
55. Hlavica, P., I. Golly and J. Mietaschk. 1983. Comparative studies on the cumene hydroperoxide- and NADPH-supported N-oxidation of 4-chloroaniline by cytochrome P-450. Biochem. J. 212:539-547.
56. Ibne-Rasa, K. M. and J. O. Edwards. 1962. The mechanism of the oxidation of some aromatic amines by peroxyacetic acid. J. Amer. Chem. Soc. 84:763-768.
57. Kalhorn, T., A. R. Becker and L. A. Sternson. 1981. Evidence for general catalysis and formation of nitrobenzene in the oxidation of phenylhydroxylamine in aqueous phosphate buffer. Bioorg. Chem. 10:144-151.
58. Kazanis, S. and R. A. McClelland. 1992. Electrophilic intermediate in the reaction of glutathione and nitrosoarenes. J. Amer. Chem. Soc. 114:3052-3059.
59. Klehr, H., P. Eyer and W. Schäfer. 1985. On the mechanism of reactions of nitrosoarenes with thiols. Biol. Chem. Hoppe-Seyler 366:755-760.
60. Knight, G. T. and B. Saville. 1973. Hydrogen transfer from N-arylhydroxylamines to nitrosoarenes: an accompaniment to azoxyarene formation. J. Chem. Soc. Perkin Trans. II 1973:1550-1553.
61. Koerber, S. C., P. Schack, A. M.-J. Au and M. F. Dunn. 1980. Investigation of a novel liver alcohol dehydrogenase catalyzed redox-elimination reaction involving arylnitroso substrate analogues. Biochemistry 19:731-738.
62. Kohnstam, G., W. A. Petch and D. L. H. Williams. 1984. Kinetic substituent and isotope effects in the acid-catalyzed rearrangement of N-phenylhydroxylamines. Are nitrenium ions involved? J. Chem. Soc. Perkin Trans. II 1984:423-427.
63. Kolanczyk, R. C., H. R. Gutmann and I. R. Rutks. 1992. Effect of bovine serum albumin on the extent of *ortho* rearrangement of N-(sulfooxy)-2-fluorenylacetamide and of enzymatically activated N-hydroxy-2-fluorenylacetamide and on the binding of reactive esters to nucleic acids. Chem. Res. Toxicol. 5:274-279.
64. Land, S. J., K. Zukowski, M.-S. Lee, C. Y. Wang and C. M. King. 1993. Purification and characterization of a rat hepatic acetyltransferase that can metabolize aromatic amine derivatives. Carcinogenesis 14:1441-1449.
65. Leskovac, V., J. Svircevic, S. Trivic, M. Popovic, and M. Radulovic. 1989. Reduction of aryl nitroso compounds by pyridine and flavin coenzymes. Int. J. Biochem. 21:825-834.
66. Lim, L. O., B. R. Corbett and M. D. Corbett. 1987. Irreversible inhibition of the cytosolic metabolism of N-hydroxy-2-acetylaminofluorene by its glycolyl analog. Cancer Lett. 37:205-211.

67. Liu, Y.-Y., A. Y. H. Lu, R. A. Stearns and S.-H. L. Chiu. 1992. In vivo covalent binding of [^{14}C]trinitrotoluene to proteins in the rat. Chem.-Biol. Interact. 82:1-19.
68. Lobo, A. M., M. M. Marques, S. Prabhakar and J. S. Rzepa. 1987. Tetrahedral intermediates formed by nitrogen and oxygen attack of aromatic hydroxylamines on acetyl cyanide. J. Org. Chem. 52:2925-2927.
69. Lutz, R. E. and M. R. Lytton. 1937. Oxidation-reduction potentials of a series of nitrosobenzene-phenylhydroxylamine systems. J. Org. Chem. 2:68-75.
70. Mangold, B. L. K., J. Erickson, C. Lohr, D. J. McCann and J. B. Mangold. 1990. Self-catalyzed irreversible inactivation of rat hepatic aryl sulfotransferase IV by N-hydroxy-2-acetylaminofluorene. Carcinogenesis 11:1563-1567.
71. Maples, K. R., P. Eyer and R. P. Mason. 1990. Aniline-, phenylhydroxylamine-, nitrosobenzene-, and nitrobenzene-induced hemoglobin thiyl free radical formation *in vivo* and *in vitro*. Molec. Pharmacol. 37:311-318.
72. Maskos, Z. and G. W. Winston. 1993. Alcohol dehydrogenase-dependent reduction of 2-nitrosofluorene and rearrangement of N-hydroxy-2-aminofluorene. Biochemistry 32:12768-12773.
73. Mason, R. P. 1982. Free-radical intermediates in the metabolism of toxic chemicals, p. 161-222. *In* W.A. Pryor (ed.), Free radicals in biology. Academic Press, New York.
74. Michels, J. and G. Gottschalk. 1994. Inhibition of the lignin peroxidase of *Phanerochaete chrysosporium* by hydroxylamino-dinitrotoluene, an early intermediate in the degradation of 2,4,6-trinitrotoluene. Appl. Environ. Microbiol. 60:187-194.
75. Minchin, R. F., K. F. Ilett, C. H. Teitel, P. T. Reeves and F. F. Kadlubar. 1992. Direct O-acetylation of N-hydroxy arylamines by acetylsalicylic acid to form carcinogen-DNA adducts. Carcinogenesis 13:663-667.
76. Nishino, S. F. and J. C. Spain. 1993. Degradation of nitrobenzene by a *Pseudomonas pseudoalcaligenes*. Appl. Environ. Microbiol. 59:2520-2525.
77. Novak, M., M. J. Kahley, E. Eiger, J. S. Helmick and H. E. Peters. 1993. Reactivity and selectivity of nitrenium ions derived from ester derivatives of carcinogenic N-(4-biphenyl)hydroxylamine and the corresponding hydroxamic acid. J. Amer. Chem. Soc. 115:9453-9460.
78. Novak, M. and R. K. Lagerman. 1988. Hydrolysis of Fe^{2+}-induced reduction of N-aryl-O-pivaloylhydroxylamines: aqueous solution chemistry of model carcinogens. J. Org. Chem. 53:4762-4769.
79. Novak, M., K. A. Martin and J. L. Heinrich. 1989. S_N2 reactions of a carbon nucleophile with N-aryl-O-pivaloylhydroxylamines: a model for in vivo reactions of carcinogenic metabolites of aromatic amines. J. Org. Chem. 54:5430-5431.
80. Novak, M., M. Pelecanou. A. K. Roy, A. F. Andronico, F. M. Plourde, T. M. Olefirowicz and T. J. Curtin. 1984. Solvolysis of N-sulfonoxyacetanilides in aqueous and alcohol solutions: generation of electrophilic species. J. Amer. Chem. Soc. 106:5623-5631.
81. Novak, M. and A. K. Roy. 1985. Hydrolysis of N-(sulfonatooxy)-*p*-acetotoluidide: solution chemistry of models for carcinogenic metabolites of aromatic amides. J. Org. Chem. 50:571-580.
82. Panda, M., M. Novak and J. Magonski. 1989. Hydrolysis kinetics of the ultimate hepatocarcinogen N-(sulfonatooxy)-2-(acetylamino)fluorene: detection of long-lived hydrolysis intermediates. J. Amer. Chem. Soc. 111:4524-4525.
83. Pearson, R. G. and J. Songstad. 1967. Application of the principle of hard and soft acids and bases to organic chemistry. J. Amer. Chem. Soc. 89:1827-1836.
84. Pelecanou, M. and M. Novak. 1985. Oxidation-reduction reactions of N-sulfonoxyacetanilides: mechanisms of the halide-induced reduction of models for the carcinogenic metabolites of aromatic amides. J. Amer. Chem. Soc. 107:4499-4503.
85. Ringer, D. P., T. R. Norton and R. R. Self. 1992. Reaction product inactivation of aryl sulfotransferase IV following electrophilic substitution by the sulfuric acid ester of N-hydroxy-2-acetylaminofluorene. Carcinogenesis 13:107-112.
86. Sakamoto, Y., T. Yoshioka and T. Uematsu. 1989. N-Arylhydroxamic acids: reaction of nitroso aromatics with α-oxo acids. J. Org. Chem. 54:4449-4453.
87. Sone, T., Y. Tokuda, T. Sakai, S. Shinkai and O. Manabe. 1981. Kinetics and mechanisms of the Bamberger rearrangement. Part 3. Rearrangement of phenylhydroxylamines to *p*-aminophenols in aqueous sulphuric acid solution. J. Chem. Soc. Perkin. Trans. II 1981:298-302.
88. Ulbrich, R., M. Famulok, F. Bosold and G. Boche. 1990. S_{N2} at nitrogen: the reaction of N-(4-cyanophenyl)-O-diphenylphosphinoylhydroxylamine with N-methylaniline. A model for the reactions of ultimate carcinogens of aromatic amines with (bio)nucleophiles. Tetrahedron Lett. 31:1689-1692.
89. Weisburger, E. K. 1978. Mechanism of chemical carcinogenesis. Ann. Rev. Pharmacol. Toxicol. 18:395-415.

90. Yoshioka, T., T. Suzuki and T. Uematsu. 1989. Biotransformation of N-substituted aromatic compounds in mammalian spermatozoa. Nonoxidative formation of N-hydroxy-N-arylacetamides from nitroso aromatic compounds. J. Biol. Chem. 264:12432-12438.
91. Yoshioka, T. and T. Uematsu. 1993. Formation of N-hydroxy-N-arylacetamides from nitrosoaromatic compounds by the mammalian pyruvate dehydrogenase complex. Biochem. J. 290:783-790.

11

TRANSFORMATION OF INORGANIC N-OXIDES BY DENITRIFYING AND NITRIFYING BACTERIA

Pathways, Mechanisms, and Relevance to Biotransformation of Nitroaromatic Compounds

Bruce A. Averill

E. C. Slater Instituut
Universiteit van Amsterdam
Plantage Muidergracht 12
1018 TV Amsterdam
The Netherlands

INTRODUCTION

The inorganic nitrogen cycle (Fig. 1) consists of several linked biological processes and one abiological process, the reaction of N_2 and O_2 in lightning discharges (and internal combustion engines) to produce NO_x and, ultimately, nitrate. All of the processes except nitrogen fixation involve reduction or oxidation of species containing N-O bonds of order greater than one, such as NO_3^-, NO_2^-, NO, and N_2O. Except for the assimilatory reduction of nitrite to ammonium, which is a complex six-electron process that is unique in nitrogen biochemistry, the reactions of inorganic N-oxides are of potential interest as relatively well-studied paradigms that may be relevant to aspects of biological degradation of nitroaromatic compounds. One might expect the existence of analogous chemistry in the reduction of NO_2^- or HNO_2 and $ArNO_2$ (where Ar is a general abbreviation for an aromatic group). Certainly both the inorganic and organic species contain nitrogen in the same oxidation state (+ 3), and they also contain two N-O bonds that possess at least partial multiple bond character.

The purpose of this paper is to explore the potential analogies between the two major biological processes, nitrification and denitrification, that interconvert inorganic nitrogen-oxygen species and the known biochemistry of degradation of nitroaromatic compounds, as developed at length in many of the other papers in this volume. To that end, the chapter will begin with a brief summary of what is known regarding the pathway of nitrification and the mechanisms of the enzymes involved, followed by a consideration of the better understood pathway of denitrification and the underlying enzymatic mechanisms. With this as a basis, the known reactions of nitroaromatic compounds will be examined for possible analogies

Biodegradation of Nitroaromatic Compounds
Edited by J. Spain, Plenum Press, New York, 1995

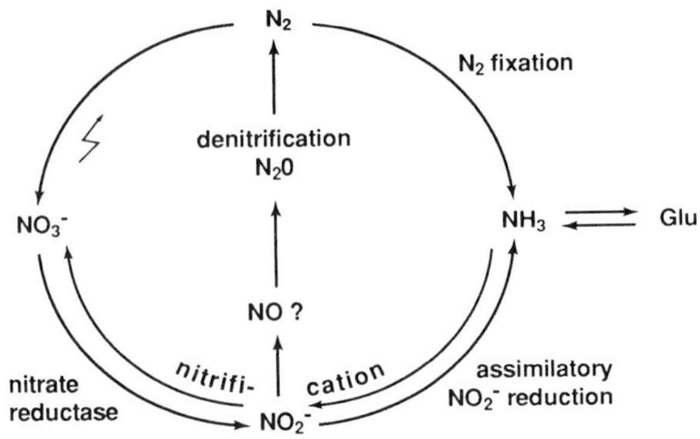

Figure 1. The inorganic nitrogen cycle.

and the likelihood of the existence of other, as yet undiscovered, reactions will be discussed. To anticipate slightly, one of the most important conclusions to be reached in this analysis is that direct analogies between reactions of inorganic and organic nitrogen species in the same oxidation state are *not* to be expected. Instead, one must focus on the arrangement of electrons in the valence shell of the nitrogen atom to begin to see at least the possibility of parallel reactivity.

NITRIFICATION

Nitrification is the process by which several genera of soil bacteria are able to grow using reduced nitrogen species as their sole energy source (32). Nitrification is a process of major agricultural importance, because it converts NH_4^+ in soils to NO_2^- and NO_3^-, the major form in which fixed nitrogen is taken up by plants. Unfortunately, however, NO_2^- and NO_3^- are both susceptible to leaching from the soil and are also substrates for denitrifying microorganisms that compete rather successfully with crops for these nutrients. As a result of the combined effects of leaching and denitrification, up to 30-40% of fertilizer ammonia can be lost from soils. Consequently, nitrapyrin, an inhibitor of ammonia oxidation by *Nitrosomonas,* has found some utility in inhibiting nitrification or in controlling the rate of nitrate production, such that nitrate is produced no faster than it can be utilized by crops. Nitrification is also potentially important for removal of toxic ammonium during bioremediation of waste water.

The pathway of nitrification is shown in Figure 2. Unlike denitrification, and most other processes found in chemolithotropic organisms, nitrification is unique in that it requires the presence of at least two distinct genera of bacteria to convert ammonium to nitrate. For example, members of the genus *Nitrosomonas* catalyze the oxidation of ammonium to nitrite via hydroxylamine, while members of the genus *Nitrobacter* carry out the oxidation of nitrite to nitrate and cannot utilize ammonium or hydroxylamine for growth. In contrast to denitrification, the enzymes involved in nitrification have proven to be difficult to isolate in

$$NH_3 \longrightarrow NH_2OH \longrightarrow NO_2^- \longrightarrow NO_3^-$$

Nitrosomonas *Nitrobacter*

Figure 2. The pathway of nitrification.

pure, active forms. Consequently, most of them remain poorly characterized, and comparatively little mechanistic information is available (16).

The first step in the nitrification process is the oxidation of ammonia to hydroxylamine by the enzyme ammonia monooxygenase (AMO) via the reaction:

$$2H^+ + NH_3 + 2e^- + O_2 \rightarrow NH_2OH + H_2O.$$

Although this appears to be a typical monooxygenase reaction of the type catalyzed by cytochromes P450, there is no evidence for the involvement of a cytochrome in the reaction. Instead, the activity is localized in the membrane fraction of *Nitrosomonas* and exhibits a sensitivity to metal chelators consistent with the presence of copper (17), suggesting a possible analogy with the copper-containing, membrane-bound methane monooxygenase (MMO) of methanotrophic bacteria (5). An additional similarity between the two systems is the broad range of substrates that can be oxidized by each. MMO can hydroxylate a range of hydrocarbons in addition to methane, while AMO can oxidize small olefins (e.g., C_2H_4), aromatics (e.g., C_6H_6), alkanes (e.g., C_6H_{12}), and phenol in addition to ammonia (16). ^{18}O labelling experiments have clearly shown that the source of oxygen in hydroxylamine is O_2 (15).

The second step in the nitrification process is the oxidation of hydroxylamine to nitrite by the enzyme hydroxylamine oxidoreductase (HAO) via the reaction:

$$NH_2OH + H_2O \rightarrow HONO + 4e^- + 4H^+.$$

HAO is a remarkably complex enzyme, containing eight covalently bound hemes per 63 kDa subunit in an *a*2 or *a*3 oligomer. Seven of these hemes appear to be variants on the cytochrome *c* heme. The eighth, which had long been known to exhibit anomalous spectral properties, including an intense band at 463nm upon reduction, and was thought to be the substrate binding site, has recently been shown to contain a covalently modified heme *c* with a bond between a *meso* carbon atom of the heme and a tyrosyl aromatic ring (3) (Fig.3). Why oxidation of hydroxylamine should require such complex enzymatic machinery is not at all clear. ^{18}O-labelling studies monitored by ^{15}N-NMR have shown that one of the oxygen atoms in 15 $^{15}NO_2^-$ produced from $^{15}NH_4^+$ originates from O_2 and the second from H_2O (2). These

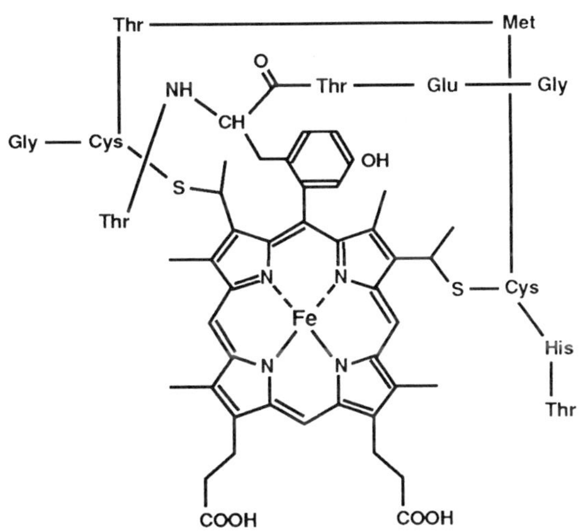

Figure 3. Proposed structure for the P-460 cofactor of hydroxylamine oxidoreductase (3).

results are consistent with sequential dehydrogenation of NH_2OH bound to the P-460 iron center to coordinated HNO and then to coordinated NO^+, followed by hydration and release as nitrite.

Cultures of *Nitrosomonas* produce significant amounts of N_2O and NO under certain conditions (16). Some of the gaseous products may be due to loss of intermediates in NH_2OH oxidation (i.e., HNO or NO) from HAO. In addition, however, a soluble copper protein that contains both Type 1 and Type 2 Cu centers (9), exhibits nitrite reductase activity (27), and is induced at low O_2 concentrations (26) has been isolated from *Nitrosomonas,* suggesting a potential similarity to the copper nitrite reductases found in denitrifying bacteria (see below). Whether the organism obtains energy by reduction of nitrite, either of exogenous origin or produced by oxidation of hydroxylamine, remains unclear.

The final step in the nitrification process is the oxidation of nitrite to nitrate by the enzyme nitrite oxidase (NoX) or, more properly nitrite dehydrogenase, via the reaction:

$$NO_2 + H_2O \rightarrow NO_3 + 2e^- + 2H^+.$$

The electrons produced are transferred to oxygen via a cytochrome αa_3-type cytochrome *c* oxidase. The NoX enzyme is membrane-bound and has proven to be extremely difficult to purify. Preparations have been reported that contain Mo as well as both heme and non-heme iron and acid-labile sulfur (10). These results, together with the simplicity of the reaction catalyzed, strongly suggest that the enzyme is a typical molybdenum-containing oxotransferase with additional redox cofactors, analogous to the nitrate reductases isolated from many organisms. This hypothesis is also consistent with the rapid exchange of ^{18}O between $^{14}N^{18}O_3^-$ and $^{15}N^{16}O_3^-$ catalyzed by *Nitrobacter* during oxidation of nitrite, without exchange of oxygen with solvent water. Such O-atom transfer reactions are typical of molybdoenzymes (see below).

DENITRIFICATION

Denitrification is the anaerobic use by bacteria of N-oxides as terminal electron acceptors in place of O_2 (35). It is important because it constitutes the only process that returns large amounts of fixed nitrogen to the atmosphere, thereby completing the terrestrial nitrogen cycle (Fig. 1). It is also important commercially, in that denitrification, by itself or in combination with nitrification, can result in the loss of up to 30-40% of fixed nitrogen fertilizer. In addition, denitrification is a "leaky" process under many conditions, resulting in the release into the atmosphere of large amounts of N_2O, a greenhouse gas that is also implicated in atmospheric ozone depletion. Denitrifiers are potentially of great importance in bioremediation efforts, since dissolved nitrate concentrations are often easier to control than are oxygen concentrations. In addition, the well-studied enzymes of denitrification provide potential structural and spectroscopic models for mammalian enzymes that produce and utilize NO in a variety of signal transduction pathways (22).

The overall pathway of denitrification is shown in Figure 4. As shown, a total of four enzymatic steps are involved in most organisms. (A few organisms appear to lack the last enzyme, and consequently produce N_2O as the major product.) Certain fungi also appear to reduce nitrite to N_2O with the involvement of a cytochrome P-450-like enzyme (29). In contrast to nitrification, the reactions of denitrification are typically all carried out by a single organism. In addition, with the exception of NO reductase, the enzymes of the denitrification pathway are generally soluble, reasonably easy to obtain in pure form, and exhibit less complexity in terms of their metal cofactor contents (14). Consequently, the mechanisms of the enzymes of denitrification are in several cases well-understood.

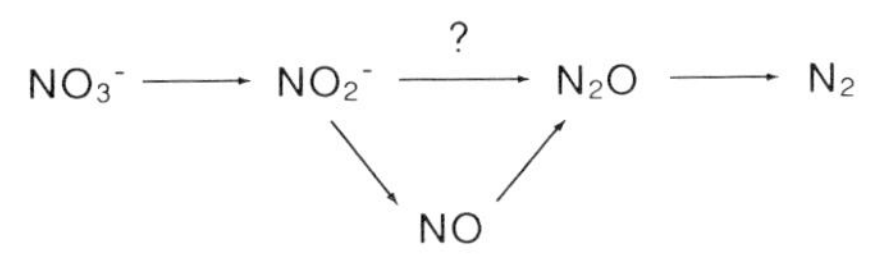

Figure 4. The pathway of denitrification.

The major point of controversy in recent years has been whether NO is a free obligatory intermediate in the denitrification pathway, or whether nitrite reductases produce N_2O directly from nitrite. The evidence for NO as an obligatory intermediate in denitrification is by now convincing: (i) purified nitrite reductases (NiR's) produce only or mostly NO from NO_2^-; (ii) denitrifiers produce and consume NO during nitrite reduction; (iii) denitrifiers produce N_2O from NO with concomitant active transport, proton translocation, and cell growth; (iv) NiR's catalyze exchange of N between NO_2^- and an exogenous NO pool; (v) nir⁻ mutants retain NO reductase activity; (vi) preparations containing high levels of NO reductase have now been obtained; and, perhaps most convincingly, (vii) deletion of the gene coding for the putative NO reductase protein is lethal in the presence of NO_2 (35,37). Nonetheless, certain points of ambiguity remain, the most important from isotope exchange experiments. It has been known for some time that many nitrite reductases (those containing heme *cd₁* chromophores - see below) are capable of catalyzing the exchange of ^{18}O from $H_2{}^{18}O$ into NO or N_2O product by a reversible dehydration/hydration reaction of bound nitrite:

$$2H^+ + E\text{-}Fe^{2+}\text{-}^{15}NO_2^- \leftrightarrow E\text{-}Fe^{2+}\text{-}^{15}NO^+ + H_2O,$$

and that the $E\text{-}Fe^{2+}\text{-}^{15}NO^+$ species could be trapped by nucleophiles such as $^{14}NH_2OH$ or $^{14}N_3^-$ to give appropriately labelled forms of N_2O (21). Unexpectedly, however, it has recently been shown that a similar exchange process occurs during reduction of NO to N_2O. If a linear pathway such as that shown at the top of Figure 4 is operative, then one would expect that in all cases the amount of ^{18}O exchange observed with NO as substrate would be less than with NO_2^- since each molecule of NO_2^- would have two opportunities to undergo ^{18}O exchange, once at the NO_2^- level and once at the NO level. Although this expectation has been confirmed for many organisms, at least two organisms are known that give precisely the opposite result: *Achromobacter cycloclastes* and *Rhodopseudomonas spheroides* forma sp. *denitrificans* were found to incorporate negligible amounts of ^{18}O into N_2O starting from NO_2^- (4 and 6%, respectively), while incorporating substantial amounts of ^{18}O (30 and 37%, respectively) starting with NO (36). It is not clear how such findings can be reconciled with NO as an obligatory intermediate in the reduction of NO_2^-. Nonetheless, it seems clear that in most organisms at least most of the time, denitrification does proceed via NO as an obligatory intermediate.

The organization of the enzymes of denitrification in gram-negative bacteria, as determined by antibody labelling and electron microscopy studies, is shown in Figure 5. The first enzyme, nitrate reductase (NaR), resides in the cytoplasmic membrane with its active site accessed from the cytoplasmic side, necessitating transport of nitrate across both the periplasmic and cytoplasmic membranes. The product nitrite is transported back into the periplasmic space, where it is reduced by the nitrite reductase (NiR). Most NiR's appear to be soluble enzymes, although there have been reports of preparations in which the activity was associated with membrane fractions. The nitric oxide reductase (NoR) is also localized in the cytoplasmic membrane, and releases its product N_2O back into the cytoplasmic space, where the soluble enzyme nitrous oxide reductase (NoS) converts it to N_2.

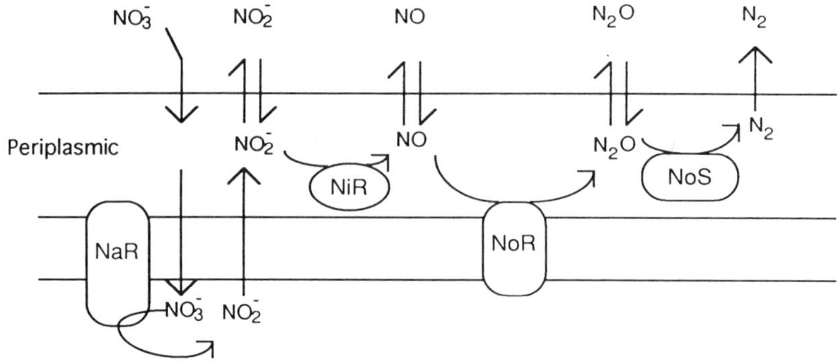

Figure 5. The arrangement of the enzymes of denitrification in gram-negative bacteria: NaR, nitrate reductase; NiR, nitrite reductase; NoR, nitric oxide reductase; NoS, nitrous oxide reductase (35).

The dissimilatory nitrate reductases appear to be fairly typical molybdenum oxotransfer enzymes, although their membrane-bound nature has made it difficult to obtain pure homogeneous preparations for detailed physical studies. All preparations appear to contain at least two types of subunit (α, 104-150 kDa; β, 52-63 kDa), together with molybdenum, both heme and non-heme iron, and acid-labile sulfur, and carry out the reaction:

$$NO_3^- + 2H^+ + 2e^- \rightarrow NO_2^- + H_2O,$$

which is simply the reverse of the nitrite oxidation reaction observed in *Nitrobacter.*

The situation with the nitrite reductases is more complex, inasmuch as two distinct types of NiR are known: those containing heme cd_1 chromophores, and those containing copper. Organisms containing the former appear to be more abundant in nature, although organisms containing the latter occupy a wider range of ecological niches and exhibit more physiological diversity (7). There is no correlation between the type of enzyme present and the genus/species of the organism; no organism has yet been identified that contains both types of NiR. The basic reaction catalyzed by both types of enzyme appears to be the same:

$$NO_2^- + e^- + 2H^+ \rightarrow NO + H_2O \text{ (but see below).}$$

The cytochrome cd_1 nitrite reductases (heme cd_1 NiR's) are typically dimers of 60 kDa subunits, each of which contain a single heme c and a heme d_1 chromophore. The heme c has typical spectroscopic properties, and appears to act as an electron donor to the heme d_1, which appears to constitute the active site of the enzyme. The heme d_1 has anomalous spectroscopic properties and a novel structure, a dioxoisobacteriochlorin (6) (Fig. 6). The mechanism of action of the heme cd_1 NiR's is reasonably well established, and is shown in Figure 7.

As mentioned earlier, a variety of work, including elegant isotope exchange and trapping studies from Hollocher's group (21) has established the existence of an electrophilic nitrosyl intermediate derived from nitrite, which is capable of undergoing nucleophilic attack by nucleophiles such as water, azide, and hydroxylamine. This species can be formulated as either the NO^+ complex of the ferrous heme d_1 or as the NO complex of ferric heme d_1. In either case, it is one electron more oxidized than the paramagnetic ferrous heme-NO complex formed by reaction of such proteins as deoxymyoglobin or deoxy-hemoglobin with NO. The Fe^{2+}-NO^+ species appears to be quite unstable, and undergoes an internal electron transfer

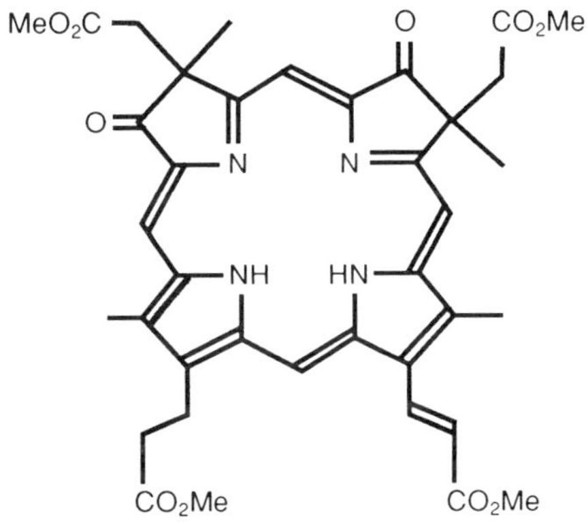

Figure 6. The structure of heme d_1(6).

reaction to produce NO and the ferric heme. Electron transfer from the ferrous heme *c* reduces the heme d_1 back to the ferrous state. Two important points should be made regarding the ferrous-NO^+ intermediate. First, although it is diamagnetic and hence not detectable by EPR, it has recently been observed directly by Fourier transform infrared spectroscopy (FTIR), where it gives an N-O stretching band at ca. 1910 cm^{-1} vs. 1925 cm^{-1} for met Hb•NO and 1615 cm^{-1} for deoxy Hb•NO (Wang and Averill, unpublished). Second, the Fe^{2+}-NO^+ species is highly reactive (28), and decomposes rapidly to NO and the Fe^{3+} heme if the NO^+ species is not trapped by a nucleophile or reduced by one electron to the very stable Fe^{2+}-NO species. EPR spectra due to the latter have been observed many times upon treatment of heme cd_1 NiR's with NO_2^-, but the extremely slow rate of dissociation for NO (days) (30) and the difficulty in reducing the Fe^{2+}-NO unit further (28) make it very unlikely that the species observed by EPR is on the catalytic pathway. Instead, it is functionally dead enzyme. (For the same reason, NO functions as a potent inhibitor of its own formation due to buildup of the Fe^{2+}-NO species.) The role of the heme *c* appears to be to rapidly reduce the ferric d_1 heme back to the ferrous form, ensuring that it is ready to interact with another molecule of substrate.

The copper-containing nitrite reductases (Cu NiR's) also contain two distinct types of chromophore. The best characterized is that from *Achromobacter cycloclastes,* for which a 2.3Å resolution X-ray structure has been reported (12). The enzyme consists of an α_3 trimer of 34.5 kDa subunits, each of which contains two copper atoms in distinct sites. One is an

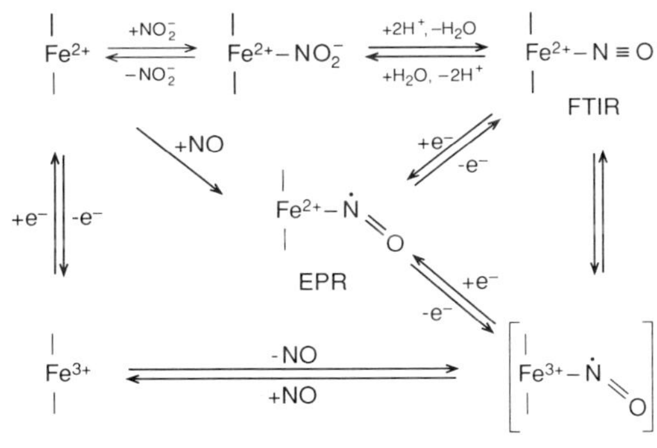

Figure 7. Proposed mechanism for the reduction of nitrite by heme cd_1-containing nitrite reductases.

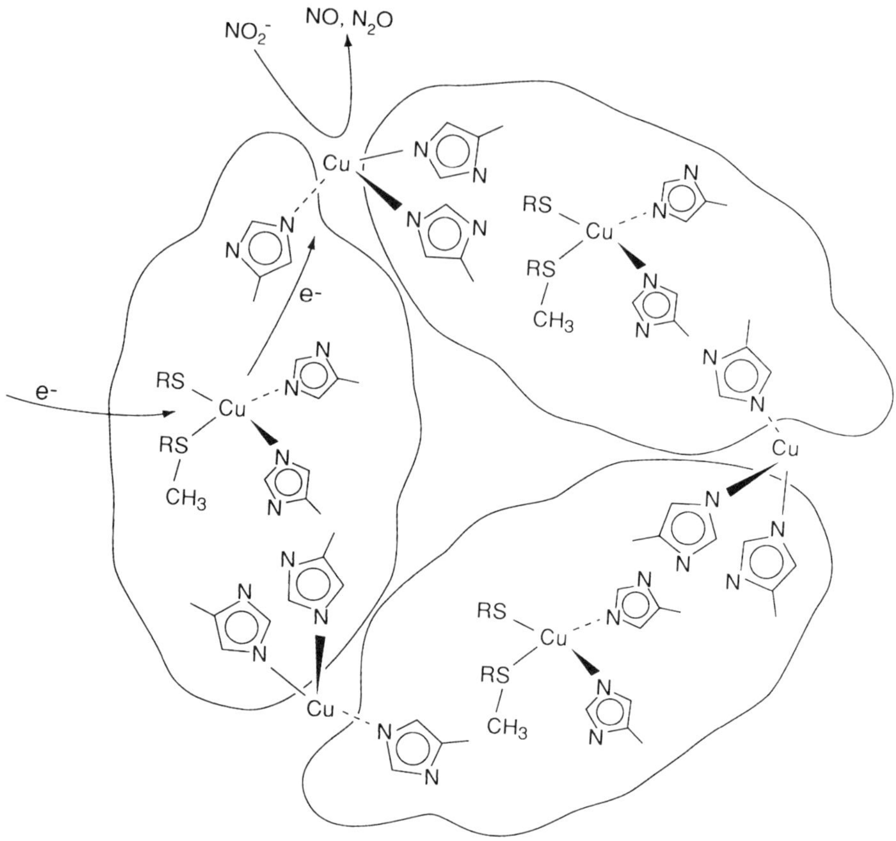

Figure 8. Schematic drawing showing the structure and arrangement of copper centers in the Cu-containing nitrite reductase from *Achromobacter cycloclastes*.

unusual green variant of the "blue" or Type 1 Cu center familiar from plastocyanin or azurin, while the other, a "non-blue" or Type 2 Cu, is coordinated in a roughly tetrahedral fashion by three histidine imidazoles and, presumably, a molecule of water (Fig. 8).

Both crystallographic studies of nitrite binding to the oxidized enzyme (12) and studies of Type 2 Cu-depleted enzyme (24), where a linear correlation between Type 2 Cu content and specific activity was observed, have shown that the Type 2 Cu center is the site at which NO_2^- binds and is reduced, analogous to the heme d_1 center in the heme cd_1 enzymes. The role of the Type 1 Cu appears to be that of an electron transfer center, analogous to the role of the heme *c* in the heme cd_1 enzymes. Although a wide variety of copper contents, colors, and quaternary structures has been reported for Cu NiR's, it seems likely that most, if not all, of the enzymes have structures and optimal copper contents similar to that of the A. *cycloclastes* enzymes. This assertion is based on: (i) extensive sequence similarities between the A. *cycloclastes* NiR and that from P. *aureofaciens*, previously reported to contain only a single Type 1 Cu per monomer (11); (ii) on studies of the Cu content of the A. *xylosoxidans* NiR, also previously reported to contain only Type 1 Cu (1); and (iii) on the extensive immunological cross-reactivity observed with antibodies to the *A. cycloclastes* enzyme (7).

A mechanism similar to that observed with the heme cd_1 NiR has been proposed for the Cu NiR's (19) (Figure 9 - lower half), but several lines of evidence suggest that there

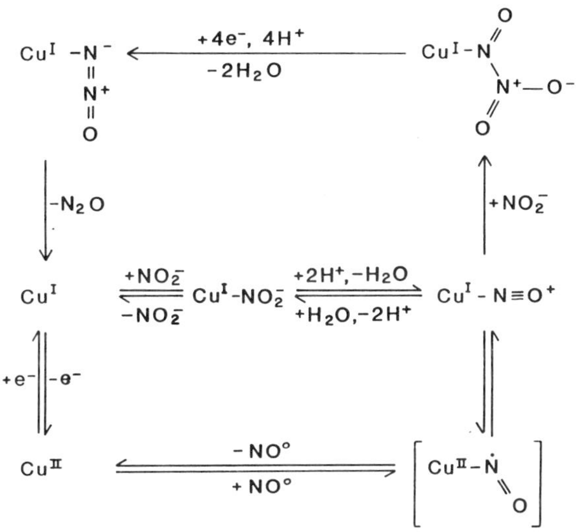

Figure 9. Proposed mechanism for the reduction of nitrite to nitric oxide (lower portion) and to nitrous oxide (upper portion) by Cu-containing nitrite reductases.

may be significant differences between the two types of enzyme. First, no direct evidence for a Cu^+-NO^+ species has been obtained via $H_2{}^{18}O$ exchange or trapping experiments with the purified enzyme. Such evidence has only been obtained with cell-free extracts in which the anomalous properties of the NO reductase might be responsible. Second, unlike the heme cd_1 enzymes, the Cu NiR's can produce a significant amount of N_2O from NO_2^-, but only after substantial levels of NO have built up (20). In addition, reaction of $^{15}NO_2^-$ with the enzyme in the presence of ^{14}NO produced large amounts of $^{14,15}N_2O$ (20). These results suggest that a labile Cu-NO species accumulates and is capable of reacting either with NO_2^- (Figure 9 - upper half) or with NO derived from NO_2^- to produce N_2O. It is not clear whether this reaction is significant under normal physiological conditions, where NO is rapidly removed, or if it may be important only under conditions where NO reductase function is impaired. Nitric oxide reductase (NoR) was the last of the enzymes of denitrification to be obtained in pure form. As indicated in Figure 5, the enzyme appears to be an integral membrane protein, and a variety of detergents have been used to solubilize it. Preparations obtained to date consist of a heterodimer with subunits of ca. 53 and 17 kDa, containing a cytochrome *b* and a cytochrome *c*, respectively. In addition, significant quantities of non-heme iron (3-6 per mol) are reported to be present in at least some preparations.

The enzyme catalyzes the reductive dimerization of NO to N_2O via the reaction:

$$2NO + 2e^- + 2H^+ \rightarrow N_2O + H_2O.$$

If the enzyme utilizes one of the heme groups as the site of NO binding and reduction, then it poses an intriguing mechanistic question: how does a heme, normally capable of binding only a single small molecule such as NO, catalyze the dimerization reaction? Hollocher (31) has suggested that the enzyme reduces NO to NO^- (HNO), which then dimerizes rapidly and spontaneously in a non-enzymatic reaction to produce N_2O and H_2O. Although initially attractive, this proposal suffers from the fact that reduction potentials for known ferrous heme-NO complexes are almost one volt more negative than those of the physiological reductants succinate or ascorbate (28). Therefore it is difficult to see how the

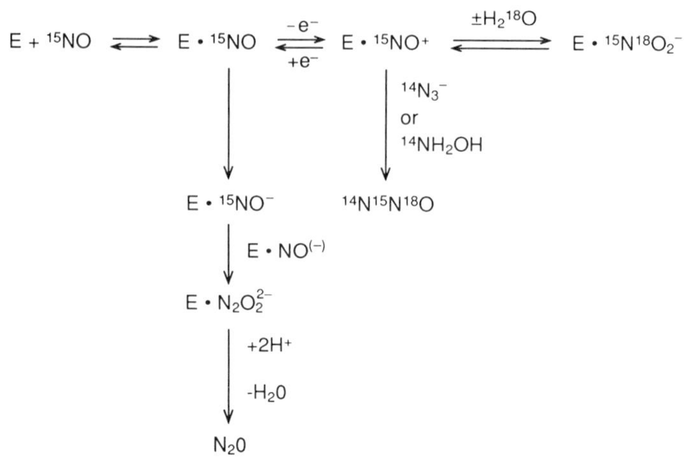

Figure 10. One possible mechanism to explain the observed $H_2{}^{18}O$ exchange during reduction of NO to N_2O (36).

enzyme could produce NO^- bound to a heme. Furthermore, $H_2{}^{18}O$ isotope exchange studies have demonstrated the existence of an electrophilic species derived from NO (believed to be bound NO^+) during catalysis of NO reduction to N_2O (36); in addition, exchange of ^{18}O into the substrate NO has also been demonstrated (Ye, Tiedje, and Averill, unpublished results). This intriguing finding suggests a number of mechanistic possibilities, one of which is shown in Figure 10. In this scheme, the electrophilic NO-derived species responsible for the observed ^{18}O exchange reaction is shown as Iying off the catalytic pathway, and arising from incidental oxidation of the E•NO complex. Alternative mechanisms can also be envisioned, in which the E-NO^+ species actually lies on the catalytic pathway (36). It would be most intriguing if *reduction* of NO to N_2O were found to proceed via initial *oxidation* to bound NO^+. An additional mechanistic possibility that has not been widely considered utilizes the non-heme iron atoms as the site at which two NO molecules are reductively coupled via a variant of the mechanisms well established for a variety of synthetic organometallic complexes (Fig. 11)

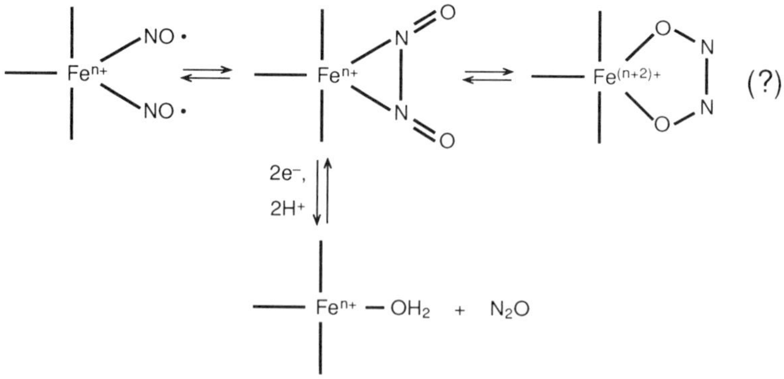

Figure 11. Possible mechanism for reductive coupling of two nitric oxide molecules to nitrous oxide at a mononuclear non-heme iron site in nitric oxide reductase (35).

Table 1. Reactions of denitrification and nitrification

Reaction	Enzyme (Abbr) Denitrification	Cofactors
Denitrification		
$NO_3^- \rightarrow NO_2^-$	Nitrate reductase (NaR)	Mo, heme & non-heme Fe sulfide
$NO_2^- \rightarrow NO$	Nitrite reductase (NiR)	Cu (green), heme cd_1, P450
$NO \rightarrow N_2O$	NO reductase (NoR)	Heme bc & non-heme Fe?, P-450?
$N_2O \rightarrow N_2$	N_2O reductase (NoS) Nitrification	Cu (purple)
Nitrification		
$NH_3 + O_2 \rightarrow NH_2OH$	Ammonia monoxygenase (AMO)	Cu?
$NH_2OH \rightarrow NO_3^-$	Hydroxylamine oxidoreductase (HAO)	heme *c*, P-460
$NO_2^- \rightarrow NO_3^-$	Nitrite oxidase (NiO)	Mo, Fe-S?

Although the non-heme iron atoms have been largely ignored and considered to be due to contamination by adventitious iron, large amounts of non-heme iron such as those reported for NoR are atypical, and the iron does not display the typical g'=4.3 EPR spectrum observed for adventitious protein-bound ferric iron (37). Clearly additional work is needed to distinguish among the various possibilities outlined here. An additional complication is that there appears to be a strong coupling between the nitrite reductase and nitric oxide reductase activities at all levels: transcriptional, translational, and even mechanistic (33, 34, 37). These findings suggest that the two enzymes may function as a multienzyme complex in at least some organisms (37).

The final step in the denitrification process is carried out by the soluble enzyme nitrous oxide reductase (NoS). This enzyme has been isolated from a number of sources, and is unusual in a number of ways. In most cases, it is a homodimer of ca. 74 kDa subunits with ca. 4 Cu/subunit, but the enzyme is bright purple or pink as isolated, depending on conditions, and becomes the typical blue color expected for copper proteins only after reduction with dithionite. A variety of spectroscopic studies strongly suggest that the enzyme contains at least one mixed-valent, thiolate-bridged Cu(I)—Cu(II) unit that may well be similar to a binuclear copper center in cytochrome *c* oxidase (23). The reaction catalyzed by the enzyme is deceptively simple:

$$N_2O + 2e^- + 2H^+ \rightarrow N_2 + H_2O,$$

but no mechanistic studies have been reported.

The reactions known to occur in the denitrification and nitrification pathways are summarized in Table 1, along with the most commonly encountered cofactors.

IMPLICATIONS FOR METABOLISM OF NITROAROMATIC COMPOUNDS

As is discussed in substantial detail elsewhere in this volume, a number of reductive and oxidative reactions of nitroaromatic compounds and their derivatives are known to occur in microbes; some of these are summarized in Table 2, along with the most analogous reactions known in inorganic N-oxide chemistry.

It is now clear that reduction of the nitro group of nitroaromatic compounds by bacteria usually proceeds in two-electron steps, as does reduction of nitrate during denitri-

Table 2. Known biological reactions of organic N-oxides compared to the most analogous reactions of inorganic N-oxides

Organic	Inorganic
$ArNO_2 \rightarrow ArNO$	$NO_3^- \rightarrow NO_2^-$
$ArO \rightarrow ArNHOH$	$NO_2^- \rightarrow [HNO?] \rightarrow N_2O$
$ArNHOH \rightarrow ArNH_2$	$NO_2^- \rightarrow NH_3$
$ArNHOH \rightarrow ArNO$	$N_2NOH \rightarrow [HNO] \rightarrow NO_2^-$
$ArNH_2 \rightarrow ArNO_2$	$NH_3 \rightarrow NO_3^-$

fication. The similarities between the reactions of denitrification and aromatic nitro group reduction become most apparent if one pairs the organo-nitrogen and inorganic compounds, not by oxidation state of nitrogen, but by the arrangement of valence electrons on the nitrogen atom, especially the presence or absence of lone pairs and/or hydrogen atoms. Thus, nitroaromatic compounds ($ArNO_2$) are more similar to nitrate (NO_3^-) than to nitrite (NO_2^-), and nitrosoaromatic compounds (ArNO) are analogous to nitrite in that both have a lone pair of electrons on the nitrogen. These similarities are most clearly seen if one compares the organonitrogen compound with the parent acid of the inorganic nitrogen compound: $ArNO_2$ vs. $HONO_2$; $Ar\ddot{N}O$ vs. $HO\ddot{N}O$. Some nitroreductases from eukaryotes are known that take advantage of the stabilization of radicals by the aromatic group to produce the one-electron reduction products, $ArNO_2^{\cdot}$ radicals (Corbett, this volume). Such a reaction is analogous to reduction of NO_3^- to $NO_2^{\cdot}$, which is not known to occur in bacteria.

There are, however, major differences in the reactions observed for the organic vs. inorganic nitrogen compounds in each pair, and it is rather easy to understand the chemical basis for these differences. First, the inorganic species exist as the anions at physiological pH, not as the free acids. This means that the susceptibility of the nitrogen atom to nucleophilic attack by a hydride equivalent is greatly reduced due to simple electrostatic effects. As a result, the nitro- and nitrosoaromatics are readily reduced by flavoenzymes that can donate the equivalent of H^- to the nitrogen, while this mode of reaction does not occur for nitrate and nitrite. Instead, the latter undergo oxygen atom transfer and protonation/dehydration reactions respectively, due to the increased nucleophilicity/basicity of the oxygen atoms, which carry a partial negative charge. Second, in the lower oxidation states (between +3 and -1), the most stable inorganic nitrogen compounds are gasses with multiple N-O and/or N-N bonds (NO, N_2O, N_2). Consequently, the analogy between reactions of inorganic and organic nitrogen compounds breaks down. At least conceptually, however, one can imagine that the two-electron reduction of NO_2^- produces HNO (analogous to the reduction of ArNO to ArNHOH), but the HNO spontaneously dimerizes to N_2O and H_2O. Third, one would clearly expect reduction of arylhydroxylamines, ArNHOH, to proceed via an entirely different mechanism than reduction of the nitro and nitroso precursors, because of lack of multiple N-O bonds to permit formation of an intermediate derived from hydride attack at N that satisfies the octet rule at nitrogen. Recent results clearly indicate that at least one purified, inducible nitroreductase produces only the hydroxylamine from nitrobenzene (Spain, this volume). Some other mechanism must be operative for arylhydroxylamine reduction, but the only parallel in inorganic nitrogen chemistry is not particularly informative: reduction of N to NH_3 by assimilatory nitrite reductase is a single six-electron process with no free intermediates, although hydroxylamine is a substrate for the enzyme (25).

Table 3. Some oxo transfer reactions catalyzed by molybdoenzymes

Sulfite Oxidase

$$SO_3^{2-} + \overset{O}{\overset{\|}{Mo^{VI}}} = O \rightarrow SO_4^{2-} + Mo^{IV} = O$$

Aldehyde Oxidase

$$RCHO + \overset{O}{\overset{\|}{Mo^{VI}}} = O \rightarrow RCO_2H + Mo^{IV} = O$$

Nitrate Reductase

$$NO_3^- + Mo^{IV} = O \rightarrow NO_2^- + \overset{O}{\overset{\|}{Mo^{VI}}} = O$$

DMSO Reductase

$$Me_2SO + Mo^{IV} = O \rightarrow Me_2S + \overset{O}{\overset{\|}{Mo^{VI}}} = O$$

Nitroaromatic Reductase

$$ArNO_2 + Mo^{IV} = O \rightarrow ArNO + \overset{O}{\overset{\|}{Mo^{VI}}} = O\ ??$$

Conceptually, however, this process is clearly related to the overall four-electron reduction of ArNO to $ArNH_2$.

Oxidative reactions of organonitrogen species that do not involve molecular oxygen are rather limited. The only case for which the evidence is at all convincing is the oxidation of arylhydroxylamines to arylnitroso species (Table 2). This reaction resembles the first half of the hydroxylamine oxidoreductase reaction found in nitrifying bacteria. The key difference is that the aryl nitroso compound is stable (although condensation with the arylhydroxylamine can occur to produce the azoxy compound, ArN(O)NAr), while the inorganic analog is nitroxyl, HNO, which if released from the enzyme would rapidly dimerize and dehydrate to form N_2O. Consequently, HAO does *not* release the HNO or NO^- intermediate, but instead oxidizes it to nitrite before any substrate-derived species are released.

What then can we learn about possible reactions of nitroaromatic compounds from the better studied reactions of inorganic N-oxide species? First, it seems highly likely that alternative modes will be discovered for biological reduction of the nitroaromatic compounds using metalloenzymes, in particular molybdenum. Table 3 summarizes only a few of the many oxygen atom transfer reactions known to be catalyzed by molybdoenzymes. Although the $ArNO_2$ substrate is clearly well set up for attack by hydride, the mechanism that appears to operate in the flavoenzyme nitroreductases discovered so far, it is also well-suited for simple direct oxygen atom transfer to produce the nitroso compound, in a reaction very similar to that catalyzed by DMSO reductase (18). One of the guiding principles of comparative biochemistry is that similar or identical reactions in different organisms may well utilize chemically equivalent but unrelated process. Compare, for example, oxygen transport in mammals (hemoglobin) vs. molluscs and arthropods (hemocyanin) vs. sipunculids (hemerythrin) or the copper vs. heme cd_1 nitrite reductases found in denitrifiers vs. the cytochrome P-450 system in fungi. An oxygen atom transfer reaction for arylnitroso compounds is, in contrast, unlikely, because the product would be a highly unstable and reactive nitrene. Second, thus far in inorganic N-oxide chemistry enzymes have taken advantage of the presence of a lone pair of electrons on nitrogen whenever it has occurred to permit an interaction with a metal ion to facilitate the catalytic reaction. It would be most surprising if this did not occur with the organonitrogen analogs, Ar*romän*NO and ArṄHOH, in at least some organisms. Coordination to a metal center is a means often utilized in biochemistry to stabilize otherwise unstable intermediates during multi-step reactions (cf. "HNO" during oxidation of NH_2OH by HAO). Third and finally, among the major products produced by microbial degradation of nitroaromatics are ammonia and nitrite, both of which have significant toxicity in their own right. Nitrification and denitrification are the only microbial processes that are likely to be able to deal with the large amounts of these inorganic nitrogen species that will be produced during bioremediation of highly contaminated sites. Consequently, workers in the field should develop at least a minimal understanding of and appreciation for the microorganisms that are likely to also be involved in successful bioremediation efforts.

ACKNOWLEDGEMENTS

Research in the author's laboratory has been supported by grants from USDA-NRICGP (91-37305-6663) and NSF (DMB-8917427, MCB-9307501), and has been a close collaboration with the group of J. M. Tiedje at Michigan State University. I thank him for educating me regarding the importance of bacterial physiology, and A. B. Hooper for providing me with recent reprints and unpublished work on nitrification.

REFERENCES

1. Abraham, Z. H. L., D. J. Lowe, and B. E. Smith. 1993. Purification and characterization of the dissimilatory nitrite reductase from *Alcaligenes xylosoxidans* subsp. *xylosoxidans* (NCIMB 11015). Evidence for the presence of both type-1 and type-2 copper centers. Biochem. J. 295:587-593.
2. Andersson, K. K., and A. B. Hooper. 1983. O_2 and H_2O are each the source of one O in NO_2^-: ^{15}N-NMR evidence. FEBS Lett. 164:236-240.
3. Arciero, D. M., A. B. Hooper, M. Cai, and R. Timkovich. 1993. Evidence for the structure of the active site heme P460 in hydroxylamino oxidoreductase of *Nitrosomonas*. Biochemistry 32:9370-9378.
4. Carr, G. J., and S. J. Ferguson. 1990. The nitric oxide reductase of *Paracoccus denitrificans*. Biochem. J. 269:423-430.
5. Chan, S. I., and H. -H. T. Nguyen. 1994. Insights into the nature of the copper sites in particulate methane monooxygenase from EPR and magnetic measurements, abstr. PHYS-49. Abstr. 208th Natl. Mtg. Amer. Chem. Soc., Washington, D.C.
6. Chang, C. K. 1986. The porphyrindione structure of heme d_1. J. Biol. Chem. 261:8593-8596.
7. Coyne, M. S., A. Arunakumari, B. A. Averill, and J. M. Tiedje. 1989. Immunological identification and distribution of dissimilatory heme cd_1 and nonheme copper nitrite reductases in denitrifying bacteria. Appl. Environ. Microbiol. 55: 2924-2931.
8. Demastia, M., T. Turk, and T. C. Hollocher. 1991. Nitric oxide reductase. Purification from *Paracoccus denitrificans* with use of a single column and some characteristics. J. Biol . Chem. 266:10899-10905.
9. Di Spirito, A. A., L. R. Taaffe, J. D. Lipscomb, and A. B. Hooper. 1985. A "blue" copper oxidase from *Nitrosomonas europaea*. Biochim. Biophys. Acta 827: 320-326.
10. Fukuoka, M., Y. Fukumori, and T. Yamanaka. 1987. *Nitrobacter winogradskyi* cytochrome a_1c_1 is an iron-sulfur molybdoenzyme having hemes *a* and *c*. J. Biochem. (Tokyo) 102:525-530.
11. Glockner, A. B., and W. G. Zumft. 1993. Copper-containing nitrite reductase from *Pseudomonas aureofaciens* is functional in a mutationally cytochrome cd_1-free background (Nir S^-) of *Pseudomonas stutzeri*. Arch. Microbiol. 160:18-26.
12. Godden, J. W ., S. Turley, D. C. Teller, E. T. Adman, M.-Y. Liu, W. J. Payne, and J. LeGall. 1991. The 2.3 angstrom X-ray structure of nitrite reductase from *Achromobacter cycloclastes* . Science 253:438-442.

13. Heiss, B., K. Frunzke, and W. G. Zumft. 1989. Formation of the N-N bond from nitric oxide by a membrane-bound cytochrome *bc* complex of nitrate-respiring (denitrifying) *Pseudomonas stutzeri*. J. Bacteriol. 171:3288-3297.
14. Hochstein, L. l., and G. A. Tomlinson. 1988. The enzymes associated with denitrification. Ann. Rev. Microbiol. 42:231-261.
15. Hollocher, T. C., M. E. Tate, and D. J. D. Nicholas. 1981. Oxidation of ammonia by *Nitrosomonas europaea:* Definitive ^{18}O-tracer evidence that hydroxylamine formation involves a monoxygenase. J. Biol. Chem. 256:10834-10836.
16. Hooper, A. B. 1989. Biochemistry of the nitrifying lithotropic bacteria, p. 239-265. *In* H. G. Schlegel and B. Bowein, (ed.) Autotrophic bacteria. Science Technology Publishers, Madison.
17. Hooper, A. B., and K. R. Terry. 1973. Specific inhibitors of ammonia oxidation in *Nitrosomonas*. J. Bacteriol. 115:480-485.
18.. Holm, R. H . 1990. The biologically relevant oxygen atom transfer of molybdenum. Coord. Chem. Rev. 100:183-222.
19. Hulse, C. H., B. A. Averill, and J. M. Tiedje. 1989. Evidence for a coppernitrosyl intermediate in denitrification by the copper-containing nitrite reductase of *Achromobacter cycloclastes*. J. Am. Chem. Soc. 111:2322-2323.
20. Jackson, M. A., J. M. Tiedje, and B. A. Averill. 1992. Evidence for an NO-rebound mechanism for production of N_2O from nitrite by the copper-containing nitrite reductase from *Achromobacter cycloclastes*. FEBS Lett. 291:41-44.
21. Kim, C. H., and T. C. Hollocher. 1984. Catalysis of nitrosyl transfer reactions by a dissimilatory nitrite reductase (cytochrome cd_1). J. Biol. Chem. 259:2092-2099.
22. Koshland, D. E., Jr. 1992. The molecule of the year. Science 258:1861.
23. Kroneck, P. M. H., W. A. Antholine, J. Riester, and W. G. Zumft. 1989. The nature of the cupric site in nitrous oxide reductase and of Cu_a in cytochrome *c* oxidase. FEBS Lett. 248:212-213.
24. Libby, E., and B. A. Averill. 1992. Evidence that the type 2 copper centers are the site of nitrite reduction by *Achromobacter cycloclastes* nitrite reductase. Biochem. Biophys. Res. Commun. 187:1529-1535.
25. Lui, S. M., W. Liang, A. Soriano, and J. A. Cowan. 1994. Enzymatic reduction of inorganic anions. Variable-temperature steady-state and presteady-state kinetics experiments to map the energy profile of an enzymatic multielectron redox reaction. Application to the dissimilatory sulfite reductase from *Desulfovibrio vulgaris* (Hildenborough). J. Am. Chem. Soc. 116:4531-4536.
26. Miller, D. J., and D. J. D. Nicholas. 1985. Characterization of a soluble cytochrome oxidase/nitrite reductase from *Nitrosomonas europaea*. J. Gen. Microbiol. 131:2851-2854.
27. Miller, D. J., and P. M. Wood. 1983. The soluble cytochrome oxidase of *Nitrosomonas europaea*. J. Gen. Microbiol. 129:1645-1650.
28. 0lson, L. W., D. Schaeper, D. Lancon, and K. W. Kadish. 1982. Characterization of several novel iron nitrosyl porphyrins. J. Am. Chem. Soc. 104:2042-2044.
29. Shoun, H., D.-H. Kim, H. Uchiyama, and J. Sugiyama. 1992. Denitrification by fungi. FEMS Microbiol. Lett. 94:277-281.
30. Traylor, T. G., and S. Sharma. 1992. Why NO? Biochemistry 31:2847-2849.
31. Turk, T., and T. C. Hollocher. 1992. Oxidation of dithiothreitol during turnover of nitric oxide reductase: evidence for generation of nitroxyl with the enzyme from *Paracoccus denitrificans*. Biochem. Biophys. Res. Commun. 183:983-988.
32. Wood. P. M. 1988. Chemolithotropy, p. 183-230. *In* C. Anthony (ed.) Bacterial energy transduction. Academic Press, New York.
33. Ye, R. W., A. Arunakumari, B. A. Averill, and J. M. Tiedje. 1992. Mutants of *Pseudomonas fluorescens* deficient in dissimilatory nitrite reduction are also altered in nitric oxide reduction. J. Bacteriol. 174:2560-2564.
34. Ye, R. W., B. A. Averill, and J. M. Tiedje. 1992. Characterization of Tn*5* mutants deficient in dissimilatory nitrite reduction in *Pseudomonas sp.* strain G179, which contains a copper nitrite reductase. J. Bacteriol. 174:6653-6658.
35. Ye, R. W., B. A. Averill, and J. M. Tiedje. 1994. Denitrification: production and consumption of nitric oxide. Appl. Environ. Microbiol. 60:1053-1058.
36. Ye, R. W., I. Toro-Suarez, J. M. Tiedje, and B. A. Averill. 1991. $H_2{}^{18}O$ isotope exchange studies on the mechanism of reduction of nitric oxide and nitrite to nitrous oxide by denitrifying bacteria: evidence for an electrophilic nitrosyl during reduction of nitric oxide. J. Biol. Chem. 266:12848-12851.
37. Zumft, W. G. 1993. The biological role of nitric oxide in bacteria. Arch. Microbiol. 160:253-264.

12

ENVIRONMENTAL PROCESSES INFLUENCING THE RATE OF ABIOTIC REDUCTION OF NITROAROMATIC COMPOUNDS IN THE SUBSURFACE

Stefan B. Haderlein and René P. Schwarzenbach

Swiss Federal Institute for Environmental Science and Technology (EAWAG)
CH-8600 Dübendorf
Switzerland
Swiss Federal Institute of Technology (ETH)
CH-8600 Dübendorf
Switzerland

1. INTRODUCTION

Numerous synthetic chemicals contain one or several nitro groups that are bound to an aromatic ring. Figure 1 shows the structures of some prominent representatives of such nitroaromatic compounds (NACs). The high toxicity of some NACs, particularly the mutagenic and carcinogenic potential of some nitrated polycyclic aromatic hydrocarbons (PAHs), has led to considerable interest in the fate of such compounds in the environment. Due to their widespread use, NACs are ubiquitous contaminants, especially in aqueous environments. In addition to contamination originating from agricultural use, from production facilities, and waste disposal sites, diffuse input into the pedosphere via the atmosphere has been documented (21, 27, 36, 37, 59, 65, 69, 80, 81). Atmospheric production of significant quantities of NACs by photochemical processes has been reported (19, 29, 42, 80). Table 1 lists some typical concentrations of NACs that have been measured in various compartments of the environment. Very high concentrations of nitroaromatic explosives (2,4,6-trinitrotoluene (TNT) and by-products) have been found especially in soil and subsurface systems. At those sites, significant concentrations of substituted aromatic amines that may have been formed from the reduction of NACs are frequently encountered.

The reduction of the aromatic nitro group(s) to the amine is the principal abiotic transformation pathway of NACs and occurs via nitroso- and hydroxylamino-intermediates:

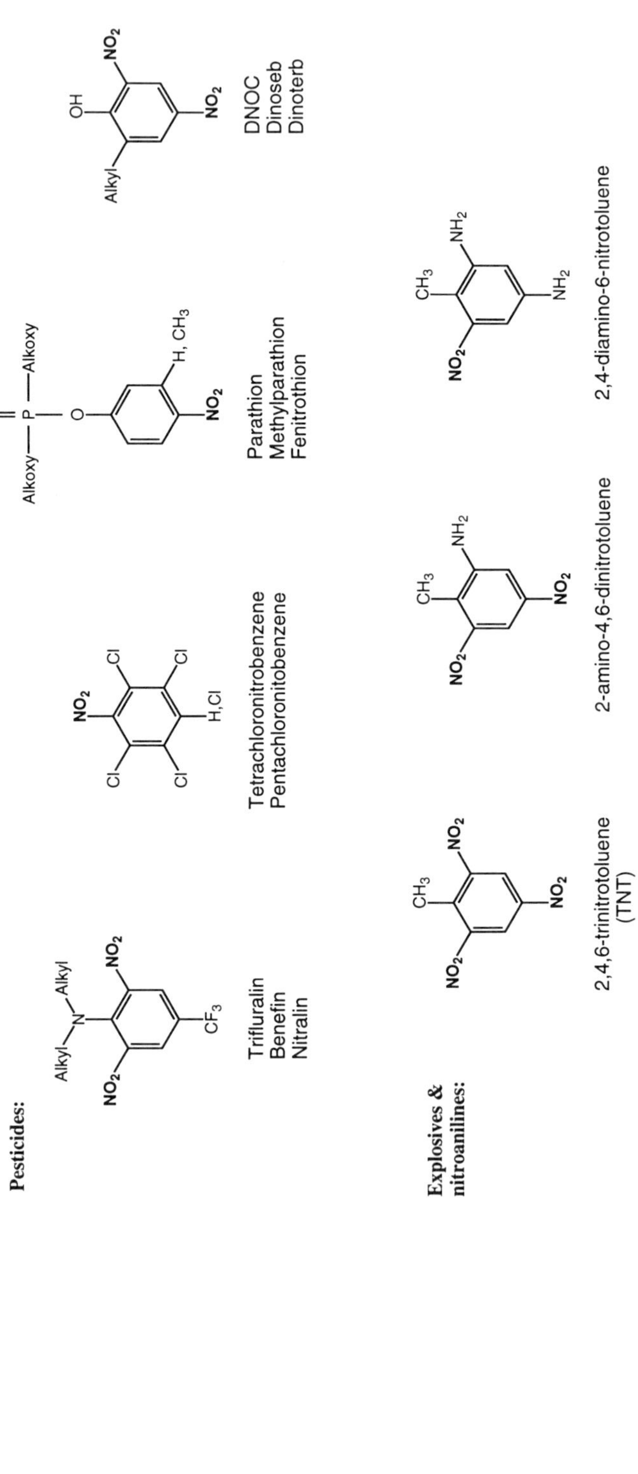

Figure 1. Examples of nitroaromatic compounds that are of environmental relevance.

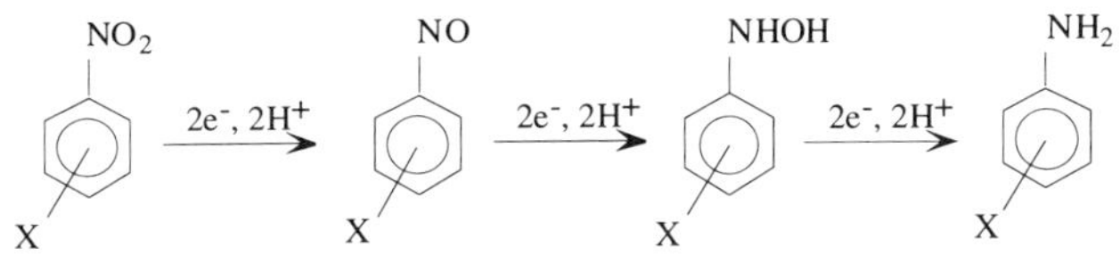

Since these compounds or other side products resulting from reactions among the intermediates are also of considerable toxicological concern, the reductive transformation of NACs in soils, aquifers and sediments is of great interest. Nitro reduction is generally postulated to be the first in a series of transformation reactions followed by either complete mineralization of NACs or by reactions with the solid matrix leading to products commonly referred to as "bound residues" (9, 28, 31, 44, 53, 55). It should be noted that even under aerobic conditions, microbial degradation of NACs may involve the reduction of the nitro group(s) as an initial step (24, 68, 93).

There are a number of studies which have found extremely fast initial reduction rates for NACs in soils and sediments (1, 84, 88). This has led to the hypothesis that reduction of NACs may occur, at least in part, by abiotic reactions. Table 2 is a survey of investigations where abiotic reduction of NACs has been postulated in the field or in laboratory experiments carried out with natural sediments. Additional information on abiotic redox reactions of organic compounds including NACs can be found in the reviews of Macalady et al. (44) and Wolfe & Macalady (89). In most of the work carried out in natural systems, representatives of one of the three major groups of nitroaromatic pesticides have been studied: dinitroanilines, organophosphorous esters, and chlorinated nitrobenzenes. Very recently, investigations on the abiotic reduction of nitroaromatic explosives have been conducted. Unfortunately most of the studies in natural systems do not provide conclusive information about the types of natural reductants that are involved in abiotic NAC reduction. Nevertheless, the scarce information about the active redox couples in these "dirty systems" has led to a series of investigations in model systems under more controlled (redox) conditions. An overview of these studies is given in Table 3. From these studies in model systems it is now clear that NACs may undergo abiotic reduction at significant rates in the environment.

The purpose of this chapter is to evaluate under which conditions abiotic reduction of NACs can be expected in soils, aquifers, and sediments. Emphasis is on the various types of environmental factors that may control the rates of reduction of NACs in natural systems. In this context it is not only important to evaluate the factors that govern the actual rate of NAC reduction but also the processes that determine the speciation of NACs (and thus their availability for reactions) in a given system.

2. DISTRIBUTION OF NACS IN HETEROGENEOUS SYSTEMS

The distribution between the aqueous (mobile) phase and the various sediment or soil (stationary) phases is one of the key parameters determining both transport and transformation of contaminants in subsurface systems. While the impact of adsorption/desorption processes on subsurface transport is obvious and has been the subject of numerous investigations, their effects on transformation reactions of organic contaminants has been widely ignored in the past. It is important to realize that both the extent and the type of distribution of a contaminant between aqueous and solid phase(s) may determine its *availability* as well as its *reactivity* towards abiotic and microbial transformation reactions. For instance, it has been shown for a series of *para*-substituted n-alkyl-nitrobenzenes that the abiotic reduction of these compounds in natural sediments is strongly affected by the

Table 1. Occurrence and typical concentrations of NACs in various environmental systems

Compartment	Description	Compound(s)	Concentrations	Reference
air	remote areas (treated grove, USA)	parathion	$7 \cdot 10^3 - 15 \cdot 10^5$ ng m^{-3}	(48)
	Mississippi Delta, USA	methyl parathion	100 - 800 (2000) ng m^{-3}	(85)
	Sacramento Valley	methyl parathion	25 ng m^{-3}	(85)
	remote areas (FRG)	substituted mono- and dinitrophenols	5 - 75 ng m^{-3}	(30)
	ambient air (USA, CA)	nitro-PAHs,		
	dust particles, diesel exhaust	nitrodibenzopyranones	0.1 - 1 ng m^{-3}	(29)
	auto-engine exhausts	Σ subst. nitrophenols	$1 - 5 \cdot 10^6$ ng m^{-3}	(80)
water	fog (CA, USA)	methyl parathion	0.04 - 0.43 μg·L^{-1}	(85)
	fog (FRG)	pendimethalin	0.1 - 1.3 μg·L^{-1}	(30)
	fog (remote areas, FRG)	substituted nitrophenols	5 - 200 μg·L^{-1}	(58)
	fog (Switzerland)	"	100 - 5000 μg·L^{-1}	(80)
	rain (Switzerland)	"	10 - 1500 μg·L^{-1}	(80)
	rain (remote areas, FRG)	"	3 - 200 μg·L^{-1}	(30)
	streams (western USA)	methyl parathion	up to 0.23 μg·L^{-1}	(85)
	rivers (NC, USA)	"	0.47 μg·L^{-1}	(85)
	Mississippi (USA)	"	0.08 - 0.46 μg·L^{-1}	(85)
	Negro river (Argentina)	"	0.04 μg·L^{-1}	(85)
	Lake (Greece)	"	0.008 - 0.032 μg·L^{-1}	(85)
	various surface waters (USA)	subst. nitrobenzenes	μg - mg·L^{-1}	(13)
	soil drain effluents (CA, USA)	methyl parathion	0.01 - 0.17 μg·L^{-1}	(85)
	groundwater beneath munitions waste disposal sites (USA)	TNT	> 0.6 mg·L^{-1}	(59)
	groundwater beneath ammunition plant (USA)	DNT isomers	1 - 10 mg·L^{-1}	(90)
	" (USA)	TNT	0.1 - ≥ 100 μg·L^{-1}	(69)
	" (FRG)	TNT and byproducts	≤ 600 μg·L^{-1}	(21)
	" (FRG)	"	up to 63 mg·L^{-1}	(61)

	waste water (production plant,USA)	methyl parathion	$2x10^3$ µg·L^{-1}	(85)
	road runoff	substituted nitrophenols	0.6 - 20 µg·L^{-1}	(71)
	munitions plant effluents (USA)	TNT, nitroanilines	< 0.005 - 150 mg·L^{-1}	(59)
	—	TNT	ca 130 mg·L^{-1}	(74)
	waste lagoon water	—	1300 mg·L^{-1}	(35)
	waste water	—	19 mg·L^{-1}	(35)
	process water, munitions plant (USA)	—	51 mg·L^{-1}	(35)
	cropland soils (USA)	methyl parathion	0.001 - 1.9 mg·kg^{-1}	(85)
	contaminated soils (FRG)	TNT, DNT-isomers, nitroanilines	0.1 - 100 g·kg^{-1}	(20)(50)
	contaminated soils (USA)	TNT	DNT isomers	12 mg·kg^{-1} 0.1 - 40 mg·kg^{-1}
sediments	waste disposal sites at ammunition plants (FRG)	TNT and byproducts	mg·kg^{-1} up to 15 g·kg^{-1}	(61)
	waste lagoon sediment	TNT	3000 mg·kg^{-1}	(59)
	waste lagoon sediments (FRG)	TNT and byproducts	mg·kg^{-1} - 50% (w/w)	(61)
	drainage systems at ammunition plants (FRG)	TNT and byproducts	up to 20% (w/w)	(61)

Table 2. Investigations in natural sediments where abiotic reduction of NACs has been observed/suggested

Type of Sediment	Redox Milieu	NACs Studied	Half Life ($t_{1/2}$)	Type of Experiment	Refs.
flooded soils	"anaerobic"	parathion	seconds - minutes	batch	(84)
flooded soils	sulfidogenic	parathion,methyl-parathion, fenitrothion	minutes	batch	(1)
sewage sludge	"anaerobic"	2,3,5,6-tetrachloro-nitrobenzene, biphenox	hours	batch	(16)
pond and stream sediments	"anaerobic"	methyl parathion	minutes	batch	(88)
various sediments	"anaerobic"	4-substituted nitrobenzenes	minutes - days	batch	(10)
various sediments	"anaerobic"	4-(n-alkyl)-nitrobenzenes	hours - days	batch	(45, 92)
contaminated soils + Fe(O)	"anaerobic"	2,4,6-trinitrotoluene (TNT)	hours	batch	(52)
river bank	ferrogenic	monosubstituted nitrobenzenes	hours (for C_o(NAC) = 10 μM; zero-order k_{obs})	columns	(25)

fraction of NAC present in the sorbed phase. No reducing activity was found in the aqueous phase (45, 92).

In the case of neutral NACs two types of adsorption mechanisms have been found to play a predominant role in most of the subsurface sediments: (i) partitioning into the organic fraction of the sediment (43, 59, 70) and (ii) adsorption due to complex formation with clay mineral surfaces that bear exchangeable NH_4^+ or K^+ cations. This type of interaction has been postulated as an electron donor-acceptor (EDA) complex where oxygen atoms at the siloxane surfaces of clay minerals act as electron donors and NACs act as electron

Table 3. Investigations on abiotic reduction of NACs in model systems.

Redox System	NACs Studied	Type of Experiments	Reference
(hydro)quinones in anaerobic aqueous solution	methyl parathion, profluralin	batch	(78)
natural dissolved organic matter / H_2S	methyl parathion, profluralin	batch	(11)
bacterial exudates / H_2S	substituted nitrobenzenes, TNT	batch	(17, 56)
protein fraction extracted from sediments in aqueous solution	2,4,6-trinitrotoluene (TNT)	batch	(86, 87)
Fe-porphyrin in organic solvents	substituted nitrobenzenes	batch	(83)
(hydro)quinones / H_2S, Fe-porphyrin / H_2S	substituted nitrobenzenes and nitrophenols	batch	(64)
magnetite suspensions $Fe(II)_{aq}$	4-chloronitrobenzene	batch	(26)
Fe(III)-(hydr)oxides, magnetite, Al-(hydr)oxides, quartz, kaolinite $Fe(II)_{aq}$	substituted nitrobenzenes	batch	(38)
sulfide, cysteine	nitrophenols	batch	(18)
sulfide minerals in anaerobic aqueous suspension	nitrobenzene	batch	(91)
metallic iron in anaerobic aqueous solutions	nitrobenzene	batch, column	(2)

acceptors (22, 23). Weakly hydrated cations, e.g., NH_4^+ or K^+, adsorbed to the negatively charged siloxane surfaces are essential for strong EDA complex formation.

Figure 2 shows the results of model calculations for the distribution of TNT, representing neutral aromatic polynitro compounds, and of 2,4-dinitro-*ortho*-cresol (DNOC), representing dissociable nitrophenols, estimated for typical aquifer (a, b) and soil (c, d) matrices. In these scenarios, the fraction of compounds present in sorbed form is shown as a function of pH and K^+-saturation of the clay minerals. In the case of TNT, for all systems and all conditions considered, a very high fraction of the total amount of compound in the system is associated with the sediment phase(s). In the aquifer the adsorption of TNT is very much dominated by the formation of EDA complexes with clay mineral surfaces. In the soil the sorption due to EDA complex formation dominates over the hydrophobic partitioning into particulate organic matter at K^+-saturation of the clay minerals above $f_{(K^+)} \approx 1.2$ %. Since $f_{(K^+)}$ values of 1 - 5 % are frequently encountered in soil (3, 60, 73) it is likely that EDA complex formation is responsible for the frequently observed high affinity of TNT to the sediment matrix during soil remediation measures. EDA complex formation may also have been the cause for the much higher retardation of TNT as compared to hexahydro-1,3,5-trinitro-1,3,5-triazine (RDX) observed in a contaminated aquifer (69).

In the case of DNOC the fraction adsorbed to the sediment is significantly less for all conditions considered as compared to TNT. This is largely because DNOC is present predominantly as the phenolate anion. For nitrophenolates relatively weak adsorption can be expected for both organic and inorganic phases under most natural conditions (33, 51). At low pH values and for significant concentrations of K^+ cations at clay mineral surfaces, however, adsorption processes may also contribute substantially to the fate of nitrophenols such as DNOC in the subsurface.

3. "ABIOTIC" REDUCTION OF NACS

Several authors have attempted to assess the relative importance of abiotic and biological processes responsible for the reduction of NACs in natural systems or in systems simulating complex natural environments such as sediment slurries or aquifer columns (44, 79, 89). Although clear evidence for the occurrence of abiotic reduction of NACs was presented in some of these studies (Table 2), the actual reductants could not be identified.

As we will demonstrate in this paper, there are numerous potential natural reductants that can reduce NACs in anaerobic environments. Since many of these reductants are involved continuously in biogeochemical reactions, it is somewhat arbitrary to classify NAC reductions into strictly abiotic and biological processes. Nevertheless, we will focus the following discussion on those reactions in which the electron transfer to the NAC occurs at reaction sites that are not part of living (micro)organisms.

Thermodynamic Aspects of NAC Reduction

Figure 3 summarizes reduction potentials for selected organic and inorganic redox couples. The $E_h^0(w)$ values indicated represent standard potentials at molar concentrations of the redox couples but at concentrations of the major dissolved species, i.e., H^+, HCO_3^-, Cl^-, Br^-, that are typical for environmental conditions (for more details see (63, 73)). As can be seen from the relative positions of the redox couples on the $E_h^0(w)$ scale in Figure 3, under standard environmental conditions, NACs as well as polyhalogenated methanes and ethanes can be reduced by a variety of reduced iron and sulfur species. In addition, reduced organic moieties present in natural organic matter (e.g., hydroquinone structures) as well as com-

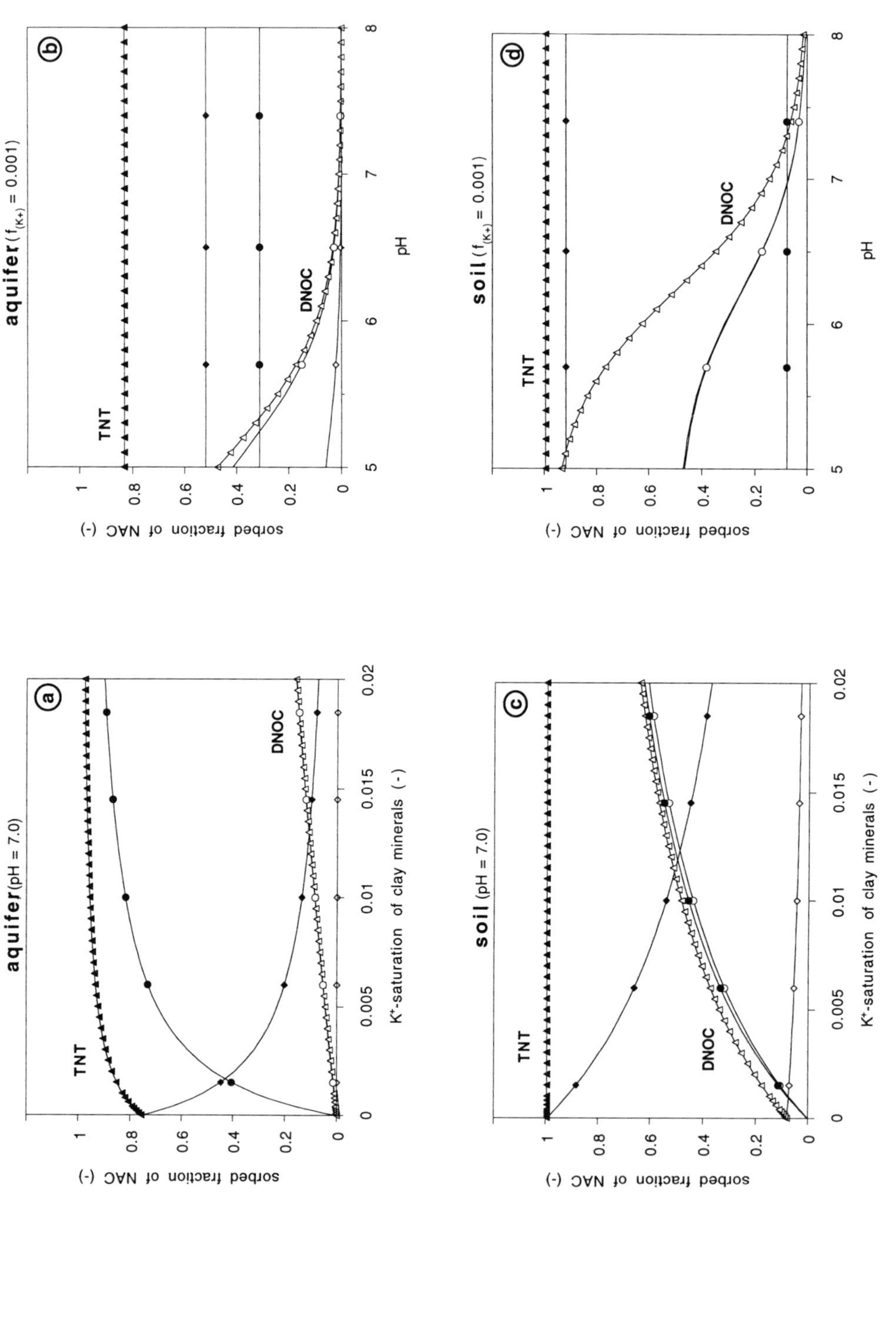
aquifer (pH = 7.0)
a
TNT
DNOC
sorbed fraction of NAC (-)
K+-saturation of clay minerals (-)
aquifer (f(K+) = 0.001)
b
TNT
DNOC
pH
soil (pH = 7.0)
c
TNT
DNOC
soil (f(K+) = 0.001)
d
TNT
DNOC

Figure 2. Distribution of DNOC (open symbols), and TNT (filled symbols) between aqueous and solid phases of typical aquifer and soil materials. Shown is the fraction of sorbed species as a function of pH and K^+-saturation of the clay minerals. The relative importance of the two dominating sorption mechanisms are compared: hydrophobic partitioning into sediment organic matter (◇, ◆) and specific EDA-complex formation with clay mineral surfaces (O, ●); triangles (Δ,▲) represent the overall speciation for DNOC and TNT respectively. Sediment parameters were chosen as follows: fraction of organic carbon: f_{oc} (soil) = 0.05; f_{oc} (aquifer) = 0.001; fraction of clay minerals: f_{clay} (soil) = 0.35; f_{clay} (aquifer) = 0.05; porosity: ε(soil) = 0.4; ε(aquifer) = 0.3; bulk density; ρ(soil) = ρ(aquifer) = 2.5 kg L^{-1}; water saturation: θ(soil) = 0.5; θ(aquifer) = 1; fraction of K^+-saturation of clay minerals: $f_{(K^+)}$ = 0.001 (when pH was used as system variable); pH = 7.0 (when $f_{(K^+)}$ was used as system variable). Linear adsorption isotherms of TNT and DNOC (neutral species) were assumed to calculate the speciation: log K_{oc}(TNT) = 2.7; log K_{oc}(DNOC) = 2.05; log K_{K^+clay}(TNT) = 3.3; log K_{K^+clay}(DNOC) = 4.2. The acidity constant of DNOC is: pK_a(DNOC) = 4.3.

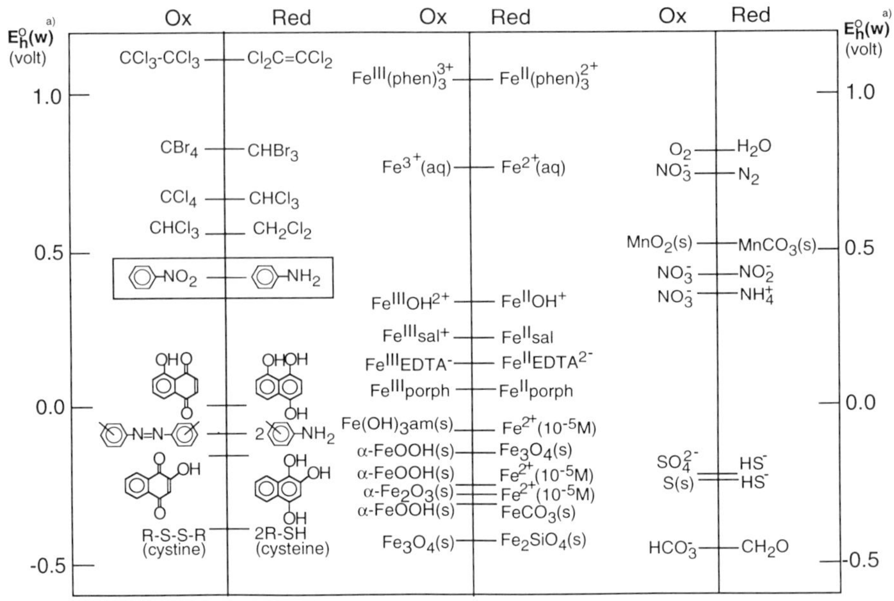

a) pH = 7; $[HCO_3^-]$; $[Cl^-] = 10^{-3}$ M; $[Br^-] = 10^{-5}$ M

Figure 3. Half-reaction reduction potentials of selected organic redox couples (left side), iron(III)/iron(II) couples (middle), and of some biogeochemically important redox couples (right side). Indicated are standard reduction potentials, $E_h^o(w)$, at environmentally relevant conditions, i.e., T = 25°C, pH = 7.0, $[Cl^-] = [HCO_3^-] = 10^{-3}M$, $[Br^-] = 10^{-5}M$. Note that the standard free-energy change, $\Delta G^o(w)$, for a given redox reaction is obtained from the difference between the $E_h^o(w)$ values of the corresponding half-reactions (see also example given by Eqs. 3-1 and 3-2): $\Delta G^o(w) = - n\, F\, \Delta E_h^o(w)$; where n is the number of electrons transferred, and F = 96.48 kJ mol^{-1} V^{-1} is the Faraday's constant. Data from Stumm (72), Schwarzenbach et al. (63), and references cited therein (am = amorphous; aq = aqueous; phen = phenanthroline; sal = salicylate; s = solid; porph = porphyrin).

plexes of other transition metals (e.g., copper, cobalt, not shown in Figure 3) exhibit reduction potentials that are low enough to reduce NACs.

As an illustrative example for calculating the redox equilibrium between a NAC and a naturally occurring reductant, we consider the reduction of nitrobenzene at 25°C in a 5 mM aqueous H_2S solution buffered at pH 7.0. We assume that nitrobenzene (NB) is reduced to aniline (AN) and that hydrogen sulfide is oxidized to elemental sulfur (S(s)). From Figure 3 we get the $E_h^o(w)$ values for the two half-reactions:

$$NB + 6H^+ + 6e^- = AN + 2H_2O; \qquad E_h^o(w) = 0.42\ V \tag{3-1a}$$

$$3S(s) + 3H^+ + 6e^- = 3HS^-; \qquad E_h^o(w) = -0.24\ V \tag{3-1b}$$

Subtraction of Eq. 3-1b from Eq. 3-1a yields the overall reaction Eq. 3-1c:

$$NB + 3HS^- + 3H^+ = AN\ 3S(s) + 2H_2O;\ \Delta E_h^o(w) = 0.66\ V \tag{3-1c}$$

The resulting $\Delta G^o(w)$ of the reaction is calculated by (for more details see 63, Chapter 12):

$$\Delta G^o(w) + -n\, F\, \Delta E_h^o(w) = -(6)\,(96.4)\,(0.66) = -382\ kJ\ mol^{-1} \tag{3-2}$$

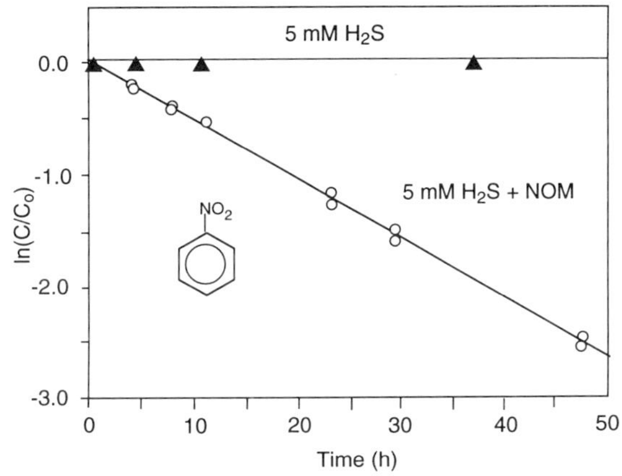

Figure 4. Reduction of nitrobenzene in 5 mM aqueous hydrogen sulfide solution in the absence and in the presence of NOM (Hyde County) at pH 7.2 and 25°C: Plot of ln (C/Co) versus time. C_o and C are the concentrations at time zero and t, respectively. Adapted from (11).

Hence, under standard environmental conditions, the reaction (Eq. 3-1c) occurs spontaneously from the left to the right. In the system considered, where H^+ (10^{-7}M), H_2O, and S(s) are present at standard concentrations, the equilibrium is described by

$$0 = \Delta G^o(w) + RT \ln \frac{[An]}{[NB][HS^-]^3} \tag{3-3}$$

Insertion of $\Delta G^o(w)$ = -382 kJmol^{-1} (Eq. 3-2) and [HS$^-$] = 2.5 x 10^3M into Eq. 3-3 yields (since pK_a (H_2S) = 7.0, [HS$^-$] = 2.5 mM at pH 7; note also that we neglect in our example calculation that about 0.5 mM HS$^-$ is consumed by oxidation during the reaction):

$$\ln\frac{[An]eq}{[NB]eq} = \frac{382 \text{ kJ mol}^{-1}}{2.48 \text{ kJ mol}^{-1}} + \ln(2.5 \times 10^{-3})^3 = 154 - 18 = 136$$

$$\text{or } \frac{[An]eq}{[NB]eq} \sim 10^{59} \tag{3-4}$$

Thus, virtually all nitrobenzene will be ultimately reduced to aniline in 5 mM aqueous $[H_2S]_{tot}$ solution at pH 7.0.

Figure 4 shows, however, that the reduction of a given NAC with H_2S in homogeneous aqueous solution is a very slow process. However, in the presence of natural organic matter (NOM) the reduction occurs at a much faster rate. In order to understand these findings, a quick look at some general kinetic aspects of redox reactions involving NACs is helpful.

Kinetic Considerations of NAC Reductions

Abiotic reductions of organic compounds in the environment usually require the transfer of *two* electrons to the compound to yield a stable product. In the majority of such reactions, the two electrons are transferred in sequential steps (12). Transfer of the first electron yields a radical species which is, in general, much more reactive than the parent compound. Hence, very often, the *rate determining step* in the overall reduction of an organic pollutant will be the formation of a radical intermediate, i.e., transfer of the first electron.

In the case of NACs, the transfer of the first electron yields at neutral pH a nitroaromatic radical anion, $ArNO_2^{\bullet-}$ (the pK_a values of $ArNO_2^{\bullet-}$ radicals are well below 5 (49)):

$$ArNO_2 + e^- = ArNO_2^{\bullet -}; \qquad E_h^{1\prime}(ArNO_2) \tag{3-5}$$

where $E_h^{1\prime}(ArNO_2)$ denotes the one-electron reduction potential of the half-reaction Eq. 3-5 at pH 7.0. Since for NACs, the formation of the radical anion is reversible, $E_h^{1\prime}$-values for NACs can be measured relatively easily by pulse radiolysis (47), and are available for a variety of such compounds (63).

Figure 5 depicts a simplified scheme of the various steps that may determine the overall rate of a one-electron transfer reaction between a reductant R and a NAC. Depending on the reductant(s) involved in the reaction, the actual transfer of the electron and/or other processes such as precursor complex formation or successor complex dissociation may be rate determining for the formation of the nitroaromatic radical anion, $ArNO_2^{\bullet -}$. Figures 5a and 5b show energy profiles of reactions where different steps of the reaction sequence are rate determining. In Figure 5a, the actual electron transfer is the rate determining step. The transition state in such a reaction is energetically closer to the radical products than to the precursor complex and the free energy change of the reaction, ΔG_1^o, can be assumed to be proportional to the activation energy, $\Delta G_1^{\neq}$. Thus, the reaction kinetics (logarithm of rate constants) of such reactions may be correlated with the free energy change, ΔG_1^o, which can be expressed by the difference in the redox potentials, $\Delta E_h^{1\prime}$, of the redox couples $ArNO_2/ArNO_2^{\bullet -}$ and R_{red}/R_{ox} (12). If other steps in the reaction sequence are rate determining (Figure 5b), other factors have to be considered for establishing quantitative structure (re)activity relationships (QSARs).

In Figure 6, such energetic considerations are schematically depicted for the reduction of nitrobenzene by hydrogen sulfide in homogeneous aqueous solution (see also Eqs. 3-1 to 3-4). In this case, it is very likely that neither the formation of the precursor complex nor the decay of the successor complex is rate determining (below). Thus, the free-energy of activation of the transfer of the first electron from HS^- to NB (i.e., $\Delta G_1^{\neq}$) can be assumed to be proportional to the standard free-energy change, $\Delta G_1^o(w)$, of the reaction:

$$ArNO_2 + HS^- = ArNO_2^{\bullet -} + HS^{\bullet}; \qquad \Delta G_1^o(w) \tag{3-6}$$

$\Delta G_1^o(w)$ can be calculated from the $E_h^{1\prime}$ values of the corresponding half-reactions:

$$\Delta G_1^o(w) = -F(E_h^{1\prime}(ArNO_2) - E_h^{1\prime}(HS^{\bullet})) = +\,154 \text{ kJ mol}^{-1} \tag{3-7}$$

Because of this large positive $\Delta G_1^o(w)$, it is not surprising that the reaction of NB with HS^- is very slow (Fig. 4). However, upon addition of natural organic matter to an H_2S solution, reduced NOM species may be formed that exhibit more negative reduction potentials than $HS^{\bullet}$ (see range of $\Delta E_h^o(w)$ values of $NOM_{red/ox}$ couples, dashed line in Figure 6). As illustrated by Figure 4 and discussed in more detail below, such NOM species (e.g., hydroquinone type moieties) may reduce NACs at much faster rates. Because such species may be re-reduced continuously and at high rates by the bulk reductant H_2S, they may act as electron transfer mediators (11).

For evaluating the reduction kinetics of NACs in a given natural system the reaction rates of a series of structurally closely related compounds can be used to probe whether the actual electron transfer or whether other steps including, for example, precursor complex formation (Fig. 5) or the regeneration of reactive sites is limiting the overall reaction rate. To this end, the rate constants observed for the reduction of NACs can be analyzed using a linear free-energy relationship of the general form

$$\log k_R = a\,\frac{E_h^{1\prime}(ArNO_2)}{RT/nF} + b \tag{3-8}$$

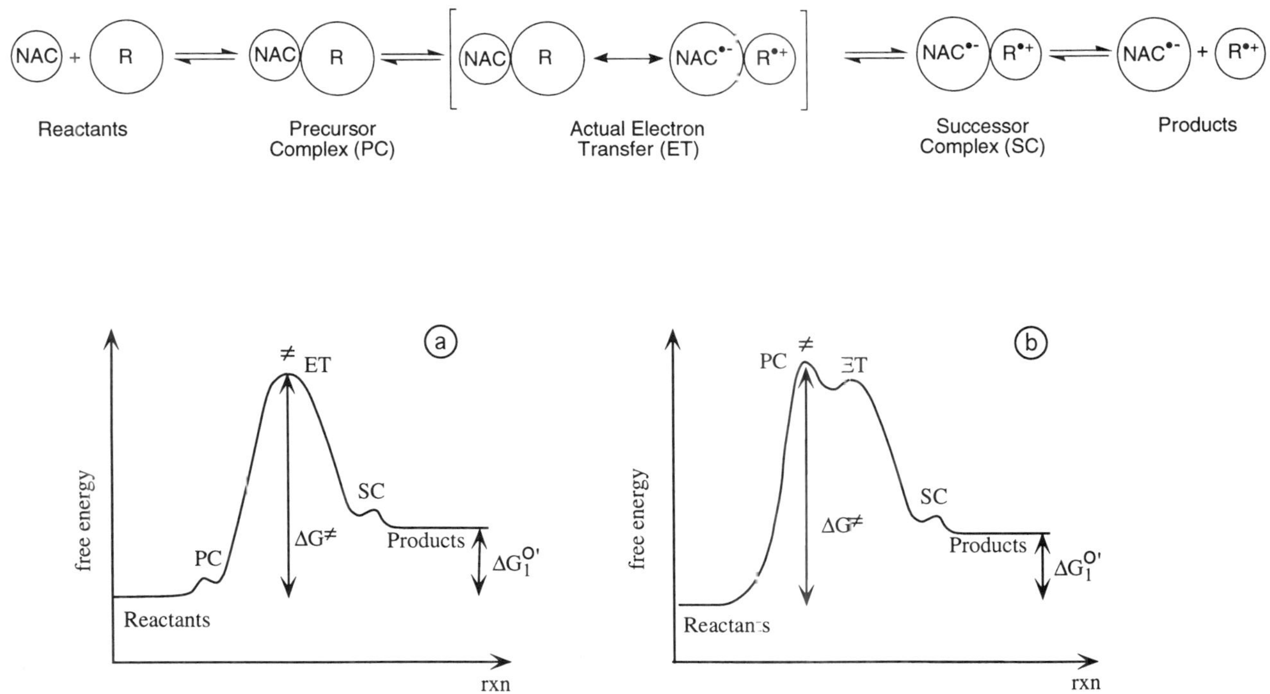

Figure 5. Simplified scheme for the transfer of the first electron from a reductant to a NAC (adapted from 12). Panels a) and b) show energy profiles of reactions where the actual electron transfer (a) or other steps such as precursor complex formation (b) are rate determining. Note that only if the actual electron transfer is the rate determining step (i.e., if the radical products are energetically closer to the transition state than the precursor complex), can a linear correlation with a slope of a = 1 be expected between the logarithm of the reduction rates and the one-electron reduction potentials ($\Delta E_h^{1'}$) of the species involved.

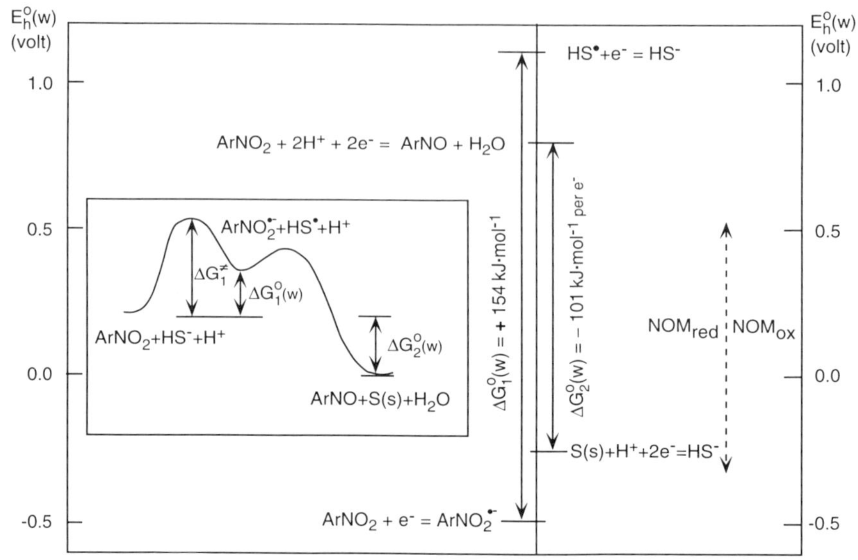

Figure 6. Energetic considerations for the reduction of nitrobenzene ($ArNO_2$) to nitrosobenzene (ArNO) with HS^- as electron donor. The standard free-energy change at pH 7 for the reduction of $ArNO_2$ to ArNO (transfer of 2 e^-) is negative [$\Delta G_2^o(w)$], whereas for the reduction of $ArNO_2$ to the nitroaromatic radical anion ($ArNO_2^{\bullet-}$; transfer of 1 e^-) a large positive standard free-energy change, $\Delta G_1^o(w)$, is obtained. If the actual transfer of the first electron from HS^- to $ArNO_2$ is rate limiting (Figure 5a), then, $\Delta G_1^{\neq}$, and thus the logarithm of the reaction rate constant will be proportional to $\Delta G_1^o(w)$ [insert]. Because the $E_h^o(w)$ values of the $NOM_{red/ox}$ couples are lower than the $E_h^o(w)$ values of the bulk reductant couple $HS^-/HS^{\bullet}$(Fig. 3), the reaction of $ArNO_2$ with NOM_{red} will be less endergonic. This explains the enhanced reduction rate of $ArNO_2$ in the presence of reduced NOM species (Fig. 4).

where k_R denotes the reaction rate constant (e.g., zero-order, (pseudo)first-order, second-order), and RT/nF has a value of 0.059V at 25°C and n = 1. If, in a given case, a significant correlation is found between log k_R and $E_h^{1'}(ArNO_2)$ with a slope $\underline{a}$ of close to 1.0, it can be concluded that, for the series of NACs considered, the actual transfer of the electron from the reductant to the NACs is rate determining (according to the Marcus relation (12)). On the other hand, a much weaker dependency of log k_R from $E_h^{1'}$ (i.e., a slope of $\underline{a} \ll 1$), or no correlation at all, indicates that other reaction steps (Figure 5) and/or other processes (below) are important. The following examples will illustrate these different cases.

Reduction of NACs in Homogeneous Aqueous Solution By NOM as ElectronTransfer Mediator and H_2S as Bulk Electron Donor

This first example is relevant for environments where reduced sulfur species (e.g., HS^-) are present (e.g., produced by microbial sulfate reduction). As already mentioned above, in the presence of hydrogen sulfide, dissolved natural organic matter (NOM) constituents may act as electron transfer mediators for NAC reduction (Figure 4). Such NOM constituents most probably include hydroquinone structures (11, 60, 64), and sulfur derivatives that result from addition reactions of quinone type structures with hydrogen sulfide (54). Dunnivant et al. (11) found that pseudo-first-order reduction rate constants, k_{obs} (Figure 4), determined for a series of substituted nitrobenzenes and nitrophenols were proportional to NOM concentrations. For a given compound and at given conditions (T, pH, $[H_2S]_{tot}$), they calculated a carbon-normalized second-order rate constant, k_{NOM}

Table 4. Model NACs: compound names, one-electron reduction potentials ($E_h^{1'}$), carbon-normalized second-order rate constants (k_{NOM}) for the reaction with Hyde County NOM, second-order rate constants for the reaction with an iron porphyrin (k_{FeP}), and apparent zero-order rate constants for reduction in an aquifer column under ferrogenic conditions (k_{obs})

Compound	Abbreviation	$E_h^{1'a}$ (V)	$k_{NOM}^{a,b}$ ($(mgC/L)^{-1}\cdot h^{-1}$)	k_{FeP}^{c} ($M^{-1}\cdot s^{-1}$)	k_{obs}^{d} ($\mu M\cdot h^{-1}$)
nitrobenzene	H	-0.485	6.4×10^{-5}	9.6×10^{-1}	0.82
2-methylnitrobenzene	2-Me	-0.590	3.2×10^{-6}	2.5×10^{-1}	0.82
3-methylnitrobenzene	3-Me	-0.475	8.8×10^{-5}	1.1	0.66
4-methylnitrobenzene	4-Me	-0.500	3.9×10^{-5}	6.7×10^{-1}	0.66
2-chloronitrobenzene	2-Cl	-0.485	2.3×10^{-4}	8.8	0.82
3-chloronitrobenzene	3 -Cl	-0.405	1.2×10^{-3}	5.3	0.66
4-chloronitrobenzene	4-Cl	-0.450	5.0×10^{-4}	2.7	0.82
2-acetylnitrobenzene	2-$COCH_3$	-0.475	5.1×10^{-4}	23.8	0.71
3-acetylnitrobenzene	3-$COCH_3$	-0.405	1.5×10^{-3}	6.3	0.90
4-acetylnitrobenzene	4-$COCH_3$	-0.360	5.9×10^{-2}	19.3	0.90
2,4-dinitrotoluene	2,4-DNT	-0.360	6.5×10^{-3}		
2-nitrophenol	2-NP	-0.492	2.3×10^{-4}		
2,4-dinitro-6-methylphenol	DNOC	-0.430	5.0×10^{-4}		

a) From (11); note that for 2-NP and DNOC, the $E_h^{1'}$ values are apparent reduction potentials at pH 7.5.
b) pH 7.5, T=25°C, $[H_2S]_{tot}$=5mM, source of NOM: Hyde County.
c) Second-order rate constant for the reaction with meso-tetrakis (N-methyl-pyridyl) iron(II) porphyrin; data from (64).
d) According to Eq. 3-10, data from (25).

$$k_{NOM} = k_{obs}/[NOM] \tag{3-9}$$

where [NOM] is the total dissolved organic carbon concentration. They found that k_{NOM} values determined for NOM derived from very different environments varied by less than one order of magnitude, and increased with increasing pH and increasing $[H_2S]_{tot}$ concentration. An illustrative set of k_{NOM} values is given in Table 4. From the k_{obs} values obtained for the different substituted nitrobenzenes it can be seen that the presence of an electron withdrawing substituent (i.e., -Cl, -$COCH_3$) leads to a decrease in the one-electron reduction potential, $E_h^{1'}$, which is paralleled by an increased rate of reduction. Substituents *ortho* to the nitro group increase the one electron reduction potential and exhibit quite drastic rate diminishing effects when compared to the *para*-substituted compound. These effects can be rationalized by considering the resonance effects of the substituents on the aromatic system. The resonance of the nitro group with the aromatic ring is hindered when it is forced out of the ring plane due to steric interactions of an *ortho*-substituent, leading to a decrease in the stabilization of the $ArNO_2^{\bullet-}$ species. This hypothesis is supported by the higher spin density at the nitrogen atom found for the 2-methylnitrobenzene radical anion as compared to the 4-methylnitrobenzene radical anion (49).

A good linear correlation was obtained between reactivity (expressed by log k_{NOM}) and one-electron reduction potential of the compounds (Figure 7). In fact, the slope $\underline{a}$ of the linear regression line is 1.0 (Eq. 3-8), indicating that the actual transfer of the electron from the reduced NOM moieties to the NACs is rate limiting. A very similar result was found for the reduction of NACs in H_2S solutions containing NOM derived from exudates of bacteria (17).

Note that in these cases, the most reactive of the nitroaromatic model compounds (4-acetylnitrobenzene) was reduced more than four orders of magnitude faster than the least

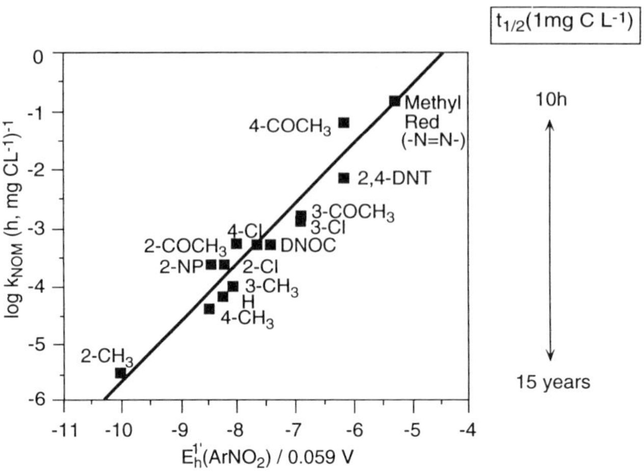

Figure 7. Plot of log k_{NOM} (Eq. 3-9) versus $[E_h^{1'}(ArNO_2)/0.059\ V]$ for the mono-substituted nitrobenzenes listed in Table 4 in aqueous solutions of Hyde County NOM at pH 7.5 and 5 mM H_2S. The regression line log $k_{NOM} = 1.0\ [E_h^{1'}(ArNO_2)/0.059\ V] + 4.4$ is the result of a linear least-squares fit of Eq. 3-8 to the data set given in Table 4. The half-lives indicated on the right-hand side are calculated for an aqueous solution containing 1 mg C/liter of Hyde County NOM; adapted from (11).

reactive NAC (2-methylnitrobenzene). As an example, the half-lives of 4-acetylnitrobenzene and 2-methylnitrobenzene in 5 mM H_2S solution (pH 7.5, 25°C) containing 1mg C/liter Hyde County NOM are predicted to be 12 hours and about 25 years, respectively (11).

Reduction of NACs by Reduced Iron Species

In abiotic reductions involving iron as an electron donor in natural systems an iron(II) species is usually oxidized to an iron(III) species. In technical applications metallic iron (Fe(0)) has been used to reduce organic pollutants including NACs (2).

Since oxidation of iron(II) to iron(III) only involves the transfer of one electron, the reduction potential of a given iron(III)/iron(II) couple (Figure 3) can be directly applied to evaluate the relative reactivities of iron(II) species in various chemical environments. Such a consideration is only applicable to reduction of organic pollutants if the actual electron transfer is at least partly rate limiting. This can not be assumed in many cases, particularly for heterogeneous reactions, where the oxidative dissolution of the iron(II) containing mineral (i.e., the regeneration of the reductant) may be rate determining in a steady-state situation. This may explain why the reduction of NACs in aqueous suspensions containing magnetite is very slow (38), although the reduction potential of the corresponding redox couple (i.e., $Fe_3O_4(s)/\alpha\text{-}FeOOH(s)$ is quite low (Figure 3). In contrast, Fe^{2+} species adsorbed to magnetite reduce NACs at very fast rates (Figure 8). In batch experiments, Fe^{2+} adsorbed to iron oxide surfaces including goethite, lepidocrocite, and amorphous iron hydroxide was found to reduced NACs very efficiently (38). Much slower reaction rates were observed when aqueous Fe^{2+} was adsorbed to other mineral surfaces such as $\gamma\text{-}Al_2O_3$, TiO_2, SiO_2 or kaolinite. In these latter systems, reduction rates increased after iron (III) coatings developed at the mineral surfaces. The surface properties of such coatings are also very likely to control the long term reactivities of metallic iron phases proposed as reductants in remediation measures (2).

Figure 8 shows an important feature that is typical for the type of results that are obtained when studying heterogeneous (abiotic) reactions of organic pollutants in batch systems. As can be seen, during the course of the reaction, the rate law of the reaction changed significantly. In the case of the reduction of nitrobenzenes in heterogeneous Fe(II)/Fe(III) systems, a pseudo-first-order rate law was obtained only for the initial phase of the experiments followed by a significant deviation from first-order behavior with slow reaction rates. Such behavior often indicates that the most reactive reductants (in this case, Fe^{2+}-spe-

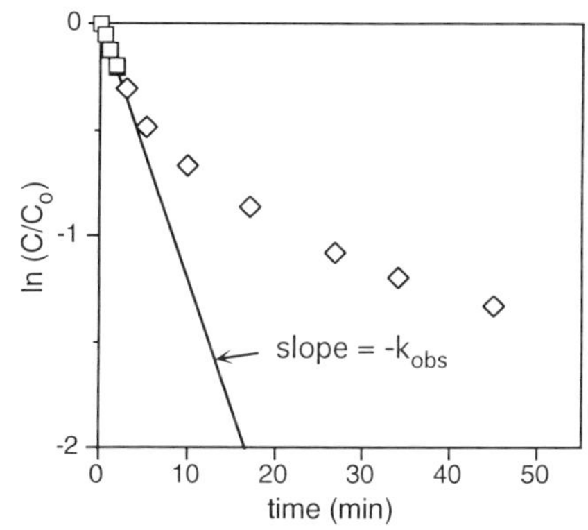

Figure 8. Reduction of nitrobenzene (NB) in aqueous suspensions containing 200 mg L^{-1} magnetite and 1.5 mM Fe^{2+} at pH 7. Plot of ln(C/C_o) versus time. C and C_o are the concentrations of NB at time zero and t, respectively. The initial pseudo-first order rate constant, k_{obs}, is obtained by a linear least-squares fit of $\ln(C/C_o) = - k_{obs} \cdot t$, using only the first few data points. Data from (38).

cies adsorbed to various surface sites of magnetite) are rapidly depleted and less reactive species and/or the changing properties of the system control the reaction rates at later stages of the experiments. In most studies on abiotic reduction reactions, only such initial rates (usually expressed by a pseudo-first-order rate constant, k_{obs}, Figure 8) derived from a few data points at the beginning of the experiment, are reported. This raises the question of how significant such initial rates are for assessing rates that may occur under typical environmental (steady-state) conditions where a prolonged exposure of a given contaminant to the reactive sites has to be assumed. However, as will be demonstrated below, despite the fact that initial rates obtained from batch experiments in model systems may not be very meaningful for predicting actual reduction rates in natural systems, they may be useful to provide insights about what types of reductants are primarily responsible for the reaction of a given class of chemicals in a complex environmental system.

In Figure 9 plots of the logarithms of the relative reduction rates (k_{rel}, relative to 4-Cl) of ten mono-substituted nitrobenzenes versus their $E_h^{1'}$ values for two different systems illustrate the usefulness of initial rates for inferring information on reaction mechanism(s).

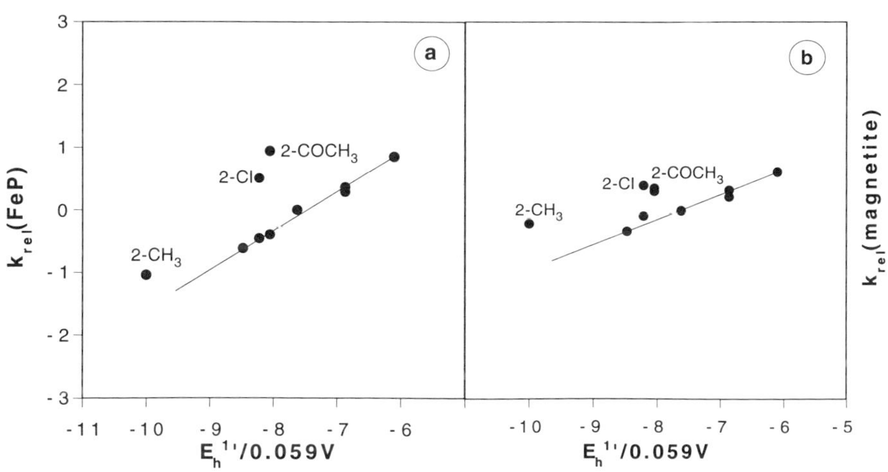

Figure 9. Plots of log k_{rel} versus [$E_h^{1'}(ArNO_2)$/0.059 V] for the mono-substituted nitrobenzenes listed in Table 4 for (a) the reaction with an iron porphyrin, and (b) the reaction with Fe^{2+} adsorbed to magnetite. Note that in case (a) k_{rel} represents the ratio of two second-order rate constants (i.e., $k_{rel} = k_{FeP}(X) / k_{FeP}(4\text{-Cl})$), whereas in case (b) k_{rel} is the ratio of two apparent pseudo-first-order rate constants, i.e., $k_{rel} = k_{obs} / k_{obs}(4\text{-Cl})$.

Figure 9a refers to a reaction in homogeneous aqueous solution containing a water soluble iron porphyrin as the electron transfer mediator and cysteine as the bulk electron donor while Figure 9b refers to a reaction in aqueous suspensions of magnetite and dissolved Fe^{2+}. Note that Figure 9a represents relative *second-order* rate constants, k_{FeP} (Table 4), for the reaction of the NACs with the iron porphyrin, whereas Figure 9b represents relative initial *first-order* rate constants, k_{obs} (see Figure 8). Qualitatively, a very similar picture is obtained for the two different iron redox systems. Similar results were also obtained for the reaction of these model NACs with other naturally occurring reduced iron species (25, 38). The common feature in these iron redox systems is the weak (linear) relationship between log k_{rel} and $E_h^{1'}(ArNO_2)$. For *meta-* and *para-*substituted nitrobenzenes the slopes $\underline{a}$ of regression lines (Eq. 3-8) are between 0.3 and 0.5. In addition, significantly higher relative reaction rates were obtained for *ortho-*substituted nitrobenzenes as would be expected from their one-electron reduction potentials. The flat slopes of the correlation lines between log k and $E_h^{1'}(ArNO_2)$ as well as the behavior of the *ortho-*substituted nitrobenzenes indicate that in contrast to the reactions with NOM constituents described above, the formation of the precursor complex is a rate limiting step in the transfer of the first electron from an iron(II) species to a given NAC. This hypothesis is corroborated by the fact that strong competition was observed between different NACs present in binary or ternary mixtures (38, 64). For a given set of isomers (e.g., the chloronitrobenzenes), a typical competition pattern favoring the reduction of *ortho-*substituted isomers was found for the reaction with the various iron(II) species investigated.

In the case of the iron redox systems presented here, the most reactive of the nitroaromatic model compounds (4-acetylnitrobenzene) was reduced by only about one order of magnitude faster than the least reactive NAC (2-methylnitrobenzene) whereas there was 4 orders of magnitude difference in the H_2S/NOM system.

In summary, this example shows that a series of structurally related NACs may exhibit a typical reaction pattern with respect to a given type of reductant. Insights obtained from experiments with model reductants may help to identify the reductants primarily responsible for NAC reduction in more complex natural systems. One example is the NOM mediated reduction of the substituted nitrobenzenes where a reduction pattern very similar to the reaction with two model quinones was found (see above and (11)). A second example involving reduced iron species follows.

Abiotic Reduction of NACs Coupled to Microbial Iron Reduction in Laboratory Aquifer Columns

Heijman et al. (25) investigated the reduction of the mono-substituted nitrobenzenes listed in Table 4 in several laboratory column systems containing aquifer materials from the banks of a river-groundwater infiltration site. All NACs were reduced quantitatively to the corresponding anilines and as indicated by the constant reactivity of the reactants over the entire column length, apparent zero-order kinetics were found (a typical example is given in Figure 10). For a quantitative comparison of the reaction rates of the various NACs and of the different column systems, Heijman et al. calculated apparent zero-order rate constants, k_{obs}

$$k_{obs} = \frac{([ArNO_2]_{in} - [ArNO_2]_{out})}{\tau_w} \tag{3-10}$$

where $[ArNO_2]_{in}$ and $[ArNO_2]_{out}$ are the NAC concentrations at the beginning and at the end of the column (or a given column segment), and τ_w is the average residence time of the

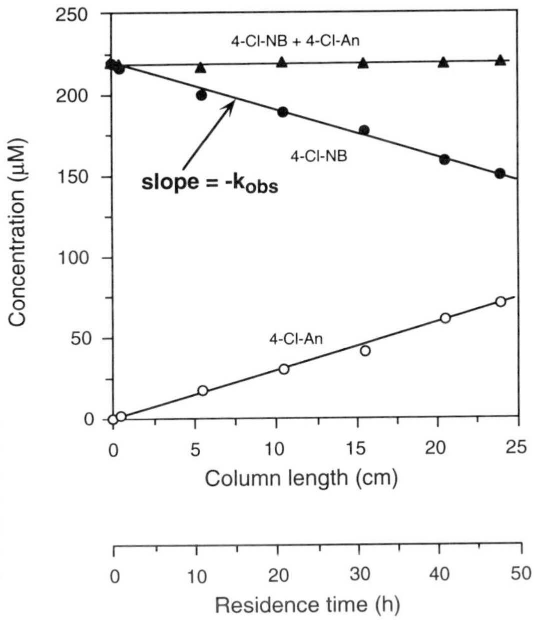

Figure 10. Example of concentration gradients typically found for the reduction of NACs in laboratory columns containing aquifer material from the banks of a river-groundwater infiltration site. Shown are the concentrations of 4-chloronitrobenzene (4-Cl-NB) and of its transformation product 4-chloroaniline (4-Cl-An) over the length of the column. Also indicated is the mass balance; adapted from (25).

water calculated as the ratio of the flow rate and the pore volume. Under the conditions chosen, sorption of the NACs to the aquifer material could be neglected, therefore, it was assumed that the average residence times of the NACs in the columns were similar to the residence time of the water.

For a given sediment column and conditions the calculated apparent zero-order rate constants k_{obs} were virtually identical for all compounds (Table 4), indicating that neither precursor complex formation nor the actual electron transfer between reductants and NACs was rate limiting. Nevertheless, with mixtures of NACs, a competition behavior similar to that described for the reduction of the NACs with iron(II) species in batch experiments was found.

The k_{obs} increased steadily with increasing temperature in the range of 5°C to 65°C. In accordance with additional evidence (25, 26) this temperature dependence of k_{obs} suggests that the reduction of NACs in the aquifer columns occurred by an abiotic process. However, prolonged heating of the column resulted in a rapid decrease in k_{obs}. Furthermore, an immediate increase in k_{obs} (up to a factor of 10) was observed, when feeding the columns with acetate, i.e., when offering an easily utilizable carbon source for iron reducing bacteria but not for fermenting bacteria. It was shown that 80-90% of the electrons produced from the oxidation of acetate were transferred to the NACs. Heijman et al. (25) propose that in these columns, reduction occurred by an abiotic reaction of NACs with surface-bound iron (II) species which served as mediators for the transfer of electrons. The reactive iron(II) species originated from microbial oxidation of organic material by dissimilatory iron reducing bacteria (Figure 11). The different NACs were all reduced at the same *relative* rates with and without addition of acetate. Therefore, it is most likely that the regeneration of the reactive sites (i.e., the surface-bound iron(II) species) by the bacteria and/or the formation of reactive Fe(III) surface coatings was the rate limiting process.

In contrast to the H_2S/NOM and batch Fe(II)/Fe(III) systems, in the sediment column systems the most reactive of the nitroaromatic model compounds (4-acetylnitrobenzene) was reduced at *almost the same rate* as the least reactive NAC (2-methylnitrobenzene). Distinct

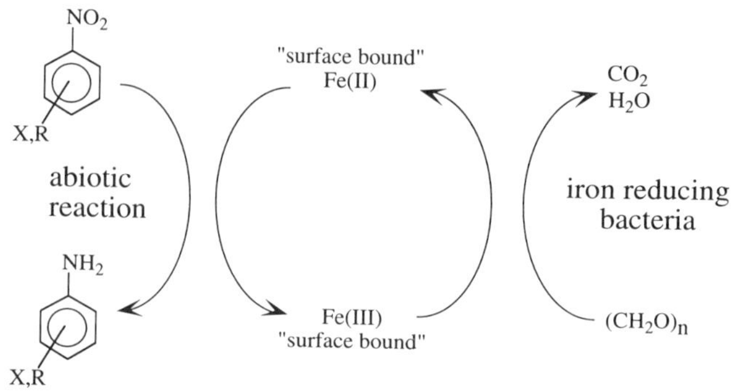

Figure 11. General reaction scheme proposed by Heijman et al. (25) for the reduction of NACs in aquifer columns under ferrogenic conditions.

competition effects similar to those in the batch iron redox systems were found for binary and ternary mixtures when *ortho*-substituted NACs were present.

Other Potential Electron Donors for Abiotic NAC Reduction in the Subsurface

Besides the reduced iron, NOM, and dissolved sulfur species discussed above, other potential electron donors present in anaerobic environments may contribute to abiotic NAC reduction. Such reductants include solid metal sulfides (e.g., MnS, FeS, FeS_2, ZnS) formed in sulfate reducing environments, and molecular hydrogen produced under methanogenic conditions. In addition, it has been postulated that redox active species present in sediments and aquifers, e.g., proteins and other biomolecules including cofactors that contain transition metal complexes, may be responsible for NAC reduction (Table 3). However, the currently available data in the literature are too scarce to allow a sound assessment of the importance of these electron donors for NAC reductions in the environment. Yu and Bailey (91) have shown that, in principle, NACs may be reduced in aqueous solution by various metal sulfides. Unfortunately, the conditions under which these reactions were studied were not very representative of natural environments. In our own laboratory, we have found that the reduction of 4-chloronitrobenzene in aqueous suspensions of aged FeS and FeS_2 phases was extremely slow (40). Because the oxidative dissolution of metal sulfides by NACs under environmental conditions is a relatively slow process, it seems rather unlikely that such reactions are of great environmental significance. In contrast, molecular hydrogen could play a more important role in NAC reduction under methanogenic conditions. In the presence of certain metal catalysts, H_2 is a very versatile reducing agent (66). Hence, NACs could be reduced either directly by H_2 in the presence of catalytic surfaces, or by electron transfer mediators that react with H_2.

4. FATE OF REDUCTION PRODUCTS OF NACS IN AQUIFERS, SOILS AND SEDIMENTS

Aromatic nitroso-, hydroxylamino-, or amine compounds exhibit significant toxicity or can combine to form toxic products (5, 14, 53). One prominent example is the occupational exposure to anilines and related compounds that has had severe impacts on human health due to the carcinogenic potential of these compounds. Therefore, the reactivity and fate of

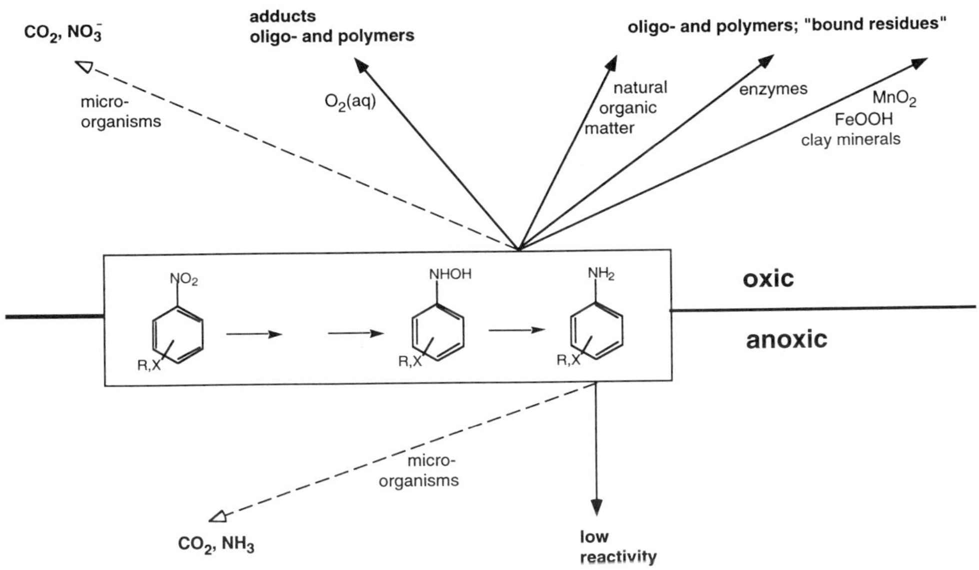

Figure 12. Examples of abiotic reactions that determine the fate of aromatic anilines and hydroxylamines under aerobic and anaerobic conditions.

these compounds in subsurface environments should be considered in the context of NAC reduction reactions. Under anaerobic conditions, anilines resulting from abiotic reduction of NACs in natural and model systems are relatively stable products (10, 11, 17, 25, 38). In aerobic and suboxic environments, however, the recovery of anilines or intermediates from NAC reduction is usually very poor due to the formation of adducts, oligo- and polymers, or "bound residues" that are difficult to analyse (15, 28, 55, 67) . Most of the extensive literature on the fate of anilines and related compounds in soil environments has been covered by comprehensive reviews (14, 53, 82).

An illustration of the abiotic reactions that are believed to dominate the fate of aromatic amines in soils and sediments is given in Figure 12. In oxic environments, numerous reactants and reaction schemes have been suggested to account for the rapid disappearance of anilines and related compounds. Many of the products obtained from the oxidation of aromatic amines can be rationalized by examination of the reactivity of the aryl-amino radical, $ArNH^{\bullet}$. Delocalization by resonance stabilizes the aryl-amino radical but makes the *ortho-* and *para-* positions of the ring subject to reactions (41, 46).

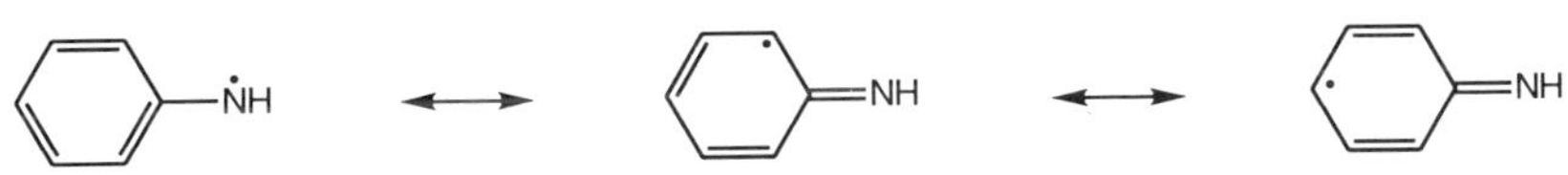

In the absence of otherreactants, $ArNH^{\bullet}$ radicals can combine with one another to form coupling products ($N^{\bullet} + N^{\bullet}$, $N^{\bullet} + C^{\bullet}$) including some compounds of high environmental concern such as substituted azobenzenes or phenazines (14, 46, 53). In soils and aquifers, however, numerous naturally occurring reactants may be present to react with aryl-amino radicals to yield poorly characterized products. Various types of mineral surfaces including clay minerals, iron (hydr)oxides, and manganese oxides have been identified as reactants or catalysts in oxidative coupling and polymerization reactions of aromatic amines (39, 82). However, in most natural soils and sediments, natural organic matter (humines, humic- and fulvic acids, lignin components) is believed to be the most important sink for aromatic amines due to covalent binding and subsequent irreversible polymerization and rearrangement reactions (8, 9, 31). The kinetics of the formation of such "bound residues" can be largely enhanced by addition of enzyme mediators and of easily oxidizable cosubstrates (7, 75, 76, 77). The persistence and the long term effects (4, 6, 21, 28, 32) of such bound residues on the subsurface ecology are still subjects of current and future research.

In reduced environments most of the potential oxidants capable of reaction with anilines are not available which causes these compounds to be quite persistent with respect to chemical transformations. Under such conditions, anaerobic biodegradation is likely to be the most important elimination process for aromatic amines (62).

SUMMARY AND CONCLUSIONS

The reduction of NACs to the corresponding amino compounds is a widely observed reaction in sediments, soils, and aquifers. As has been illustrated in this paper, there are numerous potential electron donors present in natural systems (e.g., reduced iron species, reduced sulfur species, reduced NOM constituents) that may reduce NACs in abiotic reactions. Because all these species are involved continuously in biogeochemical processes, and are thus intimately coupled to microbial activity, it is not very meaningful to separate NAC reduction in a given environment into strictly abiotic and strictly biological processes (Figure 13).

Comparison of reduction rates of NACs within a given environment or between different environments requires an understanding that a variety of different processes or reaction steps such as generation of reactive species or sites, formation of a precursor complex, or actual transfer of the electron, may be rate limiting. Thus, depending on the type of electron donor(s) available, NACs exhibiting very different one-electron reduction

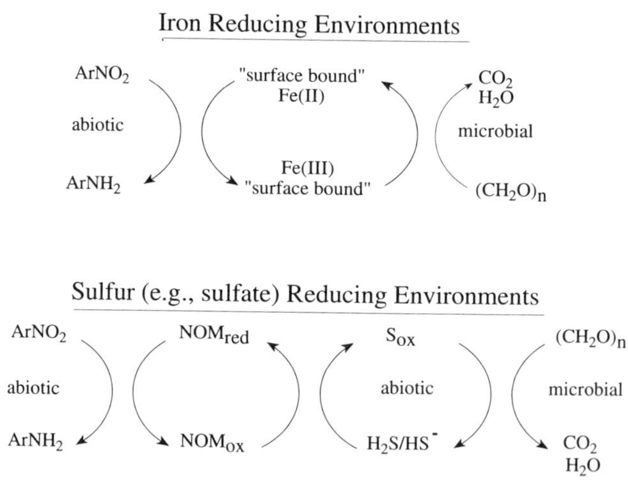

Figure 13. Example of the interdependence of abiotic and microbial processes involved in the reduction of nitroaromatic compounds in iron- and sulfur- reducing environments. In both systems microbial activity is essential for providing the bulk reductants (i.e., $Fe(II)_{aq}$ or HS^-/H_2S) that drive the "abiotic" reduction of NACs by reactive mediators (adsorbed Fe(II) or reduced NOM). Very different processes, such as actual transfer of the first electron, precursor complex formation, or regeneration of reactive sites may control the overall rate of "abiotic" reduction of NACs in such coupled systems.

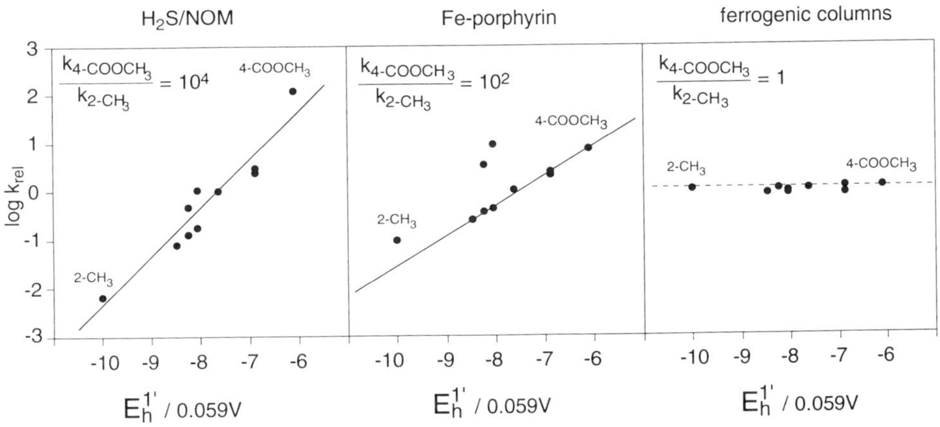

Figure 14. Comparison of the relative reduction rates (k_{rel}, the rates are relative to 4-Cl) of the mono-substituted nitrobenzenes listed in Table 4 in three different systems; (a) in aqueous solutions containing hydrogen sulfide and natural organic matter; data from (11); (b) in aqueous solutions containing cysteine and a water soluble iron porphyrin; data from (64); (c) in aquifer columns under ferrogenic conditions; data from (25) The relative rates, expressed as log k_{rel}, are plotted against the normalized one-electron reduction potentials, $[E_h^{1'}(ArNO_2)/0.059\ V]$, of the compounds.

potentials may in one system be reduced at rates that differ by orders of magnitude (Figure 14). An example is the NOM mediated reduction of NACs in the presence of hydrogen sulfide. In another system, the same compounds may react at virtually the same rate. Such is the case in reductions mediated by surface-bound iron species that are continuously recycled by iron-reducing bacteria as was illustrated in the ferrogenic aquifer columns. Furthermore, several potential electron donors as well as microorganisms can be present in a given system and different NACs can be reduced by different processes. For example under sulfate reducing conditions very reactive NACs such as TNT can be (partially) reduced very rapidly by NOM constituents, while the reduction rates of less reactive compounds such as diaminonitrotoluenes (intermediates of TNT reduction) are primarily controlled by microbially mediated reduction (56). Therefore, it may be difficult to apply quantitative structure-reactivity relationships for describing reduction rates of NACs in complex natural systems. Consequently, tests of a given environment for reactivity with respect to NAC reduction requires the use of a series of NACs covering a wide range of one-electron reduction potentials. In addition, due to hydrophobic partitioning to soil organic matter or specific adsorption to clay minerals (particularly in NH_4^+- and K^+- rich environments), certain NACs may be reduced at much slower rates than would be predicted from their one-electron reduction potentials. In fact, many very reactive NACs including compounds with more than one nitro group also exhibit very large adsorption coefficients (22, 23). Finally, it should be noted that in mixtures of NACs, competition effects may strongly influence the adsorption behavior as well as the reduction rates of the various compounds.

There are still considerable gaps in the present knowledge concerning abiotic processes that can influence the reduction of NACs in anaerobic environments. However, a few general conclusions can be made concerning the remediation of NAC contaminated sites. First, the mobility of NACs in soils and aquifers can be significantly enhanced or diminished by adding strongly hydrated (e.g., Na^+, Ca^{2+}) or weakly hydrated (e.g., K^+) cations to the system. The distribution between aqueous and solid phases also affects the susceptibility of NACs with respect to reduction reactions. Secondly, creation or stimulation of iron reducing environments appears to be a very promising approach for promoting reduction of NACs, both for *in situ* remediation as well as for the treatment of NACs on site in reactor systems. Finally, by analogy to the iron reducing system (Figure 11) for the oxidative treatment of the reduction products (i.e., the corresponding amino compounds), a system containing a manganese oxide and manganese oxidizing bacteria would be an interesting possibility.

ACKNOWLEDGMENTS

We thank Béatrice Schwertfeger for secretarial work. Valuable comments were made by W. Angst and J. Klausen.

REFERENCES

1. Adhya, T. K., B. Sudhakar, and N. Sethunathan. 1981. Fate of fenitrothion, methylparathion, and parathion in anoxic, sulfur-containing soil systems. Pestic. Biochem. Physiol. 16:14-20.
2. Agrawal, A., and P. G. Tratnyek. 1994. Abiotic remediation of nitro-aromatic groundwater contaminants by zero-valent iron, p. 492-494. *In* ACS, Extended Abstract Book. American Chemical Society, San Diego, CA.
3. Barker, J. F., J. S. Tessmann, P. E. Plotz, and M. Reinhard. 1986. The organic geochemistry of a landfill leachate plume. J. Contam. Hydrol. 1:171-189.
4. Bartha, R. 1971. Fate of herbicide-derived chloroanilines in soil. J. Agr. Food Chem. 19:385-387.
5. Bartha, R., and D. Pramer. 1967. Pesticide transformation to aniline and azo compounds in soil. Science 156:1617-1618.
6. Bollag, J. M. 1992. Decontaminating soil with enzymes. Environ. Sci. Technol. 26:1876-81.
7. Bollag, J. M. 1992. Enzymes catalyzing oxidative coupling reactions of pollutants, p. 205-17. *In* H. Siegel and A. Siegel (ed.), Metal ions in biological systems. Degradation of environmental pollutants by microorganisms and their metalloenzymes. Marcel Dekker, New York.
8. Bollag, J. M., and C. Myers. 1992. Detoxification of aquatic and terrestrial sites through binding of pollutants to humic substances. Sci. Total Environ. 117:357-66.
9. Bollag, J. M., C. J. Myers, and R. D. Minard. 1992. Biological and chemical interactions of pesticides with soil organic matter. Sci. Total Environ 123:205-17.
10. Delgado, M. C., and N. L. Wolfe. 1992. Structure-activity relationships for the reduction of *p*-substituted nitrobenzenes in anaerobic sediments. *In* ACS, Extended Abstract Books. American Chemical Society, San Francisco, CA.
11. Dunnivant, F. M., R. P. Schwarzenbach, and D. L. Macalady. 1992. Reduction of substituted nitrobenzenes in aqueous solutions containing natural organic matter. Environ. Sci. Technol. 26:2133-2141.
12. Eberson, L. 1987. Electron transfer reactions in organic chemistry. Springer Verlag, Berlin.
13. Ewing, B. B., E. S. K. Chian, J. C. Cook, C. A. Evans, P. K. Hopke, and E. G. Perkins. 1977. Monitoring to detect previously unrecognized pollutants in surface waters. US Environmental Protection Agency, Technical Report, Washington, DC.
14. Fishbein, L. 1984. Aromatic Amines, p. 1-40. *In* O. Hutzinger (ed.), The handbook of environmental chemistry - anthropogenic compounds. Springer Verlag, Berlin.
15. Gaillard, G., A. Mermoud, and S. B. Haderlein. Fate of DNOC and 1,4-dinitrobenzene in two unsaturated soils - comparison of lysimeter and batch experiments. Submitted for publication.
16. Geer, R. D. 1978. Predicting the anaerobic degradation of organic chemical pollutants in waste water treatment plants from their electrochemical behavior. Montana Univ. Joint Water Resouces Research Center, Bozeman, Montana.
17. Glaus, M. A., C. G. Heijman, R. P. Schwarzenbach, and J. Zeyer. 1992. Reduction of nitroaromatic compounds mediated by *Sreptomyces* sp. exudates. Appl. Environ. Microbiol. 58:1945-1951.
18. Gorontzy, T., J. Küver, and K.-H. Blotevogel. 1993. Microbial transformation of nitroaromatic compounds under anaerobic conditions. J. Gen. Microbiol. 139:1331-1336.
19. Grosjean, D. 1985. Reactions of *o*-cresol and nitrocresol with NO_x in sunlight and with ozone-nitrogen dioxide mixtures in the dark. Environ. Sci. Technol. 19:968-974.

20. Haas, R., and G. Stork. 1989. Konzept zur Untersuchung von Rüstungsaltlasten. Fresenius Z. Anal. Chem. 335:839-846.
21. Haas, R., and E. von Löw. 1986. Grundwasserbelastung durch eine Altlast. Die Folgen einer ehemaligen Sprengstoffproduktion für die heutige Trinkwassergewinnung. Forum Städte-Hygiene. 37:33-43.
22. Haderlein, S. B., and R. P. Schwarzenbach. 1993. Adsorption of substituted nitrobenzenes and nitrophenols to mineral surfaces. Environ. Sci. Technol. 27:316-26.
23. Haderlein, S. B., K. Weissmahr, and R. P. Schwarzenbach. Specific adsorption of nitroaromatic explosives and pesticides to clay minerals. Submitted.
24. Hallas, L. E., and M. Alexander. 1983. Microbial transformation of nitroaromatic compounds in sewage effluent. Appl. Environ. Microbiol. 45:1234-1241.
25. Heijman, C. G., E. Grieder, C. Holliger, and R. P. Schwarzenbach. Abiotic reduction of nitroaromatic compounds coupled to microbial iron reduction in laboratory aquifer columns. Environ. Sci. Technol.29:775-783.
26. Heijman, C. G., C. Holliger, M. A. Glaus, R. P. Schwarzenbach, and J. Zeyer. 1993. Abiotic reduction of 4-chloronitrobenzene to 4-chloroaniline in a dissimilatory iron-reducing enrichment culture. Appl. Environ. Microbiol. 59:4350-4353.
27. Helling, C. S. 1976. Dinitroaniline herbicides in soils. J. Environ. Qual. 5:1-15.
28. Helling, C. S., and A. E. Krivonak. 1978. Physicochemical characteristics of bound dinitroaniline herbicides in soils. J . Agric. Food Chem. 26:1156-1163.
29. Helmig, D., J. López-Cancio, J. Arey, W. P. Harger, and R. Atkinson. 1992. Quantification of ambient nitrodibenzopyranones: further evidence for atmospheric mutagen formation. Environ. Sci. Technol. 26:2207-2213.
30. Herterich, R. 1991. Vorkommen und atmosphärischer Eintrag von Herbiziden und phytotoxischen Nitrophenolen an zwei emissionsfernen Waldstandorten in Bayern. Ph.D. thesis, Univ. Bayreuth, FRG.
31. Hsu, T.-S., and R. Bartha. 1974. Interaction of pesticide-derived chloroaniline residues with soil organic matter. Soil Science 116:444-452.
32. Hsu, T.-S., and R. Bartha. 1976. Hydrolyzable and nonhydrolyzable 3,4-dichloroaniline-humus complexes and their respective rates of biodegradation. J. Agric. Food Chem. 24:118-122.
33. Jafvert, C. T. 1990. Sorption of organic acids to sediments: initial model development. Environ. Toxicol. Chem. 9:1259-1268.
34. Jenkins, T., M. E. Walsh, P. W. Schumacher, and P. H. Miyares. 1989. Liquid chromatographic method for determination of extractable nitroaromatic and nitramine residues in soil. J. Assoc. Off. Anal. Chem. 72:890-899.
35. Jenkins, T. F., D. C. Leggett, C. L. Grant, and C. F. Bauer. 1986. Reversed-phase high-performance liquid chromatographic determination of nitroorganics in munitions wastewater. Anal. Chem. 58:170-175.
36. Kearny, P. C., and D. D. Kaufman. 1975. Herbicides: chemistry, degradation, and mode of action. Dekker, New York.
37. Kjeldsen, P., J. Kjølhold, B. Schultz, T. H. Christensen, and J. C. Tjell. 1990. Sorption and degradation of chlorophenols, nitrophenols, and organophosphorous pesticides in the subsoil under landfills - laboratory studies. J. Contam. Hydrol. 6:165-184.
38. Klausen, J. W., S. Tröber, S. B. Haderlein, and R. P. Schwarzenbach. Reduction of substituted nitrobenzenes by Fe(II) in aqueous mineral suspensions. Submitted.
39. Laha, S., and R. G. Luthy. 1990. Oxidation of aniline and other primary aromatic amines by manganese dioxide. Environ. Sci. Technol. 24:363-373.
40. Lanz, K., and R. P. Schwarzenbach. 1988. Unpublished data.
41. Larson, R. A., and E. J. Weber. 1994. Reaction mechanisms in environmental organic chemistry. Lewis Publishers, Boca Raton.
42. Leuenberger, C., W. Giger, R. Coney, J. W. Graydon, and E. Molnar-Kubica. 1987. Nitrated phenols in rain: atmospheric occurence of phytotoxic pollutants. Chemosphere 17:511-515.
43. Lyman, W. J., W. F. Rheel, and D. H. Rosenblatt. 1982. Handbook of chemical property estimation methods. McGraw-Hill Inc., New York.
44. Macalady, D. L., P. G. Tratnyek, and T. J. Grundl. 1986. Abiotic reduction reactions of anthropogenic organic chemicals in anaerobic systems: a critical review. J. Contam. Hydrol. 1:1-28.
45. Macalady, D. L., and L. N. Wolfe. 1992. Hydrophobic effects in the reduction of anthropogenic organic chemicals by natural organic matter. *In* ACS, Extended Abstract Book. American Chemical Society, San Francisco, CA.
46. March, J. 1992. Advanced organic chemistry. John Wiley & Sons Inc., New York.
47. Meisel, D., and P. Neta. 1975. One-electron redox potentials of nitro compounds and radiosensitizers. J. Am. Chem. Soc. 97:5198-5203.

48. Mulla, M. S., L. S. Mian, and J. A. Kawecki. 1981. Distribution and transport of the insecticides malathion and parathion in the environment. Residue Reviews 81:1-136.

49. Neta, P., and D. Meisl. 1976. Substituent effects of nitroaromatic radical anions in aqueous solution. J. Physical. Chem. 80:519-524.

50. Neumeier, W., R. Haas, and E. v.Löw. 1989. Mikrobieller Abbau von Nitroaromaten aus einer ehemaligen Sprengstoffproduktion. Forum Städte-Hygiene 40:32-37.

51. O'Connor, G. A., J. R. Lujan, and Y. Jin. 1990. Adsorption, degradation, and plant availability of 2,4-dinitrophenol in sludge amended calcareous soils. J. Environ. Qual. 19:587-593.

52. Ou, T.-Y., L. H. Carreira, R. W. Schottman, and N. L. Wolfe. 1992. Nitroreduction of 2,4,6-trinitrotoluene (TNT) in contaminated soils. *In* ACS, Extended Abstract Book. American Chemical Society, San Francisco, CA.

53. Parris, G. E. 1980. Environmental and metabolic transformations of primary aromatic amines and related compounds. Residue Rev. 76:1-24.

54. Perlinger, J. A. 1994. Reduction of polyhalogenated alkanes by electron transfer mediators in homogeneous aqueous solution. Ph.D thesis, ETH Zurich, Switzerland.

55. Pillai, P., C. S. Helling, and J. Dragun. 1982. Soil-catalyzed oxidation of aniline. Chemosphere 11:299-317.

56. Preuss, A., J. Fimpel, and G. Diekert. 1993. Anaerobic transformation of 2,4,6-trinitrotoluene (TNT). Arch. Microbiol. 159:345-353.

57. Reitermayer, E. 1991. Identifizierung und Bestimmung von aromatischen Aminen in Wasserproben aus dem Bereich einer Rüstungsaltlast mit Hilfe der HPLC. Ph.D. thesis, Philips Univ., Marburg, FRG.

58. Rippen, G., E. Zietz, R. Frank, T. Knacker, and W. Kloepfer. 1987. Do airborne nitrophenols contribute to forest decline? Environ. Technol. Lett. 8:475-482.

59. Rosenblatt, D. H., E. P. Burrows, W. R. Mitchell, and D. L. Parmer. 1991. Organic explosives and related compounds, p. 195-237. *In* O. Hutzinger (ed.), The handbook of environmental chemistry - anthropogenic compounds. Springer Verlag, Berlin.

60. Scheffer, F., and P. Schachtschabel. 1984. Lehrbuch der Bodenkunde. Enke Verlag, Stuttgart.

61. Schneider, U. 1990. Sanierung der Rüstungsaltlast Hessisch Lichtenau-Hirschhagen - Dokumentation. Planungsgesellschaft Boden & Umwelt mbH, Kassel, FRG; Technical Report; Hessisches Ministerium für Umwelt und Reaktorsicherheit, Wiesbaden, FRG.

62. Schnell, S., and B. Schink. 1991. Anaerobic aniline degradation via reductive deamination of 4-aminobenzoyl-CoA in *Desulfobacterium anilini*. Arch. Microbiol. 155:183-190.

63. Schwarzenbach, R. P., P. M. Gschwend, and D. M. Imboden. 1993. Environmental organic chemistry. John Wiley & Sons Inc., New York.

64. Schwarzenbach, R. P., R. Stierli, K. Lanz, and J. Zeyer. 1990. Quinone and iron porphyrin mediated reduction of nitroaromatic compounds in homogeneous aqueous solution. Environ. Sci. Technol. 24:1566-1574.

65. SDC, and AATCC. 1956. Colour index, Society of Dyers and Colourists and American Association of Textile Chemists and Colorists, Badford, Yorkshire, England; Lowell, MA.

66. Smith, J. G., and M. Fieser. 1990. Collective index for volumes 1-12 of Fieser and Fieser's reagents for organic synthesis. John Wiley & Sons Inc., New York.

67. Somasundram, L., and J. R. Coats (eds.). 1991. Pesticide transformation products - fate and significance in the environment. American Chemical Society, Washington, DC.

68. Spain, J. C. Bacterial degradation of nitroaromatic compounds under aerobic conditions. (this volume)

69. Spalding, R. F., and J. W. Fulton. 1988. Groundwater munition residues and nitrate near Grand Island, Nebraska, USA. J. Contam. Hydrol. 2:139-153.

70. Spanggord, R. J., W. R. Mabey, T. W. Chou, and J. H. Smith. 1985. Environmental fate of selected nitroaromatic compounds in the aquatic environment, p. 15-33. *In* D. E. Rickert (ed.), Toxicity of nitroaromatic compounds. Hemisphere Publishing Corp., Washington, DC.

71. Striebel, T., J. Daub, and R. Hermann. 1992. Charakterisierung und Analyse der Verschmutzung des Niederschlages und des Niederschlagsabflusses, p. 113-145. *In* H. H. Hahn and C. Xanthopoulos (ed.), Niederschlagsbedingte Schmutzbelastung der Gewässer aus städtischen befestigten Flächen (Phase I) - Niederschlag. BMFT, Karlsruhe, FRG.

72. Stumm, W. 1992. Chemistry of the solid-water interface, John Wiley & Sons Inc., New York.

73. Stumm, W., and J. J. Morgan. 1981. Aquatic chemistry. John Wiley & Sons Inc., New York.

74. Sublette, K. L., P. Cho, A. Maule, S. Schwartz, and D. Pak. 1991. Microbial and biomimetic degradation of nitrobodies, p. 756-763. *In* B. Nath (ed.), Proc. Int. Conf. Environ. Pollut. Interscience Enterprises Ltd.

75. Tatsumi, K., A. Freyer, R. D. Minard, and J. M. Bollag. 1994. Enzyme-mediated coupling of 3,4-dichloroaniline and ferulic acid: a model for pollutant binding to humic materials. Environ. Sci. Technol. 28:210-15.
76. Tatsumi, K., S. Y. Liu, and J. M. Bollag. 1992. Enzyme-catalyzed complex formation of chlorinated anilines with humic substances. Water Sci. Technol. 25:57-60.
77. Tatsumi, K., S. Wada, H. Ichikawa, S. Y. Liu, and J.-M. Bollag. 1992. Cross-coupling of a chloroaniline and phenolic acids catalyzed by a fungal enzyme. Wat. Sci. Tech. 26:2157-2160.
78. Tratnyek, P. G., and D. L. Macalady. 1989. Abiotic reduction of nitroaromatic pesticides in anaerobic laboratory systems. J. Agric. Food Chem. 37:248-254.
79. Tratnyek, P. G., and N. L. Wolfe. 1990. Characterization of the reducing properties of anaerobic sediment slurries using redox indicators. Environ. Toxicol. Chem. 9:289-295.
80. Tremp, J. 1992. Sources and fate of nitrated phenols in the atmospheric environment. Ph.D thesis, ETH Zurich, Switzerland.
81. Urbansky, T. 1984. Chemistry and technology of explosives. Pergamon Press, Oxford, UK.
82. Voudrias, E. A., and M. Reinhard. 1986. Abiotic organic reactions at mineral surfaces, p. 463-486. *In* J. A. Davis and K. F. Hayes (ed.), Geochemical processes at mineral surfaces. ACS Symposium Series, 323. American Chemical Society.
83. Wade, R. S., R. Havlin, and C. E. Castro. 1969. The oxidation of iron(II) porphyrins by organic molecules. J. Am. Chem. Soc. 91:7530-7534.
84. Wahid, P. A., C. Ramakrishna, and N. Sethunathan. 1980. Instantaneous degradation of parathion in anaerobic soils. J. Environ. Qual. 9:127-130.
85. WHO. 1993. Methyl Parathion. World Health Organization, Geneva.
86. Wolfe, N. L. 1992. Abiotic transformations of pesticides in natural waters and sediments, p. 93-104. *In* J. L. Schnoor (ed.), Fate of pesticides and chemicals in the environment. John Wiley & Sons, Inc., New York.
87. Wolfe, N. L., and D. R. Burris. 1992. Abiotic redox transformations of organic contaminants in aquifer materials, p. 2-16. *In* AWMA, 85th Annual meeting. Air & Waste Management Association, Kansas City.
88. Wolfe, N. L., B. E. Kitchens, D. L. Macalady, and T. J. Grundl. 1986. Physical and chemical factors that influence the anaerobic degradation of methyl parathion in sediment systems. Environ. Toxicol. Chem. 5:1019-1026.
89. Wolfe, N. L., and D. L. Macalady. 1992. New perspectives in aquatic redox chemistry: abiotic transformations of pollutants in groundwater and sediments. J. Contam. Hydrol. 9:17-34.
90. Wujcik, W. J., W. L. Lowe, P. J. Marks, and W. E. Sisk. 1992. Granular activated carbon pilot treatment studies for explosives removal from contaminated groundwater. Environ. Prog. 11:178-89.
91. Yu, S. Y., and G. W. Bailey. 1992. Reduction of nitrobenzene by four sulfide minerals: kinetics, products, and solubility. J. Environ. Qual. 21:86-94.
92. Zepp, R. G., and L. N. Wolfe. 1987. Abiotic transformation of organic chemicals at the particle-water interface, p. 423-455. *In* W. Stumm (ed.), Aquatic surface chemistry. John Wiley & Sons Inc., New York.
93. Zeyer, J. 1988. Abbau aromatischer Nitroverbindungen. GWF Wasser Abwasser 129:25-27.

INDEX